STATISTICAL METHODS
FOR GROUNDWATER MONITORING

STATISTICAL METHODS FOR GROUNDWATER MONITORING

Robert D. Gibbons
University of Illinois at Chicago

A WILEY-INTERSCIENCE PUBLICATION

JOHN WILEY & SONS, INC.

New York • Chichester • Brisbane • Toronto • Singapore

This text is printed on acid-free paper.

Library of Congress Cataloging in Publication Data:

Gibbons, Robert D., 1955–
    Statistical methods for groundwater monitoring / Robert D.
  Gibbons.
        p.   cm.
    Includes index.
    ISBN 0-471-58707-9 (alk. paper)
    1. Groundwater—Pollution—Measurement—Statistical methods.
  I. Title.
  TD426.G52   1994
  628.1′61—dc20   94-4187

Printed in the United States of America

10 9 8 7 6 5 4 3 2

In Memory of Burton Homa
1920–1993

ACKNOWLEDGMENTS

There are a great many people who helped make this book possible.

I am particularly indebted to Dr. Charles Davis of EnviroStat (Henderson, NV) for his helpful review of several chapters and for preparing many of the most useful tables in this book. I would also like to thank Dave Dolan of Chemical Waste Management, Professor R. Darrell Bock of the University of Chicago, and Professor Donald Hedeker of the University of Illinois at Chicago for providing excellent comments on environmental and statistical aspects of the book. I am also extremely grateful to Roberta Baruch and Evelyn Murphy for helping to prepare the manuscript and providing editorial comments and to Balaraman Saravanan for helping to prepare the tables, figures, and index.

More generally, there are a great many groundwater scientists and engineers to whom I am tremendously indebted for bringing these fascinating problems to my attention and helping to provide my education in environmental science. Among these people I especially wish to thank John Baker, Frank Jarke, Gary Williams, Peter Vardy, Lin Longshore, Rob Powell, Dave Burt, Marty Sara, Mark Adams, Sara Broadbent, Doug Borro, Bill Ross, Ken Anderson, Karl Silver, Nancy Grams, Chuck White, and Lori Tagawa.

I would also like to thank Dr. Boris Astrachan for his continued support of my eclectic research interests which vary almost randomly from day to day.

Finally, I want to thank my family for their love and support.

CONTENTS

STATISTICAL METHODS
FOR GROUNDWATER MONITORING

Introduction

Protection of our natural resources from the disposal of hazardous materials continues to be a major environmental and health concern in the United States and throughout the world. Of primary concern is the impact on groundwater of leakage from waste disposal facilities. Since both municipal solid waste and hazardous industrial waste each contain constituents that are potentially harmful to human health and the environment (although typically at greatly different concentrations and likelihood of detection), this concern has given rise to regulations in the United States requiring extensive chemical testing of groundwater quality at waste disposal facilities (e.g., Resource Conservation and Recovery Act (RCRA) solid and hazardous waste landfills and Toxic Substances Control Act (TSCA) disposal landfills). Historically, only hazardous waste disposal facilities were required to provide groundwater monitoring programs; however, recent U.S. Environmental Protection Agency (EPA) regulation mandates extensive groundwater monitoring at municipal solid waste disposal facilities as well.

The purpose of this groundwater monitoring, often called groundwater detection monitoring, is to detect the earliest possible release from the disposal facility. The result of these new regulations is that thousands of previously unregulated facilities in the United States will now be required to provide extensive geochemical assessments of groundwater on a regular basis (i.e., quarterly or semiannually) and on each occasion to determine if the facility has contaminated groundwater on the basis of these analytical results.

Groundwater monitoring decisions historically have been rooted in statistical theory. Although the methods were often inappropriate, statistics has nevertheless played a critical role in environmental impact decisions, beginning with formal regulatory discussion of these issues (see RCRA regulation, EPA, 1982). The problem is technically interesting: given a new monitoring measurement for a groundwater monitoring well, drilled in a particular aquifer, and analyzed for a particular substance, what is the probability that measurement represents an effect of some unnatural source (e.g., a hazardous waste disposal facility)? Presented this way, the statistician sees this as a problem of statistical prediction. Given a collection of historical or background measurements for the substance from an area geographically removed from the potential source of contamination (e.g., groundwater monitoring wells located upgradient of a waste disposal facility), what limit or interval will contain the new measurement with a desired level of confidence

1

(e.g., 95% confidence)? Although this seems to be a natural view of the problem, it took seven years for the EPA to acknowledge its utility for groundwater monitoring applications (EPA, 1987a).

If this were the full extent of the problem, the solution would be elementary, and routine application of statistical prediction and tolerance intervals would suffice. The problem, however, is far more complicated. A key problem involves the issue of multiple comparisons. Groundwater quality is evaluated on measurements from many monitoring wells, each analyzed for multiple constituents. The result is a proliferation of statistical comparisons, often amounting to thousands of tests per sampling interval (e.g., quarterly), which leads to conclusions of contamination near certainty, regardless of whether or not contamination is actually present. The reason is that each statistical comparison has an associated probability of indicating contamination when none is present (i.e., false positive rate). Even if this probability is small (e.g., 1%), the probability of at least one of a large number of comparisons being significant by chance alone is near certainty. For example, a facility with 10 monitoring wells, each required to be tested for 10 constituents, would have an overall probability of .63 or a 63% chance of failing at least one of these tests by chance alone, even if each individual test has only a 1% chance of failure by chance alone. "Brute force" adjustment for the number of statistical comparisons (e.g., Bonferroni, adjustment to individual comparison Type I error rate; see Miller, 1966) may result in an acceptable "sitewide" false positive rate, at the expense of a proliferation of false negative results (i.e., concluding that there is no contamination when there is).

If the problem of multiple comparison were not serious enough, monitoring constituents themselves present additional problems. Many constituents (e.g., volatile organic compounds, VOCs) are rarely, if every, detected in background monitoring wells, but are detected occasionally in monitoring wells downgradient of the facility. Detection is often taken as evidence of contamination, when in many cases it is completely consistent with chance expectations. First, as an example, waste disposal facilities are constructed with numerous downgradient monitoring wells, in some cases as many as 100 or more, but often only a handful of background or upgradient monitoring wells are constructed and a fair number have only a single background well. A rare event (e.g., detection of a VOC) is more probable in a downgradient monitoring well (for which there are a greater number of available measurements). Second, detection of such compounds is based on a statistical decision rule (method detection limit, MDL) used by the analytical chemist in making the binary decision of whether or not a compound is present in the groundwater sample. The false positive rate associated with this MDL is intended to be 1% by design (it often is much larger due to numerous methodological and statistical errors; see Clayton et al., 1987); hence, on average, we would expect one detection per hundred analytical determinations even when the constituent is not present in the sample. To make matters worse, some required lists of monitoring constituents contain as many as 300 constituents, seemingly guaranteeing a conclusion of contamination regardless of the presence or absence of contaminants.

Another inherent problem in groundwater monitoring is that even when quantitative measurements are available, a proportion of these will be "nondetects." Application of traditional statistical methods to "censored" data samples often leads to biased results and incorrect tests of hypotheses and corresponding interval estimates. The problem of estimating sufficient statistics for censored distributions [e.g., mean and variance of censored normal (Cohen, 1961) and lognormal (Aitchison, 1955) distribution] has been extensively studied, but little or no attention has been paid by regulatory agencies to the effect of censoring on error rates of statistical tests applied to the site as a whole.

In each chapter in this book a general overview of the problem is presented, followed by increasingly complex solutions. In some cases statistical theory is presented that may not be accessible to all readers, however, it is included for completeness and the hope that this book may provide a foundation for further statistical research in this area. Despite the complexity, for each solution or statistical approach to a particular problem, a relevant example is provided with computational details and/or tables that can be used for routine application of the statistical results. Attention is paid to the statistical properties of alternative approaches, including the false positive and false negative rates associated with each test and the factors related to these error rates where possible. Recommendations are provided for specific problems based on characteristics such as the number of monitoring wells, number of constituents, distributional form of measurements, and detection frequency. The reader may use the book to help craft a detection monitoring program for most waste disposal facilities and other waste management units, and in some cases these techniques may also be useful in analysis of other environmental media (e.g., soils or surface water).

The book is arranged in clusters of related chapters, although all chapters in this book are related at least to some degree. Chapters 1 through 3 discuss statistical prediction intervals, perhaps the most useful general statistical tool for detection monitoring. Multiple comparisons (i.e., simultaneous statistical inference) are stressed throughout. Chapter 1 presents prediction limits that can be applied to constituents with normal distributions, and serves as a basis for the general strategy of applying prediction limits to groundwater monitoring data. The normal prediction limit is developed and then generalized to cases involving repeated application at multiple monitoring wells and for multiple constituents. The effects of verification resampling are then described and incorporated into the prediction limit equation.

Chapter 2 presents nonparametric prediction limits. Nonparametric prediction limits are similar to their parametric counterparts except that they assume no specific distributional form for the chemical constituent. These prediction limits are defined by an order statistic, in general the largest measured value in a sample of n historical measurements. The statistical challenge is to compute the confidence level associated with such a decision rule given repeated application of the test and a particular verification resampling strategy. A major advantage of the nonparametric approach is that the limit is defined even if only a single background measurement is quantifiable. Nonparametric predictions limits are therefore well suited to many of the natural complexities associated with environmental data in general and groundwater monitoring data in particular.

Chapter 3 presents prediction limits for other parametric distributions that are often characteristic of environmental data. Included is a discussion of prediction limits for lognormal and Poisson distributed measurements. Lognormal prediction limits are useful for datasets that contain a few extreme values that are characteristic of the natural variability for that constituent. In contrast, Poisson prediction limits are well suited to rare event data, such as setting limits for the frequency of detection of volatile organic constituents. Incorporating the effects of multiple comparisons and verification resampling plans into these types of prediction limits is also covered in detail.

Chapter 4 discusses the use of statistical tolerance limits in groundwater detection monitoring. In contrast to prediction limits designed to include 100% of the next k measurements with specified confidence level $(1 - \alpha)100\%$, tolerance limits are designed to contain a specified percentage $(1 - \gamma)100\%$ of all future measurements with confidence level $(1 - \alpha)100\%$. The distinction is critical for large monitoring programs in which it may not be possible to include *all* measurements within the interval or below the limit. As in Chapter 1 through 3, tolerance limits for normal, lognormal, and Poisson distributed data and nonparametric tolerance limits are described. Two-stage procedures that combine tolerance and prediction limits are described as well. These two-stage procedures are particularly well suited to large monitoring networks (i.e., large numbers of wells and constituents).

Chapter 5 and 6 present statistical methods for computing method detection limits (MDLs) and practical quantitation limits (PQLs). Analytical chemists use MDLs to determine the presence or absence of a constituent in a sample, and PQLs are used to determine if the concentration is quantifiable. At first glance, it might seem unusual that these topics in statistical chemistry occupy such a prominent position in this book. In practice, however, MDLs and PQLs are quite commonly used as statistical decision rules for constituents that are not detected or quantified in background samples. Since, over time, the number of new monitoring measurements is far greater than the number of background samples (e.g., more downgradient measurements than upgradient measurements), the probability of a "false detection" is also far greater and can lead to costly and needless site assessments and corrective action. For this reason, it is critically important that MDLs and PQLs be clearly defined and statistically rigorous estimators used for their determination. In Chapter 5 the historical literature on MDLs is reviewed, and the strengths and weaknesses of various estimators are compared and contrasted. Estimators based on the entire calibration function and not a single concentration are shown to clearly be the methods of choice. In Chapter 6 the ideas developed in Chapter 5 are expanded to the case of estimating the PQL. This area is far less developed than the literature on MDLs with the most common historical approach being to arbitrarily define the PQL as a multiple of the MDL. In Chapter 6 it is shown that while both MDLs and PQLs can be estimated from similar analytical data, the estimators have little to do with one another, the former being a test of the null hypothesis that the concentration is zero and the latter being a point estimate of analytical precision. In practice, MDLs and PQLs are used as statistical methods for testing constituents with a detection frequency of zero; hence their important role and coverage in this book.

Chapter 7 introduces the concept of contaminant source analysis. In previous chapters a single source of contamination is assumed (i.e., the landfill). In Chapter 7 this assumption is extended to consider the possibility of multiple possible sources of contamination and to differentiate among them in terms of geochemical impact. Applications include detection or assessment monitoring of waste disposal facilities located in areas in which groundwater may be affected by (1) more than one type of facility (e.g., a landfill and a steel plant), (2) waste disposal and a naturally occurring source not represented in background such as a surface water channel from an ocean, or (3) multiple types of disposal activities (e.g., disposal of hazardous industrial liquids versus disposal of municipal solid waste) at the same facility. The statistical theory introduced in this chapter has its basis in statistical classification and pattern recognition and has been used extensively in chemometrics. Specifically, parametric and nonparametric approaches to discriminant function analysis are described and compared and contrasted using a series of relevant environmental illustrations.

Chapters 8 discusses the use of control charts and prediction limits for intrawell comparisons in which new measurements are compared to historical values from the same well. The advantage of the intrawell comparison strategy is the elimination of spatial variation that often confounds upgradient versus downgradient comparisons. In many practical applications involving traditional upgradient versus downgradient comparisons, the small number of upgradient wells may not be sufficient to characterize the spatial variability observed in the larger number of downgradient monitoring wells. In some cases the hydrogeology upgradient may simply not be representative of the hydrogeology downgradient, regardless of the number of monitoring wells. In both cases the traditional upgradient versus downgradient comparison strategy does not apply, and the only viable alternative is to compare the new monitoring measurements to historical measurements from that same well. The danger, of course, is that previous site impacts may go undetected. To this end, this chapter focuses on the development of combined Shewart–CUSUM control charts which are sensitive to both gradual and immediate releases. To further decrease the possibility of "masking" previous site impacts, adjustments for historical trends and outliers are also described. Finally, use of simultaneous prediction limits for intrawell comparisons are presented as well, and new tables for computing these intrawell prediction limits are provided.

Chapter 9 expands on the discussion of detecting historical trend introduced in Chapter 8. This chapter focuses on several nonparametric methods of estimating trend in environmental data. Adjustment of trend estimators for systematic seasonal effects is described as well.

Chapter 10 presents a comparative study of the large literature on the analysis of censored data which occur in the present context when the background data consist of a series of samples in which the analyte is detected or quantified in only a subset. In practice, this is an enormously important problem since this condition typifies environmental data in general and groundwater monitoring data in particular. This chapter focuses on two applications: estimating the mean and variance of a censored distribution and using censored data estimators in routine application of statistical prediction limits and their effect on associated error rates. The former has received considerable attention in the statistical and environmental literatures, whereas the latter has rarely been discussed and is, in fact, the relevant problem. Several estimators are described, illustrated, and compared and contrasted in terms of their properties for both types of applications, and recommendations for routine application are provided. The methods presented in Chapter 10 are directly relevant to the application of methods in Chapters 1, 3, 4, and 8.

Chapter 11 presents much of the literature on testing distributional assumptions (i.e., normality and lognormality) in both small and large environmental datasets. Topics covered include traditional tests of normality and lognormality, joint assessment of normality and lognormality in several groups (e.g., wells), and evaluating normality and lognormality in censored datasets. The methods developed in this chapter are directly applicable to the selection of an appropriate statistical prediction interval from among the possible choices described in Chapters 1 through 3. Many of the ideas presented in this chapter also form the basis of the outlier detection tests presented in Chapter 13.

Chapter 12 presents an introduction to the area of variance component models. To this point, it has been assumed that the background data consist of a sample of n independent and identically distributed measurements. When the background data consist of a pool of measurements from different wells, this assumption is demonstrably false, since measurements within a well will generally be more similar than measurements between different wells. The traditional estimator of the sample standard deviation ignores this association and is therefore overly conservative and will lead to detection monitoring programs with false positive rates that are higher than expected. Similar problems can result when there are systematic temporal or seasonal trends in the data. Chapter 12 presents two general statistical approaches to this problem: the traditional least squares approach and a more recent development based on the combination of maximum likelihood estimation and empirical Bayes estimation. Complete computational details are provided, and the methods are illustrated using relevant environmental examples.

Chapter 13 expands on the discussion of detecting outliers that was first introduced in Chapter 8. Preliminary screening of anomalous measurements is critical to groundwater detection monitoring data analysis, particularly for programs based on nonparametric prediction limits or intrawell comparisons. The presence of outliers in a background database can make it impossible to detect contamination when it is present. Chapter 13 examines numerous methods that have been proposed for detecting outliers and compares and contrasts them. Primary attention is devoted to methods that are not limited in terms of the number of outliers that can be detected. Comparisons are made between those methods that adjust for the total number of outliers tested versus those methods that do not. Recommendations for routine application are then provided.

Chapter 14 covers common mistakes and methods to avoid in groundwater monitoring applications. These include the original errors in the RCRA regulation based on Cochran's approximation to the Behrens–Fisher t-statistic and new regulations that continue to advocate the use of parametric and nonparametric analysis of variance for both hazardous and municipal solid waste facilities. Numerous other proposed statistical methods for groundwater monitoring applications that can lead to disaster are discussed. It is unfortunate how often these methods are found to have a prominent role in both regulation and guidance.

The overall purpose of this book is to examine multiple problems inherent in the analysis of groundwater monitoring data and to illustrate their application and interconnection. Issues relevant for practical application and current areas of statistical research are highlighted. The level of statistical detail is commensurate with these two objectives.

The methods presented in this book are, of course, relevant to environmental statistics in general. The focus on groundwater monitoring applications in particular is due to the importance of this problem and the limited literature and poor guidance that has historically characterized this area. I hope this book will contain information useful to a wide variety of practitioners and researchers in environmental science.

1 Normal Prediction Intervals

1.1 OVERVIEW

The fundamental problem in groundwater monitoring is the prediction of future measurements based on a background sample of historical measurements (Gibbons, 1987a; Davis and McNichols, 1987). In some cases the background sample may consist of repeated measurements from a collection of wells located upgradient of the facility. In other cases the background sample may consist of repeated measurements from a single monitoring well to which the consistency of future measurements will be compared (i.e., intrawell comparison). In either case, if the number of future comparisons is finite and known, we may wish to compute an interval that will contain all future measurements with a given level of confidence (e.g., 95% confidence). The most critical problem is correctly defining the number of future comparisons and constructing the corresponding statistical decision rule so that the confidence level pertains to the site as a whole. As will be shown, the number of future comparisons includes the total number of constituents and monitoring wells for which a statistical test is to be performed. To assume any less will result in countless site assessments and possible corrective action due to chance alone.

To provide a statistical foundation, denote the number of future comparisons as k and the confidence level $1 - \alpha$, where α represents the false positive rate or Type I error rate of the decision rule. The false positive rate is the rate at which we would reject a new value if in fact it came from the same distribution as the background measurements. The appropriate statistical interval for this application is known as a prediction interval (Gibbons, 1987a; Davis and McNichols, 1987; Gibbons and Baker, 1991d). A synonym for prediction interval is a beta-expectation tolerance interval, in that, on average, the new measurements will be contained with confidence level $1 - \alpha$.

In the context of groundwater monitoring, prediction intervals play an important role because we often know the number of statistical comparisons made on each monitoring event, and, for regulatory purposes, we must include all measurements or risk a potentially costly site assessment. It is not always a trivial question, however, of what constitutes the number of future measurements. Is it the number of monitoring wells, the number of constituents at a particular monitoring well, or a combination of both? Should the number of future comparisons be restricted to those performed on the next single monitoring event or should it include all future monitoring events? To answer these questions, it is important to understand the consequences of a false positive decision and the impact the choice of k has on the false positive and negative rates of the statistical test. The false negative rate describes the frequency of failure to reject a new measurement when it has come from a different distribution than the background measurements. To better understand the answers to these questions, let us begin with the simplest form of prediction limit: a prediction limit for the next single measurement from a normal distribution.

1.2   PREDICTION INTERVALS FOR THE NEXT SINGLE MEASUREMENT FROM A NORMAL DISTRIBUTION

Assume we have collected $n = 8$ background water quality measurements for total organic carbon (TOC) levels, denoted $x_1, \ldots, x_8$. The sample mean and sample standard deviation of these eight measurements are given by

$$\bar{x} = \sum_{i=1}^{n} \frac{x_i}{n} \tag{1.1}$$

and

$$s = \sqrt{\sum_{i=1}^{n} \frac{(x_i - \bar{x})^2}{n - 1}} \tag{1.2}$$

On the next quarterly monitoring event, we intend to collect a new TOC measurement from the same well or a corresponding compliance well located downgradient of the facility. Based on the previous eight samples, what interval will contain the next single TOC measurement with $(1 - \alpha)100\%$ confidence?

To construct such an interval, we must begin by examining the sources of uncertainty in this problem. First, note that $\bar{x}$ and s are merely sample-based estimates of the true population mean and standard deviation μ and σ. If we had measured all groundwater in the area for that time period, $\bar{x}$ would equal μ and s would equal σ. However, we only have eight available measurements; hence we will have considerable uncertainty in our estimates of μ and σ. Fortunately, we can quantify our uncertainty in σ by noting that the sample mean $\bar{x}$ is distributed normally with mean μ and standard deviation $\sigma/\sqrt{n}$ (i.e., $\bar{x} \sim N[\mu, \sigma^2/n]$). Second, note that the new measurement x_{new} also has an associated measurement error σ for which we have a sample-based estimate s and is independent of the prior measurements. Combining these two sources of uncertainty and selecting the $(1 - \alpha/2)100\%$

point of Student's t-distribution with $n - 1$ degrees of freedom yields the interval

$$\bar{x} \pm t_{[n-1, 1-\alpha/2]}\sqrt{s^2 + \frac{s^2}{n}} \tag{1.3}$$

which can be expressed in the more familiar form

$$\bar{x} \pm t_{[n-1, 1-\alpha/2]}s\sqrt{1 + \frac{1}{n}} \tag{1.4}$$

This interval will provide $(1 - \alpha)100\%$ confidence of including the next future measurement from the normal distribution for which we have a sample of n previous measurements.

Frequently, however, we are most interested in providing an upper limit for the new measurement, since, for example, a TOC measurement that is too low poses no environmental threat. In this case we compute the one-sided normal prediction limit as

$$\bar{x} + t_{[n-1, 1-\alpha]}s\sqrt{1 + \frac{1}{n}} \tag{1.5}$$

This prediction limit provides $(1 - \alpha)100\%$ confidence of not being exceeded by the next single measurement.

Example 1.1

Consider the data in Table 1.1 for TOC measurements from a single well over 2 years of quarterly monitoring. Inspection of the data reveals no obvious trends, and these data have mean $\bar{x} = 11.0$ and standard deviation $s = 0.61$. The upper 95% point of Student's t-distribution with $n - 1 = 8 -$

TABLE 1.1   Eight Quarterly TOC Measurements

Year	Quarter	TOC in mg/l
1992	1	10.0
1992	2	11.5
1992	3	11.0
1992	4	10.6
1993	1	10.9
1993	2	12.0
1993	3	11.3
1993	4	10.7

$1 = 7$ degrees of freedom is $t_{[7, 1-.05]} = 1.895$. Therefore the upper 95% confidence normal prediction limit is given by

$$11.0 + 1.895(0.61)\sqrt{1 + \tfrac{1}{8}} = 12.23 \text{ mg/l}$$

which is larger than any of the observed values. Had we required 99% confidence of including the next single measurement, the upper 99% point of Student's t-distribution on 7 degrees of freedom would be $t_{[7, .01]} = 2.998$. Therefore the upper 99% confidence normal prediction limit would be given by

$$11.0 + 2.998(0.61)\sqrt{1 + \tfrac{1}{8}} = 12.94 \text{ mg/l}$$

These limits (i.e., 12.23 mg/l and 12.94 mg/l) provide 95% and 99% confidence, respectively, of including the next single observation from a normal distribution for which eight previous measurements have been obtained with an observed mean 11.0 mg/l and a standard deviation 0.61 mg/l.

1.3 PREDICTION LIMITS FOR THE NEXT k MEASUREMENTS FROM A NORMAL DISTRIBUTION

In practice, it is rare to have an application in which only a single future measurement requires evaluation. Typically, TOC measurements are obtained from a series of downgradient or compliance wells and must be simultaneously evaluated. The simplest approach is to assume independence. Under independence if the probability of a false positive result for a single comparison is α, the probability of at least one of k comparisons being significant by chance alone is

$$\alpha^* = 1 - (1 - \alpha)^k \tag{1.6}$$

Here, α^* is the sitewide false positive rate since it simultaneously considers all k comparisons being performed on a given monitoring event. The value of k reflects the total number of statistical tests which is the product of the number of monitoring wells and the number of constituents. For example, with 95% confidence for an individual comparison (i.e., $\alpha = .05$) and $k = 10$ comparisons, the probability of at least one significant result by chance alone is

$$\alpha^* = 1 - (1 - .05)^{10} = .40$$

or a 40% chance of a statistically significant exceedance by chance alone. With 100 comparisons $\alpha^* = .99$ or a 99% chance of a statistically significant exceedance by chance alone. Since it is not uncommon for detection monitoring programs to have 20 to 30 monitoring wells, each monitored quarterly for 10 or 20 constituents (in some cases far more), the effect of these multiple comparisons on the sitewide false positive rate is considerable. The likelihood of chance failure is near certainty. A facility with 25 wells, each monitored for 20 constituents, will be performing 500 statistical tests per sampling event. Even setting $\alpha = .01$ will produce a probability of $\alpha^* = .99$ or a 99% chance of failing any one of those tests by chance alone. Since most state and federal regulations require costly site assessments that may lead to corrective action on the basis of any significant elevation of any constituent in any point of compliance well, the impact of an inflated sitewide false positive rate is enormous.

One solution to this problem is to compute a prediction limit that will provide $(1 - \alpha^*)100\%$ confidence of including all k future measurements. The simplest approach to this problem is through use of the Bonferroni

TABLE 1.2   One-Sided Values of Student's t-Statistic (95% Overall Confidence for Background $n = 4$ to 100 and $k = 4$ to 50 Future Measurements)

n	\multicolumn{10}{c}{k = Number of Future Comparisons}
	5	10	15	20	25	30	35	40	45	50
4	4.54	5.84	6.74	7.45	8.05	8.57	9.04	9.46	9.85	10.21
8	3.00	3.50	3.81	4.03	4.21	4.35	4.48	4.59	4.69	4.78
12	2.71	3.10	3.32	3.48	3.60	3.71	3.79	3.87	3.93	3.99
16	2.60	2.94	3.14	3.28	3.39	3.48	3.55	3.61	3.67	3.72
20	2.54	2.86	3.04	3.17	3.27	3.35	3.42	3.48	3.53	3.57
24	2.50	2.81	2.98	3.10	3.20	3.27	3.34	3.39	3.44	3.48
28	2.47	2.77	2.94	3.06	3.15	3.22	3.28	3.33	3.38	3.42
32	2.45	2.74	2.91	3.02	3.11	3.18	3.24	3.29	3.33	3.37
36	2.44	2.72	2.88	3.00	3.08	3.15	3.21	3.26	3.30	3.34
40	2.43	2.71	2.87	2.98	3.06	3.13	3.18	3.23	3.27	3.31
44	2.42	2.69	2.85	2.96	3.04	3.11	3.16	3.21	3.25	3.29
48	2.41	2.68	2.84	2.95	3.03	3.09	3.15	3.19	3.24	3.27
52	2.40	2.68	2.83	2.93	3.01	3.08	3.13	3.18	3.22	3.26
56	2.40	2.67	2.82	2.92	3.00	3.07	3.12	3.17	3.21	3.24
60	2.39	2.66	2.81	2.92	3.00	3.06	3.11	3.16	3.20	3.23
64	2.39	2.66	2.81	2.91	2.99	3.05	3.10	3.15	3.19	3.22
68	2.38	2.65	2.80	2.90	2.98	3.04	3.10	3.14	3.18	3.22
72	2.38	2.65	2.80	2.90	2.98	3.04	3.09	3.13	3.17	3.21
76	2.38	2.64	2.79	2.89	2.97	3.03	3.08	3.13	3.17	3.20
80	2.37	2.64	2.79	2.89	2.97	3.03	3.08	3.12	3.16	3.20
84	2.37	2.64	2.78	2.88	2.96	3.02	3.07	3.12	3.16	3.19
88	2.37	2.63	2.78	2.88	2.96	3.02	3.07	3.11	3.15	3.19
92	2.37	2.63	2.78	2.88	2.95	3.01	3.07	3.11	3.15	3.18
96	2.37	2.63	2.77	2.87	2.95	3.01	3.06	3.11	3.14	3.18
100	2.36	2.63	2.77	2.87	2.95	3.01	3.06	3.10	3.14	3.17

inequality (see Miller, 1966; Chew, 1968), noting that from (1.6)

$$\alpha = \frac{\alpha^*}{k} \qquad (1.7)$$

Application of (1.7) reveals that in order to have a sitewide error rate at $\alpha^* = .05$ when $k = 10$ comparisons are made requires we test each comparison at the $\alpha = .005\%$ level. The $(1 - \alpha)100\%$ prediction limit for the next k measurements from a normal distribution is therefore

$$\bar{x} + t_{[n-1,1-\alpha^*/k]} s \sqrt{1 + \frac{1}{n}} \qquad (1.8)$$

Table 1.2 displays one-sided values of $t_{[n-1,\alpha^*/k]}$, for $n = 4$ to 100 and $k = 5$ to 50, and Table 1.3 displays corresponding two-sided values.

TABLE 1.3   Two-Sided Values of Student's t-Statistic (95% Overall Confidence for Background $n = 4$ to 100 and $k = 4$ to 50 Future Measurements)

	k = Number of Future Comparisons									
n	5	10	15	20	25	30	35	40	45	50
4	5.84	7.45	8.57	9.46	10.21	10.87	11.45	11.98	12.47	12.92
8	3.50	4.03	4.35	4.59	4.78	4.94	5.08	5.20	5.31	5.41
12	3.10	3.48	3.71	3.87	3.99	4.10	4.19	4.26	4.33	4.39
16	2.94	3.28	3.48	3.61	3.72	3.81	3.88	3.95	4.00	4.06
20	2.86	3.17	3.35	3.48	3.57	3.65	3.72	3.78	3.83	3.88
24	2.81	3.10	3.27	3.39	3.48	3.56	3.62	3.67	3.72	3.76
28	2.77	3.06	3.22	3.33	3.42	3.49	3.55	3.60	3.65	3.69
32	2.74	3.02	3.18	3.29	3.37	3.44	3.50	3.55	3.59	3.63
36	2.72	3.00	3.15	3.26	3.34	3.41	3.46	3.51	3.55	3.59
40	2.71	2.98	3.13	3.23	3.31	3.38	3.43	3.48	3.52	3.56
44	2.69	2.96	3.11	3.21	3.29	3.35	3.41	3.45	3.49	3.53
48	2.68	2.95	3.09	3.19	3.27	3.34	3.39	3.43	3.47	3.51
52	2.68	2.93	3.08	3.18	3.26	3.32	3.37	3.42	3.46	3.49
56	2.67	2.92	3.07	3.17	3.24	3.31	3.36	3.40	3.44	3.48
60	2.66	2.92	3.06	3.16	3.23	3.30	3.35	3.39	3.43	3.46
64	2.66	2.91	3.05	3.15	3.22	3.29	3.34	3.38	3.42	3.45
68	2.65	2.90	3.04	3.14	3.22	3.28	3.33	3.37	3.41	3.44
72	2.65	2.90	3.04	3.13	3.21	3.27	3.32	3.36	3.40	3.43
76	2.64	2.89	3.03	3.13	3.20	3.26	3.31	3.35	3.39	3.42
80	2.64	2.89	3.03	3.12	3.20	3.26	3.31	3.35	3.38	3.42
84	2.64	2.88	3.02	3.12	3.19	3.25	3.30	3.34	3.38	3.41
88	2.63	2.88	3.02	3.11	3.19	3.25	3.29	3.34	3.37	3.41
92	2.63	2.88	3.01	3.11	3.18	3.24	3.29	3.33	3.37	3.40
96	2.63	2.87	3.01	3.11	3.18	3.24	3.29	3.33	3.36	3.40
100	2.63	2.87	3.01	3.10	3.17	3.23	3.28	3.32	3.36	3.39

Although the prediction limit in (1.8) limits probability of any one of k future measurements exceeding the limit by chance alone to α^*, it does so at the expense of the false negative rate. To illustrate this point, Figure 1.1 displays statistical power curves for prediction limits for the next $k = 1, 10,$ and 50 comparisons based on a background sample of $n = 8$ measurements and setting the individual comparison false positive rate to $\alpha = .05/k$. In Figure 1.1 contamination was introduced into a single monitoring well for a single constituent; hence only one of 1, 10, or 50 comparisons was contaminated. The power curves in Figure 1.1 therefore display the probability of detecting a very localized release that impacts only one of k future measurements. In practice, we would expect contamination to impact several wells and constituents; therefore the probability estimates in Figure 1.1 represent a lower bound. Inspection of Figure 1.1 reveals that the false positive rates for $k = 1, 10,$ and 50 future comparisons all approach the nominal level of 5%. However, false negative rates are dramatically affected by adjusting for larger numbers of future comparisons. For a difference of four standard deviation units and eight background samples, the false negative rates are 4%, 39%,

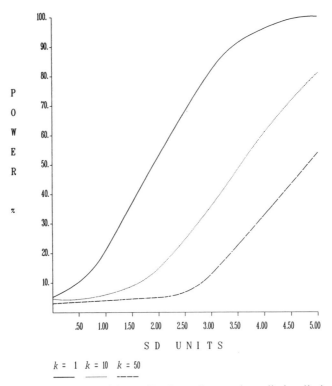

Figure 1.1 Power of 95% confidence Bonferroni normal prediction limits for 1, 10, and 50 future comparisons.

and 57% for $k = 1$, 10, and 50, respectively (i.e., one minus the probability of a significant result at the x axis $= 4$ sd units). These results indicate that by simply performing a statistical adjustment to the prediction limit to provide an overall sitewide false positive rate not greater than 5%, we sacrifice the false negative rate (i.e., failure to detect contamination when present), an unacceptable outcome. Control of the false negative rate at the expense of the false positive rate is also unacceptable. For example, the original RCRA regulation required that quadruplicate samples be obtained (i.e., a single sample split into four aliquots), and replicated measurements were to be treated as if they were independent. Of course, the intrasample correlation (i.e., correlation among the measurements for the four aliquots) was typically near unity and not zero as assumed by Cochran's approximation to the Behrens–Fisher (CABF) t-statistic. As such, the false positive rate approached 100%. A test with a false positive rate of 100% has a false negative rate of zero since it will trigger a site assessment regardless of the data.

Example 1.2

Returning to the TOC example dataset in Table 1.1, we may wish to apply the computed prediction limit to new TOC measurements from each of 10 downgradient monitoring wells, limiting the overall probability of *any* false positive result to 5%. To do this, we note that $\alpha/k = .05/10 = .005$ and the 99.5% upper percentage point of Student's t-distribution with 7 degrees of freedom is $t_{[7, .005]} = 3.499$. The upper prediction limit is therefore

$$11.0 + 3.499(0.61)\sqrt{1 + \tfrac{1}{8}} = 13.26 \text{ mg/l}$$

in contrast to 12.23 mg/l for a single future comparison.

1.4   NORMAL PREDICTION LIMITS FOR THE NEXT *r* OF *m* FUTURE SAMPLES AT EACH OF *k* MONITORING WELLS

The best currently available approach to balancing false positive and false negative rates in groundwater monitoring applications is through the use of verification resampling. Here, in the event of an initial exceedance, one or more verification resamples are obtained, and a statistical exceedance is declared if some number of the resampled values also exceeds the limit. In small monitoring programs it is sometimes possible to declare an exceedance if any of the resampled values exceed the limit (see Gibbons, 1991a). Alternatively, when background sample sizes are small and the number of future comparisons is large, a reasonable balance between false positive and false negative rates may require that statistical exceedance be declared only if all resampled values exceed the limit (see Davis and McNichols, 1987;

Gibbons, 1990). For this reason it is critical that the number of monitoring wells and constituents (i.e., k) be carefully selected and kept to a minimum.

To illustrate the effects of verification resampling on the false positive rate of a test in which the individual testwise false positive rate is set at $\alpha = .01$, consider a site with $k = 50$ future comparisons and one verification resample. Assuming independence among the k future comparisons,

$$\alpha^* = 1 - \Pr(\text{all wells okay})$$

$$= 1 - (\Pr(\text{one well okay}))^k$$

$$= 1 - (1 - \alpha + \alpha(1 - \alpha))^k$$

$$= 1 - (1 - .01 + .01(1 - .01))^{50}$$

$$= .005 \tag{1.9}$$

In this equation the first $1 - \alpha$ is for the initial sample being in bounds and the $\alpha(1 - \alpha)$ is for the initial sample out of bounds but the resample in bounds. In this case the verification resample has allowed us to use a 99% confidence prediction limit for 50 future measurements. Without verification resampling we could have only provided a sitewide 95% confidence level for $k = 5$ future monitoring measurements (i.e., $\alpha = \alpha^*/k = .05/5 = .01$), using exactly the same individual test level false positive rate (i.e., $\alpha = .01$) and corresponding prediction limit.

Now, consider a monitoring program in which, in the event of an initial significant increase, two verification resamples are to be obtained and a significant result recorded only if both verification resamples exceed the limit. In this case the sitewide false positive rate is

$$\alpha = 1 - (1 - \alpha^3)^k$$

$$= 1 - (1 - .01^3)^{50}$$

$$= .00005 \tag{1.10}$$

which is the probability of failing at least one of the initial samples and both of the verification resamples. This result suggests that for this example we have gone too far in that the sitewide false positive rate is now well below the nominal 5% level.

As a more conservative alternative, consider a monitoring program in which, in the event of an initial exceedance, two verification resamples are obtained and a significant exceedance is recorded if either resampled value exceeds the limit. In this case the sitewide false positive rate is given by

$$
\alpha^* = 1 - \left(1 - \alpha + \alpha(1 - \alpha)^2\right)^k
$$
$$
= 1 - \left(1 - .01 + .01(1 - .01)^2\right)^{50}
$$
$$
= .01 \tag{1.11}
$$

that is, the probability of failing an initial sample and at least one of the two verification resamples.

In any of these cases, we should select the most powerful solution that provides a reasonable sitewide false positive rate (i.e., $\alpha^* \sim .05$) within budgetary, legislative, and independence constraints. To do this, select (1.9), (1.10), or (1.11), such that $\alpha^* \sim .05$ for $\alpha \sim .01$ (i.e., sitewide false positive rate of approximately 5% and individual test false positive rate of approximately 1%). In this way a reasonable balance between false positive and false negative rates is achieved. Note, however, that these computations require the monitoring samples and resamples to be (adequately) stochastically independent. This implies a certain minimum time between samples. For quarterly monitoring, at most two resamples are reasonable.

Example 1.3

Returning to the TOC example dataset in Table 1.1, we may wish to apply the computed prediction limit to new TOC measurements from each of 10 downgradient monitoring wells for each of 5 monitoring constituents for a total of 50 future comparisons. Assuming that the five constituents are reasonably independent, the upper 99% confidence normal prediction limit for a single new measurement

$$
11.0 + 2.998(0.61)\sqrt{1 + \tfrac{1}{8}} = 12.94 \text{ mg/l}
$$

will provide an overall sitewide false positive rate of

$$
1 - (1 - .01)^{50} = .39
$$

or 39% without verification resampling,

$$
1 - (1 - .01 + .01(1 - .01))^{50} = .005
$$

or 0.5% with a single verification resample,

$$
1 - \left(1 - .01 + .01(1 - .01)^2\right)^{50} = .01
$$

or 1.0% with failure indicated if *either* of two verification resamples fails (i.e., exceeds 12.94),

$$1 - (1 - .01^3)^{50} = .00005$$

or 0.005% with failure indicated if *both* of two verification resamples fail.

To illustrate the effects of resampling on false negative rates, Figure 1.2 displays power curves for the four previously described alternatives (i.e., no resampling, one resample, failing the first and either of two resamples, or failing the first and both of two resamples) using a 99% confidence normal prediction limit.

Figure 1.2 reveals that the plan without resampling has an unacceptably high false positive rate; however, the rate is slightly less than predicted (i.e., 34% versus 39%). The reason for this discrepancy is that the multiple comparisons are not independent as assumed by the independence based computations. This problem is discussed in detail in the following section.

Figure 1.2 also reveals that for 10 monitoring wells and 5 constituents and a fixed prediction limit, the best balance between false positive and false

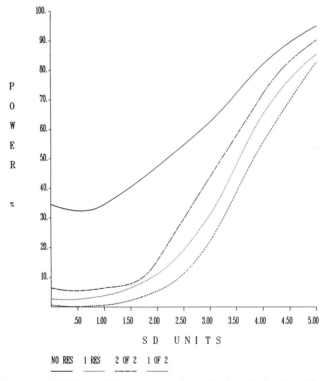

Figure 1.2 Power of 99% confidence normal prediction limits for 10 wells and 5 constituents and 4 resampling plans.

negative results is achieved for the plan in which two verification resamples are taken and failure is indicated if either exceeds the limit. Note that the false positive rates obtained via simulation for the various resampling plans were considerably higher than predicted via the Bonferroni inequality. For example, with a single verification resample we predicted 0.4% false positives but observed 2.7%. For two verification resamples, Plan A produced 6.4% false positives whereas only 0.8% were predicted, and Plan B produced 0.6% whereas only 0.004% were predicted. These discrepancies are opposite of what we observed in the case without resampling. The reason for the increased observed false positive rate is that by chance alone background levels may be particularly low for one well and constituent, but it is this same low background being compared to the verification resample(s). As such, the probability of two successive failures is not the simple product of the individual probabilities as assumed by the Bonferroni adjustment. Davis and McNichols (1987) have proposed an alternative approach to this problem that overcomes these limitations, which is described in a following section. Caution must be taken in comparing false negative rates for tests with different false positive rates. For example, one might conclude from Figure 1.2 that the strategy without resampling has the lowest false negative rate and that this is more important than the fact that it fails by chance alone on one-third of the monitoring events. False negative rates are only meaningful for tests that achieve their intended false positive rates, and comparisons between tests are only appropriate for a fixed false positive rate. The purpose of Figure 1.2 (which clearly violates this advice) is that, for a fixed prediction limit (e.g., 99% confidence normal prediction limit for the next single measurement) applied to different monitoring scenarios (e.g., multiple wells, constituents, and resampling strategies), both false positive and false negative rates can vary dramatically. In those cases where the limit is fixed, perhaps by unwise regulation, then the verification resampling strategy and adequacy of the monitoring program as a whole must be based on achieving a balance between false positive and false negative rates. Sacrifice of one for the other is unacceptable.

1.5   SIMULTANEOUS NORMAL PREDICTION LIMITS FOR THE NEXT k SAMPLES

The previous prediction limits for multiple future comparisons are simultaneous in the sense that they control the overall sitewide comparison rate (e.g., $\alpha^* = .05$) assuming that the multiple comparisons are independent. While this is true for a series of intrawell comparisons on a given sampling event (i.e., each well is compared to its own background for a single constituent), it is not true in the context of upgradient versus downgradient comparisons where each new monitoring measurement is compared to the same pooled upgradient background limit. This type of comparison strategy introduces a correlation among the k comparisons for each constituent of magnitude $r = 1/(n + 1)$. An analogous situation occurs in the context of comparing multiple treatment group means to a single control (i.e., Dunnett's test, Dunnett and Sobel, 1955). The correlation among repeated comparisons makes use of the simple probability product for α^* too conservative. Obtaining the correct joint probability of failure on any one of the k comparisons requires evaluation of the equicorrelated multivariate normal distribution, and these probabilities must be integrated over the distribution of s, the sample standard deviation.

Suppose there is interest in comparing k groups with a common background in terms of the means $\overline{X}_0, \overline{X}_1, \ldots, \overline{X}_k$ (and common standard deviation s) of $k + 1$ sets of observations which are assumed to be independently and normally distributed; $\overline{X}_0$ referring to the background and $\overline{X}_i$ to the ith comparison group ($i = 1, \ldots, k$) mean. In this case Dunnett (1955) has provided a procedure for making confidence statements about expected values of k differences $\overline{X}_i - \overline{X}_0$, the procedure having the property that the probability of *all* k statements being simultaneously correct is equal to a specified probability level $1 - \alpha$. Dunnett's procedure and the associated tables were worked out for the case of equal sample sizes in all groups. Here, we will expand the procedure for the case when sample sizes are not equal.

Suppose there are n_0 background measurements, n_1 measurements on the first well, $\ldots, n_k$ measurements on the kth well, and denote these observations by X_{ij} ($i = 0, 1, \ldots, k$; $j = 1, 2, \ldots, n_i$) and the corresponding ith well mean as $\overline{X}_i$. We assume that the X_{ij} are independent and normally distributed with common variance σ^2 and means μ_i, and that there is an estimate of σ^2 available (denoted s^2) based on ν degrees of freedom. Let

$$
z_i = \frac{\overline{X}_i - \overline{X}_0 - (\mu_i - \mu_0)}{\sqrt{\dfrac{1}{n_i} + \dfrac{1}{n_0}}}
\tag{1.12}
$$

and $t_i = z_i/s$ for $i = 1, 2, \ldots, k$. As Dunnett (1955) notes, the lower confidence limits with joint confidence coefficient $1 - \alpha$ for the k comparison group effects $\mu_i - \mu_0$ are given by

$$
\overline{X}_i - \overline{X}_0 - d_i s \sqrt{\frac{1}{n_i} + \frac{1}{n_0}}
\tag{1.13}
$$

if the k constants d_i are chosen so that

$$
\Pr(t_1 < d_1, t_2 < d_2, \ldots, t_k < d_k) = 1 - \alpha
\tag{1.14}
$$

To find the k constants d_i that satisfy these equations, joint distribution of the t_i is required, which is the multivariate analog of Student's t-distribution defined by Dunnett and Sobel (1955). Dunnett (1955) has shown how the problem of tabulating the multivariate t-distribution can be reduced to the problem of tabulating the corresponding multivariate normal (MVN) distribution. For this, note that the joint distribution of the z_i is an MVN distribution with means 0 and variances σ^2. The correlation between z_i and z_j is given by

$$\rho_{ij} = 1 \Big/ \sqrt{\left(\frac{n_0}{n_i} + 1\right)\left(\frac{n_0}{n_j} + 1\right)} \tag{1.15}$$

Notice that the joint probability statement given previously can be written in the following way:

$$
\begin{aligned}
1 - \alpha &= \Pr(t_1 < d_1, t_2 < d_2, \ldots, t_k < d_k) \\
&= \Pr(z_1 < d_1 s, z_2 < d_2 s, \ldots, z_k < d_k s) \\
&= \int_{-\infty}^{+\infty} F(d_1 s, d_2 s, \ldots, d_k s) f(s) \, ds
\end{aligned}
\tag{1.16}
$$

where $F(d_1 s, d_2 s, \ldots, d_k s)$ is the MVN cdf of the z_i and $f(s)$ is the probability density function of s. Thus, with probability values for $F(\cdot)$, the preceding equation can be evaluated using numerical integration over the distribution of s. For this, note that the density function of s is given by Pearson and Hartley (1976) as

$$f(s) = \frac{\nu^{\nu/2}}{\Gamma(\nu/2)2^{(\nu/2)-1}} \sigma^{-\nu} s^{\nu-1} \exp\left(-\frac{\nu s^2}{2\sigma^2}\right) \tag{1.17}$$

Since $s^2/\sigma^2 = \chi^2/\nu$ we can rewrite the equation for $1 - \alpha$ in terms of integration over the distribution of $y = s/\sigma$ (which is defined on 0 to $+\infty$) as

$$
\begin{aligned}
1 - \alpha &= \int_0^{+\infty} F(d_1 y, d_2 y, \ldots, d_k y) f(y) \frac{ds}{dy} \, dy \\
&= \int_0^{+\infty} F(d_1 y, d_2 y, \ldots, d_k y) \frac{\nu^{\nu/2}}{\Gamma(\nu/2)2^{(\nu/2)-1}} y^{\nu-1} \exp\left(-\frac{\nu y^2}{2}\right) dy
\end{aligned}
\tag{1.18}
$$

Numerical integration over the distribution of y can then be performed to yield the associated probability $1 - \alpha$ for selected values of d, k, and ν.

In the present context we are interested in comparing k new individual measurements (e.g., TOC levels in 10 downgradient monitoring wells) with collection of n background measurements, perhaps obtained from monitoring wells upgradient of the facility. In this case $n_i = 1$, $i = 1, \ldots, k$, $n_0 = n$, and the constant correlation is $\rho_{ij} = 1/(n + 1)$. As a result, the need for this correction decreases with increasing background sample size n since as n increases ρ_{ij} goes to zero. In this special case the probability integral simplifies to

$$F_k(ds, ds, \ldots, ds; \rho) = \int_{-\infty}^{\infty} \left[F^k \left(\frac{ds + \rho^{1/2} y}{\sqrt{1 - \rho}} \right) \right] f(y) \, dy \qquad (1.19)$$

where $f(\tau) = \exp(-\frac{1}{2}\tau^2)/(2\pi)^{1/2}$ and $F(\tau) = \int_{-\infty}^{ds} f(\tau) \, d\tau$; see Gupta (1963).

To aid in application, Table 1.4 provides the constants d for background sample sizes n from 4 to 50 and number of future comparisons k from 2 to

TABLE 1.4   Dunnett-Type Multivariate t-Statistics 95% Overall Confidence for Background $n = 4$ to 100 and $k = 4$ to 50 Future Measurements)

	k = Number of Future Comparisons									
n	5	10	15	20	25	30	35	40	45	50
4	4.00	4.72	5.14	5.42	5.64	5.82	5.96	6.09	6.20	6.29
8	2.90	3.31	3.56	3.72	3.85	3.95	4.04	4.12	4.18	4.24
12	2.67	3.02	3.22	3.36	3.47	3.56	3.63	3.69	3.75	3.80
16	2.57	2.89	3.08	3.21	3.30	3.38	3.45	3.51	3.56	3.60
20	2.51	2.82	3.00	3.12	3.21	3.29	3.35	3.40	3.45	3.49
24	2.48	2.78	2.94	3.06	3.15	3.22	3.28	3.33	3.38	3.42
28	2.45	2.74	2.91	3.02	3.11	3.18	3.24	3.29	3.33	3.37
32	2.44	2.72	2.88	2.99	3.08	3.14	3.20	3.25	3.29	3.33
36	2.42	2.70	2.86	2.97	3.05	3.12	3.18	3.22	3.27	3.30
40	2.41	2.69	2.84	2.95	3.03	3.10	3.15	3.20	3.24	3.28
44	2.40	2.68	2.83	2.94	3.02	3.08	3.14	3.18	3.23	3.26
48	2.39	2.67	2.82	2.93	3.01	3.07	3.12	3.17	3.21	3.25
52	2.39	2.66	2.81	2.92	2.99	3.06	3.11	3.16	3.20	3.23
56	2.38	2.65	2.80	2.91	2.99	3.05	3.10	3.15	3.19	3.22
60	2.38	2.65	2.80	2.90	2.98	3.04	3.09	3.14	3.18	3.21
64	2.38	2.64	2.79	2.89	2.97	3.03	3.09	3.13	3.17	3.20
68	2.37	2.64	2.79	2.89	2.96	3.03	3.08	3.12	3.16	3.20
72	2.37	2.63	2.78	2.88	2.96	3.02	3.07	3.12	3.16	3.19
76	2.37	2.63	2.78	2.88	2.95	3.02	3.07	3.11	3.15	3.19
80	2.36	2.63	2.77	2.87	2.95	3.01	3.06	3.11	3.15	3.18
84	2.36	2.62	2.77	2.87	2.95	3.01	3.06	3.10	3.14	3.18
88	2.36	2.62	2.77	2.87	2.94	3.00	3.05	3.10	3.14	3.17
92	2.36	2.62	2.76	2.86	2.94	3.00	3.05	3.09	3.13	3.17
96	2.36	2.62	2.76	2.86	2.94	3.00	3.05	3.09	3.13	3.16
100	2.35	2.61	2.76	2.86	2.93	2.99	3.05	3.09	3.13	3.16

50 for $1 - \alpha = .95$. These coefficients may be used in deriving prediction limits of the form

$$\bar{x} + d_{(n,k)}s\sqrt{1 + \frac{1}{n}} \tag{1.20}$$

Example 1.4

Returning to the TOC example dataset in Table 1.1 and Example 1.2, we may wish to apply the prediction limit for TOC to new TOC measurements from each of 10 downgradient monitoring wells, limiting the overall probability of *any* false positive result to 5%. Using the Bonferroni adjustment which assumes independence, the value in Table 1.2 that corresponds to $n = 8$ and

TABLE 1.5   Simultaneous Normal Prediction Limit Factors for $\alpha = .05$ and One of Two Samples in Bounds (Factors K Where Prediction Limit Is $\bar{x} + Ks$)

n	k = Number of Future Comparisons									
	5	10	15	20	25	30	35	40	45	50
4	2.47	3.00	3.31	3.52	3.69	3.82	3.94	4.03	4.12	4.19
8	1.72	2.03	2.21	2.33	2.43	2.51	2.57	2.63	2.67	2.72
12	1.56	1.82	1.97	2.07	2.15	2.21	2.27	2.31	2.36	2.39
16	1.48	1.72	1.86	1.95	2.03	2.08	2.13	2.18	2.21	2.25
20	1.44	1.67	1.80	1.89	1.95	2.01	2.06	2.10	2.13	2.16
24	1.41	1.63	1.76	1.84	1.91	1.96	2.01	2.05	2.08	2.11
28	1.39	1.61	1.73	1.81	1.88	1.93	1.97	2.01	2.04	2.07
32	1.38	1.59	1.71	1.79	1.85	1.90	1.95	1.98	2.02	2.04
36	1.37	1.58	1.69	1.77	1.84	1.89	1.93	1.96	1.99	2.02
40	1.36	1.56	1.68	1.76	1.82	1.87	1.91	1.95	1.98	2.00
44	1.35	1.56	1.67	1.75	1.81	1.86	1.90	1.93	1.96	1.99
48	1.34	1.55	1.66	1.74	1.80	1.85	1.89	1.92	1.95	1.98
52	1.34	1.54	1.66	1.73	1.79	1.84	1.88	1.91	1.94	1.97
56	1.33	1.54	1.65	1.73	1.79	1.83	1.87	1.91	1.94	1.96
60	1.33	1.53	1.64	1.72	1.78	1.83	1.87	1.90	1.93	1.95
64	1.33	1.53	1.64	1.72	1.77	1.82	1.86	1.89	1.92	1.95
68	1.32	1.52	1.64	1.71	1.77	1.82	1.85	1.89	1.92	1.94
72	1.32	1.52	1.63	1.71	1.77	1.81	1.85	1.88	1.91	1.94
76	1.32	1.52	1.63	1.70	1.76	1.81	1.85	1.88	1.91	1.93
80	1.32	1.52	1.63	1.70	1.76	1.80	1.84	1.88	1.90	1.93
84	1.32	1.51	1.62	1.70	1.76	1.80	1.84	1.87	1.90	1.93
88	1.31	1.51	1.62	1.70	1.75	1.80	1.84	1.87	1.90	1.92
92	1.31	1.51	1.62	1.69	1.75	1.80	1.83	1.87	1.89	1.92
96	1.31	1.51	1.62	1.69	1.75	1.79	1.83	1.86	1.89	1.92
100	1.31	1.51	1.62	1.69	1.75	1.79	1.83	1.86	1.89	1.91

Prepared by Charles Davis based on results in Davis and McNichols (1987).

$k = 10$ is 3.5, leading to the upper prediction limit

$$11.0 + 3.50(0.61)\sqrt{1 + \frac{1}{8}} = 13.26 \text{ mg/l}$$

In contrast, using the Dunnett-type multivariate t-statistic from Table 1.4 leads to the upper prediction limit

$$11.0 + 3.31(0.61)\sqrt{1 + \tfrac{1}{8}} = 13.14 \text{ mg/l}$$

which is slightly more conservative. Comparison of Tables 1.2 and 1.4 reveals that the multivariate t-statistic and Bonferroni-adjusted t-statistic become

TABLE 1.6   Simultaneous Normal Prediction Limit Factors for $\alpha = .05$ and One of Three Samples in Bounds (Factors K Where Prediction Limit Is $\bar{x} + Ks$)

n	\multicolumn{10}{c}{k = Number of Future Comparisons}
	5	10	15	20	25	30	35	40	45	50
4	1.62	2.02	2.25	2.42	2.55	2.65	2.74	2.82	2.88	2.94
8	1.12	1.37	1.51	1.61	1.69	1.75	1.80	1.84	1.88	1.92
12	1.00	1.21	1.34	1.42	1.49	1.54	1.58	1.62	1.65	1.68
16	0.94	1.14	1.26	1.33	1.39	1.44	1.48	1.52	1.55	1.58
20	0.91	1.10	1.21	1.28	1.34	1.39	1.43	1.46	1.49	1.51
24	0.89	1.08	1.18	1.25	1.31	1.35	1.39	1.42	1.45	1.47
28	0.87	1.06	1.16	1.23	1.28	1.33	1.36	1.39	1.42	1.45
32	0.86	1.04	1.14	1.21	1.27	1.31	1.34	1.37	1.40	1.42
36	0.85	1.03	1.13	1.20	1.25	1.29	1.33	1.36	1.38	1.41
40	0.85	1.02	1.12	1.19	1.24	1.28	1.32	1.35	1.37	1.39
44	0.84	1.02	1.11	1.18	1.23	1.27	1.31	1.34	1.36	1.38
48	0.84	1.01	1.11	1.17	1.22	1.27	1.30	1.33	1.35	1.38
52	0.83	1.01	1.10	1.17	1.22	1.26	1.29	1.32	1.35	1.37
56	0.83	1.00	1.10	1.16	1.21	1.25	1.29	1.32	1.34	1.36
60	0.83	1.00	1.09	1.16	1.21	1.25	1.28	1.31	1.33	1.36
64	0.83	1.00	1.09	1.16	1.20	1.24	1.28	1.31	1.33	1.35
68	0.82	0.99	1.09	1.15	1.20	1.24	1.27	1.30	1.33	1.35
72	0.82	0.99	1.08	1.15	1.20	1.24	1.27	1.30	1.32	1.34
76	0.82	0.99	1.08	1.15	1.20	1.23	1.27	1.29	1.32	1.34
80	0.82	0.99	1.08	1.14	1.19	1.23	1.26	1.29	1.32	1.34
84	0.82	0.99	1.08	1.14	1.19	1.23	1.26	1.29	1.31	1.34
88	0.81	0.98	1.08	1.14	1.19	1.23	1.26	1.29	1.31	1.33
92	0.81	0.98	1.07	1.14	1.19	1.23	1.26	1.28	1.31	1.33
96	0.81	0.98	1.07	1.14	1.18	1.22	1.26	1.28	1.31	1.33
100	0.81	0.98	1.07	1.14	1.18	1.22	1.25	1.28	1.31	1.33

Prepared by Charles Davis based on results in Davis and McNichols (1987).

relatively indistinguishable for background sample sizes of $n = 20$ or more, but the difference is more pronounced for small background sample size.

1.6 SIMULTANEOUS NORMAL PREDICTION LIMITS FOR THE NEXT *r* OF *m* MEASUREMENTS AT EACH OF *k* MONITORING WELLS

Davis and McNichols (1987) have generalized the solution given in the previous section to the case in which r out of m samples in each of k future monitoring wells are required in bounds. For example, a detection monitoring program that requires passage of both of two resamples in the event of an initial exceedance is similar to a prediction limit for two of three samples in bounds in each of k wells. Strictly speaking, however, we are concerned with

TABLE 1.7   Simultaneous Normal Prediction Limit Factors for $\alpha = .05$ and First or Next Two Samples in Bounds (Factors K Where Prediction Limit Is $\bar{x} + Ks$)

	$k =$ Number of Future Comparisons									
n	5	10	15	20	25	30	35	40	45	50
4	2.92	3.47	3.78	3.99	4.16	4.29	4.40	4.50	4.58	4.65
8	2.00	2.31	2.48	2.61	2.70	2.78	2.84	2.89	2.94	2.98
12	1.79	2.05	2.20	2.30	2.38	2.44	2.50	2.54	2.58	2.62
16	1.70	1.94	2.07	2.16	2.24	2.29	2.34	2.38	2.42	2.45
20	1.65	1.87	2.00	2.09	2.15	2.21	2.25	2.29	2.33	2.36
24	1.62	1.83	1.95	2.04	2.10	2.15	2.20	2.23	2.27	2.30
28	1.59	1.80	1.92	2.00	2.06	2.11	2.16	2.19	2.22	2.25
32	1.58	1.78	1.90	1.98	2.04	2.09	2.13	2.16	2.19	2.22
36	1.56	1.76	1.88	1.96	2.02	2.06	2.10	2.14	2.17	2.20
40	1.55	1.75	1.86	1.94	2.00	2.05	2.09	2.12	2.15	2.18
44	1.54	1.74	1.85	1.93	1.99	2.03	2.07	2.10	2.13	2.16
48	1.54	1.73	1.84	1.92	1.97	2.02	2.06	2.09	2.12	2.15
52	1.53	1.72	1.83	1.91	1.96	2.01	2.05	2.08	2.11	2.14
56	1.53	1.72	1.83	1.90	1.96	2.00	2.04	2.07	2.10	2.13
60	1.52	1.71	1.82	1.89	1.95	1.99	2.03	2.06	2.09	2.12
64	1.52	1.71	1.81	1.89	1.94	1.99	2.03	2.06	2.09	2.11
68	1.51	1.70	1.81	1.88	1.94	1.98	2.02	2.05	2.08	2.10
72	1.51	1.70	1.81	1.88	1.93	1.98	2.01	2.05	2.07	2.10
76	1.51	1.70	1.80	1.87	1.93	1.97	2.01	2.04	2.07	2.09
80	1.51	1.69	1.80	1.87	1.92	1.97	2.01	2.04	2.06	2.09
84	1.50	1.69	1.80	1.87	1.92	1.97	2.00	2.03	2.06	2.08
88	1.50	1.69	1.79	1.86	1.92	1.96	2.00	2.03	2.06	2.08
92	1.50	1.69	1.79	1.86	1.92	1.96	2.00	2.03	2.05	2.08
96	1.50	1.68	1.79	1.86	1.91	1.96	1.99	2.02	2.05	2.07
100	1.50	1.68	1.79	1.86	1.91	1.95	1.99	2.02	2.05	2.07

Prepared by Charles Davis based on results in Davis and McNichols (1987).

the case in which the first or next two samples are in bounds, which requires a slight modification of their original work. In fact, the multivariate t-statistics in the previous section (i.e., Dunnett's test) represent a special case in which $r = m = 1$.

The derivation is somewhat complicated, but a few key features are described. As in the previous derivations, we assume that the background observations and new monitoring measurements are drawn from the same normal distribution $N(\mu, \sigma^2)$. Expressing $y_{ij} = x_{ij} - \bar{x}$ (i.e., a mean deviation) for $i = 1, \ldots, k$ wells and $j = 1, \ldots, m$ samples and letting $y_{i(r)}$ denote the rth smallest of the y_{ij} for well i, and $y^* = \max_i(y_{i(r)})$, then having at least r of m future observations below $\bar{x} + Ks$ is equivalent to $y^* < Ks$, where K is the multiplier we seek. Davis and McNichols (1987) have shown

TABLE 1.8   Simultaneous Normal Prediction Limit Factors for $\alpha = .005$ and One of Two Samples in Bounds (Factors K Where Prediction Limit Is $\bar{x} + Ks$)

				$k =$ Number of Future Comparisons						
n	5	10	15	20	25	30	35	40	45	50
4	5.86	6.96	7.60	8.06	8.42	8.71	8.95	9.15	9.34	9.50
8	2.98	3.36	3.59	3.75	3.87	3.97	4.06	4.14	4.20	4.26
12	2.51	2.78	2.94	3.06	3.15	3.22	3.28	3.34	3.38	3.43
16	2.31	2.55	2.69	2.78	2.86	2.92	2.97	3.02	3.06	3.09
20	2.21	2.42	2.55	2.64	2.70	2.76	2.81	2.85	2.88	2.91
24	2.14	2.35	2.46	2.55	2.61	2.66	2.70	2.74	2.77	2.80
28	2.10	2.29	2.40	2.48	2.54	2.59	2.63	2.67	2.70	2.73
32	2.07	2.25	2.36	2.44	2.50	2.54	2.58	2.62	2.65	2.67
36	2.04	2.23	2.33	2.40	2.46	2.50	2.54	2.58	2.60	2.63
40	2.02	2.20	2.30	2.38	2.43	2.48	2.51	2.54	2.57	2.60
44	2.01	2.18	2.28	2.35	2.41	2.45	2.49	2.52	2.55	2.57
48	1.99	2.17	2.27	2.34	2.39	2.43	2.47	2.50	2.53	2.55
52	1.98	2.16	2.25	2.32	2.37	2.42	2.45	2.48	2.51	2.53
56	1.98	2.15	2.24	2.31	2.36	2.40	2.44	2.47	2.49	2.52
60	1.97	2.14	2.23	2.30	2.35	2.39	2.42	2.45	2.48	2.50
64	1.96	2.13	2.22	2.29	2.34	2.38	2.41	2.44	2.47	2.49
68	1.95	2.12	2.21	2.28	2.33	2.37	2.40	2.43	2.46	2.48
72	1.95	2.11	2.21	2.27	2.32	2.36	2.40	2.43	2.45	2.47
76	1.94	2.11	2.20	2.27	2.32	2.36	2.39	2.42	2.44	2.47
80	1.94	2.10	2.20	2.26	2.31	2.35	2.38	2.41	2.44	2.46
84	1.94	2.10	2.19	2.26	2.30	2.34	2.38	2.41	2.43	2.45
88	1.93	2.09	2.19	2.25	2.30	2.34	2.37	2.40	2.42	2.45
92	1.93	2.09	2.18	2.25	2.29	2.33	2.37	2.39	2.42	2.44
96	1.93	2.09	2.18	2.24	2.29	2.33	2.36	2.39	2.41	2.44
100	1.92	2.08	2.18	2.24	2.29	2.33	2.36	2.39	2.41	2.43

Prepared by Charles Davis based on results in Davis and McNichols (1987).

that

$$\Pr(y^* < Ks) = \int_{-\infty}^{\infty} T_{n-1, \sqrt{n}z^*}(\sqrt{n}\,K)$$

$$\times k \left[\int_{-\infty}^{z^*} m \binom{m-1}{r-1} \Phi^{r-1}(t)\phi(t)\left[1 - \Phi(t)^{m-r}\right] dt \right]^{k-1}$$

$$\times m \binom{m-1}{r-1} \Phi^{r-1}(z^*)\phi(z^*)\left[1 - \Phi(z^*)\right]^{m-r} dz^* \quad (1.21)$$

where $T_{\nu,\delta}(\cdot)$ is the cumulative density function of the noncentral t distribution, z^* is the maximum rth order statistic across all k wells, and ϕ and Φ

TABLE 1.9   Simultaneous Normal Prediction Limit Factors for $\alpha = .005$ and One of Three Samples in Bounds (Factors K Where Prediction Limit Is $\bar{x} + Ks$)

	k = Number of Future Comparisons									
n	5	10	15	20	25	30	35	40	45	50
4	4.02	4.83	5.31	5.66	5.93	6.15	6.34	6.50	6.64	6.76
8	2.10	2.39	2.57	2.70	2.79	2.87	2.94	3.00	3.05	3.10
12	1.76	1.98	2.11	2.20	2.28	2.33	2.38	2.43	2.46	2.50
16	1.62	1.82	1.93	2.00	2.07	2.11	2.16	2.19	2.22	2.25
20	1.55	1.72	1.83	1.90	1.95	2.00	2.03	2.07	2.10	2.12
24	1.50	1.67	1.76	1.83	1.88	1.92	1.96	1.99	2.02	2.04
28	1.47	1.63	1.72	1.78	1.83	1.87	1.91	1.93	1.96	1.98
32	1.44	1.60	1.69	1.75	1.80	1.84	1.87	1.90	1.92	1.94
36	1.42	1.58	1.66	1.72	1.77	1.81	1.84	1.87	1.89	1.91
40	1.41	1.56	1.64	1.70	1.75	1.79	1.82	1.84	1.87	1.89
44	1.40	1.54	1.63	1.69	1.73	1.77	1.80	1.82	1.85	1.87
48	1.39	1.53	1.62	1.67	1.72	1.75	1.78	1.81	1.83	1.85
52	1.38	1.52	1.61	1.66	1.71	1.74	1.77	1.80	1.82	1.84
56	1.37	1.52	1.60	1.65	1.70	1.73	1.76	1.78	1.81	1.83
60	1.37	1.51	1.59	1.64	1.69	1.72	1.75	1.77	1.80	1.82
64	1.36	1.50	1.58	1.64	1.68	1.71	1.74	1.77	1.79	1.81
68	1.36	1.50	1.58	1.63	1.67	1.71	1.74	1.76	1.78	1.80
72	1.35	1.49	1.57	1.63	1.67	1.70	1.73	1.75	1.77	1.79
76	1.35	1.49	1.57	1.62	1.66	1.70	1.72	1.75	1.77	1.79
80	1.35	1.48	1.56	1.62	1.66	1.69	1.72	1.74	1.76	1.78
84	1.34	1.48	1.56	1.61	1.65	1.69	1.71	1.74	1.76	1.78
88	1.34	1.48	1.55	1.61	1.65	1.68	1.71	1.73	1.75	1.77
92	1.34	1.47	1.55	1.60	1.65	1.68	1.71	1.73	1.75	1.77
96	1.34	1.47	1.55	1.60	1.64	1.68	1.70	1.73	1.75	1.77
100	1.33	1.47	1.55	1.60	1.64	1.67	1.70	1.72	1.74	1.76

Prepared by Charles Davis based on results in Davis and McNichols (1987).

TABLE 1.10   Simultaneous Normal Prediction Limit Factors for $\alpha = .005$ and First or Next Two Samples in Bounds (Factors K Where Prediction Limit Is $\bar{x} + Ks$)

	k = Number of Future Comparisons									
n	5	10	15	20	25	30	35	40	45	50
4	6.81	7.95	8.61	9.07	9.42	9.71	9.94	10.15	10.33	10.49
8	3.33	3.72	3.95	4.11	4.24	4.34	4.43	4.50	4.57	4.62
12	2.77	3.05	3.21	3.33	3.42	3.49	3.55	3.60	3.65	3.69
16	2.54	2.78	2.91	3.01	3.09	3.15	3.20	3.24	3.28	3.32
20	2.42	2.63	2.75	2.84	2.91	2.96	3.01	3.05	3.08	3.12
24	2.34	2.54	2.65	2.74	2.80	2.85	2.89	2.93	2.96	2.99
28	2.29	2.48	2.59	2.66	2.72	2.77	2.81	2.85	2.88	2.91
32	2.25	2.43	2.54	2.61	2.67	2.71	2.75	2.79	2.82	2.84
36	2.22	2.40	2.50	2.57	2.63	2.67	2.71	2.74	2.77	2.80
40	2.20	2.37	2.47	2.54	2.60	2.64	2.68	2.71	2.73	2.76
44	2.18	2.35	2.45	2.52	2.57	2.61	2.65	2.68	2.71	2.73
48	2.16	2.33	2.43	2.50	2.55	2.59	2.62	2.66	2.68	2.71
52	2.15	2.32	2.41	2.48	2.53	2.57	2.61	2.64	2.66	2.68
56	2.14	2.31	2.40	2.46	2.51	2.56	2.59	2.62	2.64	2.67
60	2.13	2.29	2.39	2.45	2.50	2.54	2.58	2.60	2.63	2.65
64	2.12	2.29	2.38	2.44	2.49	2.53	2.56	2.59	2.62	2.64
68	2.12	2.28	2.37	2.43	2.48	2.52	2.55	2.58	2.61	2.63
72	2.11	2.27	2.36	2.42	2.47	2.51	2.54	2.57	2.60	2.62
76	2.11	2.26	2.35	2.42	2.46	2.50	2.54	2.56	2.59	2.61
80	2.10	2.26	2.35	2.41	2.46	2.50	2.53	2.56	2.58	2.60
84	2.10	2.25	2.34	2.40	2.45	2.49	2.52	2.55	2.57	2.59
88	2.09	2.25	2.34	2.40	2.44	2.48	2.52	2.54	2.57	2.59
92	2.09	2.24	2.33	2.39	2.44	2.48	2.51	2.54	2.56	2.58
96	2.08	2.24	2.33	2.39	2.44	2.47	2.50	2.53	2.56	2.58
100	2.08	2.24	2.32	2.38	2.43	2.47	2.50	2.53	2.55	2.57

Prepared by Charles Davis based on results in Davis and McNichols (1987).

are the standard normal probability density and cumulative distribution functions, respectively. The equation is then solved for K such that the right-hand side is equal to $1 - \alpha$. Davis and McNichols (1987) describe a numerical algorithm for obtaining values of K conditional on values of k, r, and m, such that the overall confidence level is $1 - \alpha$. The original publication, however, was limited in terms of combinations of k, r, m, and α, so routine application of this methodology was generally not possible. Tables 1.5 to 1.13 (prepared by C. Davis) overcome this limitation in three ways. First, the tables include three levels of α which correspond to Bonferroni-adjusted sitewide confidence levels for monitoring programs consisting of 1 ($\alpha = .05$), 10 ($\alpha = .005$), and 20 ($\alpha = .0025$) constituents. Second, the tables provide values of K for background sample sizes of $n = 4$ to 100 and $k = 5$ to 50

TABLE 1.11   Simultaneous Normal Prediction Limit Factors for α = .0025 and One of Two Samples in Bounds (Factors *K* Where Prediction Limit Is $\bar{x} + Ks$)

				k = Number of Future Comparisons						
n	5	10	15	20	25	30	35	40	45	50
4	7.45	8.82	9.64	10.22	10.66	11.03	11.33	11.59	11.82	12.02
8	3.40	3.82	4.06	4.24	4.37	4.49	4.58	4.66	4.73	4.80
12	2.80	3.08	3.25	3.37	3.46	3.54	3.60	3.66	3.71	3.75
16	2.56	2.79	2.93	3.03	3.11	3.17	3.23	3.27	3.31	3.35
20	2.43	2.64	2.77	2.86	2.92	2.98	3.03	3.07	3.10	3.14
24	2.35	2.55	2.66	2.75	2.81	2.86	2.90	2.94	2.97	3.00
28	2.30	2.49	2.60	2.67	2.73	2.78	2.82	2.86	2.89	2.91
32	2.26	2.44	2.54	2.62	2.68	2.72	2.76	2.79	2.82	2.85
36	2.23	2.40	2.51	2.58	2.63	2.68	2.72	2.75	2.78	2.80
40	2.20	2.38	2.48	2.55	2.60	2.64	2.68	2.71	2.74	2.76
44	2.19	2.36	2.45	2.52	2.57	2.62	2.65	2.68	2.71	2.73
48	2.17	2.34	2.43	2.50	2.55	2.59	2.63	2.66	2.69	2.71
52	2.16	2.32	2.42	2.48	2.53	2.57	2.61	2.64	2.67	2.69
56	2.15	2.31	2.40	2.47	2.52	2.56	2.59	2.62	2.65	2.67
60	2.14	2.30	2.39	2.46	2.50	2.55	2.58	2.61	2.63	2.66
64	2.13	2.29	2.38	2.44	2.49	2.53	2.57	2.60	2.62	2.64
68	2.12	2.28	2.37	2.43	2.48	2.52	2.56	2.58	2.61	2.63
72	2.11	2.27	2.36	2.43	2.47	2.51	2.55	2.57	2.60	2.62
76	2.11	2.27	2.36	2.42	2.47	2.51	2.54	2.57	2.59	2.61
80	2.10	2.26	2.35	2.41	2.46	2.50	2.53	2.56	2.58	2.60
84	2.10	2.26	2.34	2.41	2.45	2.49	2.52	2.55	2.58	2.60
88	2.10	2.25	2.34	2.40	2.45	2.49	2.52	2.54	2.57	2.59
92	2.09	2.25	2.33	2.40	2.44	2.48	2.51	2.54	2.56	2.58
96	2.09	2.24	2.33	2.39	2.44	2.47	2.51	2.53	2.56	2.58
100	2.09	2.24	2.33	2.39	2.43	2.47	2.50	2.53	2.55	2.57

Prepared by Charles Davis based on results in Davis and McNichols (1987).

monitoring wells. Third, separate tables are prepared for one resample and two resamples under Plans A and B. In this way these important results can be practically applied to the widest variety of monitoring programs. Tables 1.5 to 1.7 provide factors of *K* for the three verification resampling strategies for α = .05, Tables 1.8 to 1.10 for α = .005, and Tables 1.11 to 1.13 for α = .0025.

To illustrate the strength of this approach, consider a monitoring program with a single monitoring constituent, *k* = 50 monitoring wells, and *n* = 8 background measurements. For a single verification resample, the limit is given by $\bar{x} + 2.72s$ (see Table 1.5). For two verification resamples in which an exceedance is recorded if either is exceeded (i.e., Plan A), the limit is given by $\bar{x} + 2.98s$ (see Table 1.6). For two verification resamples in which

TABLE 1.12   Simultaneous Normal Prediction Limit Factors for α = .0025 and One of Three Samples in Bounds (Factors K Where Prediction Limit Is $\bar{x} + Ks$)

				k =	Number	of Future	Comparisons			
n	5	10	15	20	25	30	35	40	45	50
4	5.14	6.14	6.75	7.19	7.53	7.80	8.04	8.24	8.42	8.58
8	2.42	2.74	2.93	3.07	3.17	3.26	3.33	3.40	3.45	3.50
12	1.99	2.22	2.35	2.45	2.52	2.58	2.63	2.68	2.72	2.75
16	1.82	2.01	2.12	2.20	2.26	2.31	2.36	2.39	2.43	2.46
20	1.73	1.90	2.00	2.07	2.13	2.17	2.21	2.24	2.27	2.30
24	1.67	1.83	1.93	1.99	2.04	2.09	2.12	2.15	2.18	2.20
28	1.63	1.78	1.87	1.94	1.99	2.03	2.06	2.09	2.11	2.14
32	1.60	1.75	1.84	1.90	1.94	1.98	2.01	2.04	2.07	2.09
36	1.58	1.72	1.81	1.87	1.91	1.95	1.98	2.01	2.03	2.05
40	1.56	1.70	1.79	1.84	1.89	1.92	1.95	1.98	2.00	2.02
44	1.55	1.69	1.77	1.83	1.87	1.90	1.93	1.96	1.98	2.00
48	1.53	1.67	1.75	1.81	1.85	1.89	1.92	1.94	1.96	1.98
52	1.52	1.66	1.74	1.80	1.84	1.87	1.90	1.93	1.95	1.97
56	1.52	1.65	1.73	1.79	1.83	1.86	1.89	1.91	1.93	1.95
60	1.51	1.64	1.72	1.78	1.82	1.85	1.88	1.90	1.92	1.94
64	1.50	1.64	1.71	1.77	1.81	1.84	1.87	1.89	1.91	1.93
68	1.50	1.63	1.71	1.76	1.80	1.83	1.86	1.88	1.91	1.92
72	1.49	1.63	1.70	1.75	1.79	1.83	1.85	1.88	1.90	1.92
76	1.49	1.62	1.70	1.75	1.79	1.82	1.85	1.87	1.89	1.91
80	1.48	1.62	1.69	1.74	1.78	1.81	1.84	1.87	1.89	1.90
84	1.48	1.61	1.69	1.74	1.78	1.81	1.84	1.86	1.88	1.90
88	1.48	1.61	1.68	1.73	1.77	1.81	1.83	1.86	1.88	1.89
92	1.48	1.61	1.68	1.73	1.77	1.80	1.83	1.85	1.87	1.89
96	1.47	1.60	1.68	1.73	1.77	1.80	1.82	1.85	1.87	1.88
100	1.47	1.60	1.67	1.72	1.76	1.79	1.82	1.84	1.86	1.88

Prepared by Charles Davis based on results in Davis and McNichols (1987).

an exceedance is recorded only if both are exceeded (i.e., Plan B), the limit is given by $\bar{x} + 1.92s$ (see Table 1.7). Note that by requiring both resamples be exceeded, we can use a smaller value of K and hence have greater power and lower false negative rate. This result is evident in the power curves displayed in Figure 1.3. This result is contrary to guidance and previous regulations which suggest that passing multiple resamples in the presence of an initial exceedance leads to a more conservative monitoring program. When the correct statistical model is used, in fact, the reverse is true.

Finally, also note that despite the difference in multipliers for the three verification resampling plans, the sitewide false positive rate continues to be 5% as intended. For these reasons simultaneous prediction limits of the form derived by Davis and McNichols (1987) are the optimal choice for normally distributed measurements, or those that can be suitably transformed.

TABLE 1.13   Simultaneous Normal Prediction Limit Factors for $\alpha = .0025$ and First or Next Two Samples in Bounds (Factors K Where Prediction Limit is $\bar{x} + Ks$)

					k = Number of Future Comparisons					
n	5	10	15	20	25	30	35	40	45	50
4	8.65	10.08	10.91	11.48	11.93	12.29	12.59	12.84	13.07	13.27
8	3.78	4.21	4.46	4.64	4.78	4.89	4.98	5.06	5.13	5.20
12	3.07	3.36	3.53	3.65	3.74	3.82	3.88	3.94	3.99	4.03
16	2.78	3.03	3.17	3.26	3.34	3.40	3.46	3.50	3.54	3.58
20	2.64	2.85	2.97	3.06	3.13	3.19	3.23	3.27	3.31	3.34
24	2.54	2.74	2.86	2.94	3.00	3.05	3.09	3.13	3.16	3.19
28	2.48	2.67	2.78	2.85	2.91	2.96	3.00	3.03	3.07	3.09
32	2.44	2.62	2.72	2.79	2.85	2.89	2.93	2.97	2.99	3.02
36	2.40	2.58	2.68	2.75	2.80	2.84	2.88	2.91	2.94	2.97
40	2.37	2.54	2.64	2.71	2.76	2.80	2.84	2.87	2.90	2.92
44	2.35	2.52	2.61	2.68	2.73	2.77	2.81	2.84	2.87	2.89
48	2.33	2.50	2.59	2.66	2.71	2.75	2.78	2.81	2.84	2.86
52	2.32	2.48	2.57	2.64	2.69	2.73	2.76	2.79	2.82	2.84
56	2.31	2.47	2.56	2.62	2.67	2.71	2.74	2.77	2.80	2.82
60	2.30	2.45	2.54	2.61	2.65	2.69	2.73	2.75	2.78	2.80
64	2.29	2.44	2.53	2.59	2.64	2.68	2.71	2.74	2.77	2.79
68	2.28	2.43	2.52	2.58	2.63	2.67	2.70	2.73	2.75	2.77
72	2.27	2.42	2.51	2.57	2.62	2.66	2.68	2.72	2.73	2.76
76	2.26	2.42	2.50	2.56	2.61	2.65	2.68	2.71	2.73	2.75
80	2.26	2.41	2.50	2.56	2.60	2.64	2.67	2.70	2.72	2.74
84	2.25	2.40	2.49	2.55	2.60	2.63	2.66	2.69	2.71	2.74
88	2.25	2.40	2.48	2.54	2.59	2.63	2.66	2.68	2.71	2.73
92	2.24	2.39	2.48	2.54	2.58	2.62	2.65	2.68	2.70	2.72
96	2.24	2.39	2.47	2.53	2.58	2.61	2.64	2.67	2.69	2.71
100	2.24	2.38	2.47	2.53	2.57	2.61	2.64	2.67	2.69	2.71

Prepared by Charles Davis based on results in Davis and McNichols (1987).

1.7   SUMMARY

A variety of statistical prediction limits and intervals have been presented for normally distributed measurements in order of increasing statistical sophistication required for increasingly complex applications. The final prediction limits presented, based on the work of Davis and McNichols (1987), are the most appealing since they consider both multiple comparisons and verification resampling in the most statistically rigorous way. The reader should note, however, that in the case of $k > 1$ we are invariably describing an upgradient versus downgradient monitoring plan. The assumption here is that either no spatial variability exists or that spatial variability in *all* downgradient wells is adequately described and can be statistically modeled by the spatial variability in the small number of upgradient wells available at

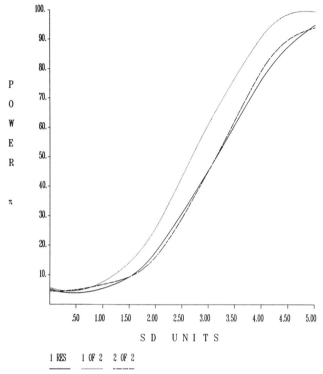

Figure 1.3 Power of 95% simultaneous prediction limits for 10 wells and 5 constituents and 3 resampling plans.

the facility (a topic to be discussed in a following chapter). Interestingly, there are no current regulatory requirements on the number of upgradient wells, and, to minimize expenses, owner/operators typically keep the number of upgradient wells to fewer than three or four. Often, there is only a single upgradient well, and potential contamination is completely confounded with spatial variability yielding the upgradient versus downgradient comparison strategy meaningless. Even with three or four upgradient wells, it is extremely unlikely that spatial variability across the site as a whole will be adequately characterized. The methods described here, and indeed any other approach, will not work in the presence of such spatial variability. If predisposal data exist, then estimates of the sitewide spatial variability are available and can be incorporated into prediction limits using components of variance models to be described later. Unfortunately, predisposal data are rarely available at most sites despite the fact that their benefits far outweigh their cost. In these cases intrawell comparisons may be the only viable alternative. Intrawell comparisons are always more powerful if they are justified.

2 Nonparametric Prediction Intervals

2.1 OVERVIEW

The distribution of a particular constituent in background groundwater quality samples is not always normal nor can the data always be suitably transformed to have approximately a normal distribution (e.g., taking natural logarithms of a lognormally distributed constituent to produce normally distributed measurements). In these cases the methods described in Chapter 1 are inappropriate and two general alternatives are available. First, we can attempt to model these data using an alternate distribution such as the Poisson (see Chapter 3). Second, we can assume that the distribution in background and monitoring wells is continuous and the same in both background and monitoring wells in the absence of contamination, but unknown, and proceed nonparametrically. Certainly, the most convenient approach is the nonparametric alternative; however, as will be shown, the number of background samples required for large monitoring programs may be unacceptably large. As with normally distributed measurements, we must be sensitive to the effects of multiple comparisons as well as verification resampling strategies in deriving these methods.

The nonparametric approach provides a natural way of dealing with nondetects with minimal distributional assumptions, but the same strong independence assumptions that are common to both the parametric and nonparametric methods still apply. Note that both temporal and spatial correlations violate the independence assumption. In addition, both the parametric and nonparametric methods assume that the uncertainty associated with each measurement is identical (i.e., homoscedastic). In the present context the nonparametric prediction limit is defined as the maximum value out of n background samples. Confidence associated with the prediction limit is a function of n, the resampling plan used (e.g., pass one of two resamples versus pass two of two resamples), and the number of future comparisons k. No information regarding the quantitative value of the $n - 1$ smallest values in the background dataset is used; hence imputing or adjusting for nondetects is not an issue. This is not at all true for parametric approaches in which a suitable value must be imputed for the nondetects (e.g., one-half the method detection limit) or the mean and variance of a censored distribution must be computed. The fact that the censoring point is not always constant [i.e., varying method detection limits (MDLs) commonly observed for metals in groundwater] further complicates the parametric solution, but does not complicate the nonparametric approach as long as the maximum value was detected. A disadvantage of the nonparametric approach is that for large k or small α, a large number of background samples is required to provide a reasonable sitewide false positive rate, often requiring several years of sampling in multiple background wells. This drawback is dependent on the verification resampling requirements. For example, smaller background sample sizes are required if failure is indicated by prediction limit exceedance of both of two verification resamples than requiring failure if either of two resamples exceeds the limit. In the following sections a variety of applications involving nonparametric prediction limits are presented along with the fundamentals of the relevant statistical theory.

2.2   PASS ONE OF m SAMPLES

The first resampling plan to be considered requires that, in the presence of an exceedance, $m - 1$ resamples are to be obtained and failure is indicated if *all* resamples exceed the limit. The m measurements refer to the initial sample plus the $m - 1$ resamples. Nonparametric prediction limits for this case have been described by Gibbons (1990). To control the facility-wide false positive rate, we need the probability that at least one of m future measurements in each of k monitoring wells will exceed the maximum of n previous samples. For example, consider a hypothetical facility with two upgradient wells and five downgradient wells for which quarterly monitoring produces relatively independent groundwater measurements. Two years of quarterly monitoring has taken place yielding 16 upgradient background measurements. Using calculations to be described and assuming that (1) distribution of the indicator parameter is continuous, (2) distribution of water quality should be the same in background and monitoring locations in the absence of a release, and (3) all measurements are independent and identically distributed, the probability that the five new monitoring values (i.e., one at each of the five downgradient wells) will be less than the maximum of the 16 background measurements is .762. This decision rule does not have a reasonable false positive rate (i.e., $1 - .762 = .238$). However, assume that in the 23.8% of the cases in which a false positive result is obtained, the owner/operator is permitted to resample the well, and if the new measurement is below the maximum of the 16 background values, the facility could return to normal detection monitoring. With a single resampling we are now interested in the probability that at least one out of two measurements at each of the five monitoring wells will be less than the maximum of the 16 background measurements. This probability is given in Table 2.1 as .969 or a false positive rate of only 3.1%. This appears to be a reasonable decision rule given this background sample size and number of monitoring wells, at least for a program using only one constituent.

It is important to note that resampling is required only for those wells that exceed the prediction limit on the initial sample. A well that produces a measurement below the prediction limit on the initial sample can either exceed or not exceed the limit on a potential resample, and still fulfill the requirement that *at least* one of m measurements did not exceed the limit. In practice, sampling at any particular well ceases as soon as one observation "in bounds" is obtained or the m chances are used up. Probability estimates computed here do not depend on the order of the m results, and only require that the m samples are independent and are drawn from a common but unspecified distribution. Note that the testing is sequential in that verification

TABLE 2.1   Probability That at Least One of Two Samples Will Be below the Maximum of n Background Measurements at Each of k Monitoring Wells

Previous	Number of Monitoring Wells (k)														
n	1	2	3	4	5	6	7	8	9	10	11	12	13	14	15
4	.933	.881	.838	.802	.771	.744	.720	.698	.679	.661	.645	.630	.617	.604	.592
5	.952	.913	.879	.849	.823	.800	.779	.760	.742	.726	.711	.697	.684	.672	.661
6	.964	.933	.906	.882	.860	.840	.822	.805	.789	.774	.761	.748	.736	.725	.714
7	.972	.947	.925	.905	.886	.869	.852	.838	.825	.812	.799	.788	.777	.766	.757
8	.978	.958	.939	.922	.906	.891	.878	.864	.852	.841	.830	.819	.809	.800	.791
9	.982	.965	.949	.935	.921	.908	.896	.885	.874	.864	.854	.844	.835	.827	.818
10	.985	.971	.957	.945	.933	.922	.911	.901	.891	.882	.873	.865	.857	.849	.841
11	.987	.975	.964	.953	.942	.933	.923	.914	.906	.897	.889	.882	.874	.867	.860
12	.989	.979	.969	.959	.950	.941	.933	.925	.917	.910	.902	.896	.889	.882	.876
13	.990	.981	.973	.964	.956	.948	.941	.934	.927	.920	.914	.907	.901	.895	.889
14	.992	.984	.976	.969	.961	.954	.948	.941	.935	.929	.923	.917	.912	.906	.901
15	.993	.986	.979	.972	.966	.959	.953	.947	.942	.936	.931	.926	.920	.915	.910
16	.993	.987	.981	.975	.969	.964	.958	.953	.948	.943	.938	.933	.928	.923	.919
17	.994	.988	.983	.978	.972	.967	.962	.957	.953	.948	.943	.939	.935	.930	.926
18	.995	.990	.985	.980	.975	.970	.966	.961	.957	.953	.949	.944	.940	.937	.933
19	.995	.991	.986	.982	.977	.973	.969	.965	.961	.957	.953	.949	.946	.942	.938
20	.996	.991	.987	.983	.979	.975	.972	.968	.964	.960	.957	.953	.950	.947	.943
25	.997	.994	.992	.989	.986	.984	.981	.978	.976	.973	.971	.968	.966	.974	.961
30	.998	.996	.994	.992	.990	.988	.986	.984	.983	.981	.979	.977	.975	.964	.972
35	.998	.997	.996	.994	.993	.991	.990	.988	.987	.986	.984	.983	.981	.980	.979
40	.999	.998	.997	.995	.994	.993	.992	.991	.990	.989	.988	.987	.985	.984	.983
45	.999	.998	.997	.996	.995	.995	.994	.993	.992	.991	.990	.989	.988	.987	.987
50	.999	.998	.998	.997	.996	.996	.995	.994	.993	.993	.992	.991	.990	.990	.989
60	.999	.999	.998	.998	.997	.997	.996	.996	.995	.995	.994	.994	.993	.993	.992
70	1.00	.999	.999	.998	.998	.998	.997	.997	.997	.996	.996	.995	.995	.995	.994
80	1.00	.999	.999	.999	.998	.998	.998	.998	.997	.997	.997	.996	.996	.996	.996
90	1.00	1.00	.999	.999	.999	.999	.998	.998	.998	.998	.997	.997	.997	.997	.996
100	1.00	1.00	.999	.999	.999	.999	.999	.998	.998	.998	.998	.998	.997	.997	.997

TABLE 2.1   (*Continued*)

Previous	Number of Monitoring Wells (k)														
n	20	25	30	35	40	45	50	55	60	65	70	75	80	90	100
4	.542	.504	.474	.449	.428	.410	.394	.380	.367	.356	.345	.336	.327	.312	.299
5	.612	.574	.543	.517	.495	.476	.459	.443	.430	.417	.406	.396	.386	.369	.355
6	.668	.631	.600	.574	.552	.532	.514	.499	.484	.472	.460	.449	.439	.420	.405
7	.713	.678	.648	.623	.600	.580	.563	.547	.532	.519	.507	.496	.485	.466	.450
8	.750	.717	.688	.664	.642	.622	.605	.589	.574	.561	.549	.537	.527	.507	.490
9	.781	.750	.723	.699	.678	.659	.642	.626	.612	.598	.586	.574	.564	.544	.527
10	.807	.777	.752	.729	.709	.691	.674	.659	.644	.631	.619	.608	.597	.578	.560
11	.828	.801	.777	.755	.736	.718	.702	.687	.674	.661	.649	.638	.627	.608	.590
12	.847	.821	.799	.778	.760	.743	.727	.713	.700	.687	.675	.664	.654	.635	.618
13	.862	.839	.817	.798	.781	.764	.750	.736	.723	.711	.699	.689	.678	.660	.643
14	.876	.854	.834	.816	.799	.784	.769	.756	.744	.732	.721	.710	.701	.682	.666
15	.888	.867	.848	.831	.815	.801	.787	.774	.762	.751	.740	.730	.721	.703	.686
16	.898	.879	.861	.845	.830	.816	.803	.791	.779	.768	.758	.748	.739	.722	.706
17	.907	.889	.872	.857	.843	.830	.817	.806	.794	.784	.774	.765	.756	.739	.723
18	.914	.898	.882	.868	.855	.842	.830	.819	.808	.798	.789	.780	.771	.754	.739
19	.921	.906	.891	.878	.865	.853	.842	.831	.821	.811	.802	.793	.785	.769	.754
20	.928	.913	.899	.886	.874	.863	.852	.842	.832	.823	.814	.806	.798	.782	.768
25	.950	.939	.929	.919	.910.	.901	.892	.884	.876	.869	.862	.855	.848	.835	.823
30	.963	.955	.947	.940	.932	.925	.919	.912	.906	.900	.894	.888	.882	.872	.861
35	.972	.966	.959	.954	.948	.942	.937	.931	.926	.921	.916	.911	.907	.898	.889
40	.978	.973	.968	.963	.958	.954	.949	.945	.941	.936	.932	.928	.924	.917	.909
45	.982	.978	.974	.970	.966	.962	.959	.955	.951	.948	.944	.941	.938	.931	.925
50	.985	.982	.979	.975	.972	.969	.966	.963	.959	.956	.954	.951	.948	.942	.937
60	.990	.987	.985	.982	.980	.978	.975	.973	.971	.968	.966	.964	.962	.958	.954
70	.992	.990	.989	.987	.985	.983	.981	.980	.978	.976	.974	.973	.971	.968	.965
80	.994	.993	.991	.990	.988	.987	.986	.984	.983	.981	.980	.979	.977	.975	.972
90	.995	.994	.993	.992	.991	.990	.988	.987	.986	.985	.984	.983	.982	.980	.978
100	.996	.995	.994	.993	.992	.991	.991	.990	.989	.988	.987	.986	.985	.983	.982

samples are only obtained in the presence of an initial exceedance. This also prevents the chance of all samples being analyzed on the same day.

To construct these nonparametric prediction limits, we begin by assuming that the n background measurements were drawn from a continuous distribution, but the exact form of the probability distribution is unknown. Furthermore, on the next round of sampling, a series of k monitoring wells, presumably downgradient of the facility, are to be sampled, and we assume that in the absence of impact from the facility, they are also drawn from the same population as the background measurements. If the spatial variability in the background wells does not reflect the spatial variability in the monitoring wells, then we already cannot assume that these measurements were drawn from the same population and neither parametric nor nonparametric statistical comparisons have great meaning. Even if upgradient spatial variability does reflect overall spatial variability, the very presence of spatial variability violates the independence assumption. If a particular downgradient well has an initial exceedance, a resample in that well is more likely to be high than under the independence model.

At each downgradient well we have the possibility of up to $m - 1$ resamples (where m is the total number of measurements including the original) to rule out laboratory error and other effects that might lead to false positive results. At any monitoring well we continue resampling until a value less than the maximum of the n background measurements is obtained or the number of samples is m. If all m samples result in values that exceed the maximum of the n background measurements, the result is deemed significant, otherwise it is not. In this context the $\gamma 100\%$ upper prediction limit is usually defined as the maximum of n background measurements, where the confidence γ is a function of n, k, and m. Our objective therefore is to select combinations of n, k, and m that satisfy the condition $\gamma \geq 1 - .05/$ number of constituents.

Let $X_{(\max, n)}$ represent the maximum value obtained out of a sample of size n and $Y_{(\min, m)}$ represent the minimum value out of a sample of size m. In light of the previous discussion, the confidence level for the simultaneous upper prediction limit is

$$\Pr\left(Y_{1(\min, m)} \leq X_{(\max, n)}, Y_{2(\min, m)} \leq X_{(\max, n)}, \ldots, Y_{k(\min, m)} \leq X_{(\max, n)}\right) = \gamma$$
(2.1)

In order to achieve a desired confidence level, say $\gamma = .95$ for a fixed number of background measurements, we must adjust m; the more resamples the greater the confidence.

Mathematically, this probability can be evaluated using a variant of the multivariate hypergeometric distribution function described by Hall, Prarie, and Motlagh (1975) and Chou and Owen (1986) as

$$\gamma = \frac{n}{km + n} \sum_{j_1 = 1}^{m} \sum_{j_2 = 1}^{m} \cdots \sum_{j_k = 1}^{m} \frac{\binom{m}{j_1}\binom{m}{j_2} \cdots \binom{m}{j_k}}{\binom{km + n - 1}{\sum_{i=1}^{k} j_i + n - 1}}$$
(2.2)

where the notation $\binom{m}{j}$ denotes the number of combinations of m things taken j at a time, where

$$\binom{m}{j} = \frac{m!}{j!(m - j)!}$$

For example, the number of ways in which 3 things can be selected from 33 things is

$$\binom{3}{33} = \frac{33!}{3!(30)!} = \frac{1 \cdot 2 \cdots 32 \cdot 33}{(1 \cdot 2 \cdot 3)(1 \cdot 2 \cdots 29 \cdot 30)} = \frac{(31)(32)(33)}{6} = 5456$$

A sketch of the derivation of (2.2) is as follows. The distribution of the rth order statistic from a sample of size N is

$$\Psi(x)\,dx = \frac{N!}{(r-1)!(N-r)!}[F(x)]^{r-1}[1-F(x)]^{N-r}f(x)\,dx \quad (2.3)$$

(see Sarhan and Greenberg, 1962, page 12). Since

$$\int \Psi(x)\,dx = 1$$

it follows that

$$\int [F(x)]^{r-1}[1-F(x)]^{N-r}f(x)\,dx = \left[\frac{N!}{(r-1)!(N-r)!}\right]^{-1}$$

$$= \left[\frac{N(N-1)!}{(r-1)!(N-r)!}\right]^{-1}$$

$$= \left[N\binom{N-1}{r-1}\right]^{-1} \quad (2.4)$$

Now, for an individual monitoring well,

$$\Pr(Y_{\min,m} \le x_{\max,n}) = \sum_{j=1}^{m}\binom{m}{j}[F(x)]^{j}[1-F(x)]^{m-j} \quad (2.5)$$

The joint probability in (2.1) therefore may be written as

$$n\sum_{j_1=1}^{m}\binom{m}{j_1}\sum_{j_2=1}^{m}\binom{m}{j_2}\cdots\sum_{j_k=1}^{m}\binom{m}{j_k}$$

$$\times \int [F(x)]^{\sum_{i=1}^{k}j_i+n-1}[1-F(x)]^{\sum_{i=1}^{k}(m-j)}f(x)\,dx \quad (2.6)$$

Note the term $\binom{n-1}{r-1}$ is equal to one since r is selected as the maximum of the n background samples (i.e., $r = n$). Given that the total number of observations in the background sample and the k monitoring wells is $km + n$, we may rewrite the joint probability in (2.1) as

$$n \sum_{j_1=1}^{m} \binom{m}{j_1} \sum_{j_2=1}^{m} \binom{m}{j_2} \cdots \sum_{j_k=1}^{m} \binom{m}{j_k} \left[(km+n) \binom{km+n-1}{k} \Big/ \left(\sum_{i=1}^{k} j_i + n - 1 \right) \right]^{-1} \geq \gamma$$

$$(2.7)$$

Chou and Owen (1986) express this probability in terms of the multivariate hypergeometric distribution function (see Johnson and Kotz, 1969, page 300), which is the form of (2.2).

Returning to the original example of a site with five downgradient monitoring wells and 16 background observations and no resampling, we have $n = 16$, $k = 5$, and $m = 1$. Application of (2.2) yields

$$\gamma = \left(\frac{16}{(5)(1)+16} \right) \left[\frac{\binom{1}{1}\binom{1}{1}\binom{1}{1}\binom{1}{1}\binom{1}{1}}{\binom{(5)(1)+16-1}{5+16-1}} \right] = .762$$

as previously stated. Alternatively, with one resampling (i.e., $m = 2$), the solution of (2.2) involves the sum of 32 terms corresponding to the 2^5 possible combinations of j_1, j_2, j_3, j_4, and j_5 equal to 1 or 2. The resulting probability is $\gamma = .969$.

Inspection of (2.2) reveals that, as the number of monitoring wells (i.e., k) gets larger, the number of terms in the probability sum gets extremely large. For example, with only a single resampling (i.e., $m = 2$) and a facility with 20 monitoring wells, (2.2) consists of a sum of $2^{20} = 1,048,576$ terms. With two resamples we have $3^{20} = 3,486,784,401$ terms. Thus the required probabilities are computationally restricted to cases in which the required number of resamplings and number of monitoring wells are rather small. Gibbons (1990) suggested an approximation by assuming independence of comparisons across wells. In this context, if we construct a limit assuming $k = 1$, but actually apply it to $k = t$ monitoring wells, where $t > 1$, then the simultaneous probability that all t measurements will be included in the limit is approximately γ^t. Therefore we can easily evaluate (2.2) for $k = 1$, but report $\gamma = \gamma^t$. The approximated probability is therefore

$$\gamma \cong \left[\frac{n}{m+n} \sum_{j=1}^{m} \frac{\binom{m}{j}}{\binom{m+n-1}{j+n-1}} \right]^{k}$$

$$(2.8)$$

Returning to our example of $n = 16$, $k = 5$, and $m = 2$, we find

$$\gamma \cong \left[\frac{16}{2+16} \left(\frac{\binom{2}{1}}{\left(\frac{2+16-1}{1+16-1}\right)} + \frac{\binom{2}{2}}{\left(\frac{2+16-1}{2+16-1}\right)} \right) \right]^5$$

$$\cong \left[.8889\left(\frac{2}{17} + 1 \right) \right]^5 = .968$$

which is very close to the value of .969 in Table 2.1. This approximation ignores dependence introduced by repeated comparison of each monitoring

TABLE 2.2   Probability That at Least One of Three Samples Will Be below the Maximum of n Background Measurements at Each of k Monitoring Wells

| Previous | Number of Monitoring Wells (k) | | | | | | | | | | | | | | |
n	1	2	3	4	5	6	7	8	9	10	11	12	13	14	15
4	.971	.944	.917	.891	.865	.840	.816	.793	.770	.748	.727	.706	.686	.666	.647
5	.982	.965	.947	.930	.914	.898	.882	.866	.850	.835	.820	.806	.791	.777	.763
6	.988	.976	.965	.953	.942	.931	.920	.909	.898	.887	.877	.866	.856	.846	.836
7	.992	.983	.975	.967	.959	.951	.943	.935	.927	.920	.912	.904	.897	.889	.882
8	.994	.988	.982	.976	.970	.964	.958	.953	.947	.941	.935	.930	.924	.918	.913
9	.995	.991	.986	.982	.977	.973	.969	.964	.960	.955	.951	.947	.942	.938	.934
10	.997	.993	.990	.986	.983	.979	.976	.972	.969	.966	.962	.959	.955	.952	.949
11	.997	.995	.992	.989	.986	.984	.981	.978	.976	.973	.970	.968	.965	.962	.960
12	.998	.996	.993	.991	.989	.987	.985	.983	.980	.978	.976	.974	.972	.970	.968
13	.998	.996	.995	.993	.991	.989	.988	.986	.984	.982	.981	.979	.977	.975	.974
14	.999	.997	.996	.994	.993	.991	.990	.988	.987	.985	.984	.982	.981	.980	.978
15	.999	.998	.996	.995	.994	.993	.991	.990	.989	.988	.987	.985	.984	.983	.982
16	.999	.998	.997	.996	.995	.994	.993	.992	.991	.990	.989	.988	.987	.986	.985
17	.999	.998	.997	.996	.996	.995	.994	.993	.992	.991	.990	.990	.989	.988	.987
18	.999	.998	.998	.997	.996	.995	.995	.994	.993	.993	.992	.991	.990	.990	.989
19	.999	.999	.998	.997	.997	.996	.995	.995	.994	.994	.993	.992	.992	.991	.990
20	.999	.999	.998	.998	.997	.997	.996	.995	.995	.994	.994	.993	.993	.992	.992
25	1.00	.999	.999	.999	.998	.998	.998	.998	.997	.997	.997	.996	.996	.996	.995
30	1.00	1.00	.999	.999	.999	.999	.999	.999	.998	.998	.998	.998	.998	.997	.997
35	1.00	1.00	1.00	1.00	.999	.999	.999	.999	.999	.999	.999	.999	.998	.998	.998
40	1.00	1.00	1.00	1.00	1.00	1.00	.999	.999	.999	.999	.999	.999	.999	.999	.999
45	1.00	1.00	1.00	1.00	1.00	1.00	1.00	1.00	.999	.999	.999	.999	.999	.999	.999
50	1.00	1.00	1.00	1.00	1.00	1.00	1.00	1.00	1.00	1.00	1.00	.999	.999	.999	.999
60	1.00	1.00	1.00	1.00	1.00	1.00	1.00	1.00	1.00	1.00	1.00	1.00	1.00	1.00	
70	1.00	1.00	1.00	1.00	1.00	1.00	1.00	1.00	1.00	1.00	1.00	1.00	1.00	1.00	
80	1.00	1.00	1.00	1.00	1.00	1.00	1.00	1.00	1.00	1.00	1.00	1.00	1.00	1.00	
90	1.00	1.00	1.00	1.00	1.00	1.00	1.00	1.00	1.00	1.00	1.00	1.00	1.00	1.00	
100	1.00	1.00	1.00	1.00	1.00	1.00	1.00	1.00	1.00	1.00	1.00	1.00	1.00	1.00	

TABLE 2.2   (*Continued*)

Previous n	20	25	30	35	40	45	50	55	60	65	70	75	80	90	100
							Number of Monitoring Wells (k)								
4	.560	.484	.419	.363	.314	.271	.235	.203	.176	.152	.131	.114	.098	.074	.055
5	.697	.637	.582	.532	.486	.444	.406	.371	.339	.310	.283	.259	.237	.198	.165
6	.787	.741	.698	.658	.619	.583	.549	.518	.487	.459	.432	.407	.384	.340	.302
7	.846	.811	.778	.746	.716	.686	.658	.631	.605	.580	.557	.534	.512	.471	.433
8	.886	.859	.833	.808	.784	.761	.738	.716	.694	.674	.653	.634	.615	.579	.544
9	.913	.892	.872	.853	.833	.815	.796	.778	.761	.744	.727	.711	.695	.664	.634
10	.932	.916	.900	.885	.869	.854	.839	.825	.810	.796	.783	.769	.756	.730	.705
11	.946	.934	.921	.908	.896	.884	.871	.860	.848	.836	.825	.814	.802	.781	.759
12	.957	.946	.936	.926	.916	.906	.896	.886	.876	.867	.857	.848	.839	.820	.803
13	.965	.956	.948	.939	.931	.923	.915	.906	.898	.890	.882	.875	.867	.851	.836
14	.971	.964	.957	.950	.943	.936	.929	.922	.915	.909	.902	.895	.889	.876	.863
15	.976	.970	.964	.958	.952	.946	.941	.935	.929	.923	.918	.912	.907	.896	.885
16	.980	.975	.969	.965	.960	.955	.950	.945	.940	.935	.930	.925	.921	.911	.902
17	.983	.978	.974	.970	.966	.961	.957	.953	.949	.945	.940	.936	.932	.924	.916
18	.985	.981	.978	.974	.970	.967	.963	.959	.956	.952	.949	.945	.942	.935	.928
19	.987	.984	.981	.978	.974	.971	.968	.965	.962	.959	.956	.952	.949	.943	.937
20	.989	.986	.983	.980	.978	.975	.972	.969	.967	.964	.961	.959	.956	.950	.945
25	.994	.992	.991	.989	.988	.986	.985	.983	.982	.980	.979	.977	.976	.973	.970
30	.996	.995	.995	.994	.993	.992	.991	.990	.989	.988	.987	.986	.985	.984	.982
35	.998	.997	.996	.996	.995	.995	.994	.994	.993	.992	.992	.991	.991	.989	.988
40	.998	.998	998	.997	.997	.996	.996	.996	.995	.995	.994	.994	.994	.993	.992
45	.999	.999	.998	.998	.998	.997	.997	.997	.997	.996	.996	.996	.995	.995	.994
50	.999	.999	.999	.999	.999	.998	.998	.998	.997	.997	.997	.997	.997	.996	.996
60	.999	.999	.999	.999	.999	.999	.999	.999	.998	.998	.998	.998	.998	.998	.997
70	1.00	1.00	1.00	.999	.999	.999	.999	.999	.999	.999	.999	.999	.999	.999	.998
80	1.00	1.00	1.00	1.00	1.00	1.00	.999	.999	.999	.999	.999	.999	.999	.999	.999
90	1.00	1.00	1.00	1.00	1.00	1.00	1.00	1.00	1.00	.999	.999	.999	.999	.999	.999
100	1.00	1.00	1.00	1.00	1.00	1.00	1.00	1.00	1.00	1.00	1.00	1.00	1.00	.999	.999

well to the same pooled background as would be the case in an upgradient versus downgradient comparison strategy; however, the approximation is exact for the case of intrawell comparisons where each new monitoring measurement is compared to its own background. In the former case the approximation works well particularly for confidence values in excess of .9 (all that is of interest) because the limits require reasonably large n and the correlation goes to zero with increasing n (i.e., $r = 1/[n + 1]$).

These approximated probabilities have been computed for selected values of n ranging from 4 to 100, selected values of k ranging from 1 to 100, and m ranging from 1 to 3 (i.e., 0 through two resamplings). The results are displayed in Tables 2.1 and 2.2 (one and two resamplings, respectively). Gibbons (1990) has shown that the approximation is nearly exact for $n > 10$.

Finally, Table 2.3 presents the number of resamplings required to achieve at least a 95% confidence level for combinations of $n = 4$ to 100 and $k = 1$ to 100. Inspection of Table 2.3 reveals that without resampling the required background sample sizes are generally too large to be of much practical value; however, a single resampling decreases the required number of background samples to a reasonable range for most waste disposal facilities depending on the number of constituents included. It is important to note that for small numbers of background measurements and large numbers of monitoring wells, the required number of resamplings will be quite large, and the false negative rate will increase as well. The intent of Table 2.3 is to point out the effect of various combinations of numbers of background samples and numbers of monitoring wells on the integrity of detection monitoring decisions.

TABLE 2.3   Number of Resamples Required to Have at Least 95% Confidence That at Least One Measurement Will Be below the Maximum of n Background Measurements at Each of k Monitoring Wells

Previous n	Number of Monitoring Wells (k)														
	1	2	3	4	5	6	7	8	9	10	11	12	13	14	15
4	2	3	3	4	4	4	5	5	5	5	> 5	> 5	> 5	> 5	> 5
5	1	2	3	3	3	3	4	4	4	4	4	4	5	5	5
6	1	2	2	2	3	3	3	3	3	3	4	4	4	4	4
7	1	2	2	2	2	2	3	3	3	3	3	3	3	3	3
8	1	1	2	2	2	2	2	2	3	3	3	3	3	3	3
9	1	1	2	2	2	2	2	2	2	2	2	3	3	3	3
10	1	1	1	2	2	2	2	2	2	2	2	2	2	2	3
11	1	1	1	2	2	2	2	2	2	2	2	2	2	2	2
12	1	1	1	1	2	2	2	2	2	2	2	2	2	2	2
13	1	1	1	1	1	2	2	2	2	2	2	2	2	2	2
14	1	1	1	1	1	1	2	2	2	2	2	2	2	2	2
15	1	1	1	1	1	1	2	2	2	2	2	2	2	2	2
16	1	1	1	1	1	1	1	2	2	2	2	2	2	2	2
17	1	1	1	1	1	1	1	1	2	2	2	2	2	2	2
18	1	1	1	1	1	1	1	1	1	2	2	2	2	2	2
19	0	1	1	1	1	1	1	1	1	1	2	2	2	2	2
20	0	1	1	1	1	1	1	1	1	1	1	2	2	2	2
25	0	1	1	1	1	1	1	1	1	1	1	1	1	1	1
30	0	1	1	1	1	1	1	1	1	1	1	1	1	1	1
35	0	1	1	1	1	1	1	1	1	1	1	1	1	1	1
40	0	0	1	1	1	1	1	1	1	1	1	1	1	1	1
45	0	0	1	1	1	1	1	1	1	1	1	1	1	1	1
50	0	0	1	1	1	1	1	1	1	1	1	1	1	1	1
60	0	0	0	1	1	1	1	1	1	1	1	1	1	1	1
70	0	0	0	1	1	1	1	1	1	1	1	1	1	1	1
80	0	0	0	0	1	1	1	1	1	1	1	1	1	1	1
90	0	0	0	0	1	1	1	1	1	1	1	1	1	1	1
100	0	0	0	0	0	1	1	1	1	1	1	1	1	1	1

TABLE 2.3 (*Continued*)

Previous	Number of Monitoring Wells (k)														
---	20	25	30	35	40	45	50	55	60	65	70	75	80	90	100
n															
4	> 5	> 5	> 5	> 5	> 5	> 5	> 5	> 5	> 5	> 5	> 5	> 5	> 5	> 5	> 5
5	5	> 5	> 5	> 5	> 5	> 5	> 5	> 5	> 5	> 5	> 5	> 5	> 5	> 5	> 5
6	4	5	5	5	5	5	> 5	> 5	> 5	> 5	> 5	> 5	> 5	> 5	> 5
7	4	4	4	4	4	5	5	5	5	5	5	5	5	> 5	> 5
8	3	3	4	4	4	4	4	4	4	4	5	5	5	5	5
9	3	3	3	3	4	4	4	4	4	4	4	4	4	4	4
10	3	3	3	3	3	3	3	4	4	4	4	4	4	4	4
11	3	3	3	3	3	3	3	3	3	3	4	4	4	4	4
12	2	3	3	3	3	3	3	3	3	3	3	3	3	3	4
13	2	2	3	3	3	3	3	3	3	3	3	3	3	3	3
14	2	2	2	3	3	3	3	3	3	3	3	3	3	3	3
15	2	2	2	2	2	3	3	3	3	3	3	3	3	3	3
16	2	2	2	2	2	2	3	3	3	3	3	3	3	3	3
17	2	2	2	2	2	2	2	2	3	3	3	3	3	3	3
18	2	2	2	2	2	2	2	2	2	2	3	3	3	3	3
19	2	2	2	2	2	2	2	2	2	2	2	2	3	3	3
20	2	2	2	2	2	2	2	2	2	2	2	2	2	2	3
25	2	2	2	2	2	2	2	2	2	2	2	2	2	2	2
30	1	1	2	2	2	2	2	2	2	2	2	2	2	2	2
35	1	1	1	2	2	2	2	2	2	2	2	2	2	2	2
40	1	1	1	1	1	2	2	2	2	2	2	2	2	2	2
45	1	1	1	1	1	1	1	1	2	2	2	2	2	2	2
50	1	1	1	1	1	1	1	1	1	1	2	2	2	2	2
60	1	1	1	1	1	1	1	1	1	1	1	1	1	1	2
70	1	1	1	1	1	1	1	1	1	1	1	1	1	1	1
80	1	1	1	1	1	1	1	1	1	1	1	1	1	1	1
90	1	1	1	1	1	1	1	1	1	1	1	1	1	1	1
100	1	1	1	1	1	1	1	1	1	1	1	1	1	1	1

Example 2.1

A facility obtained seven background measurements prior to operation and wants to identify the optimum detection monitoring plan for its three monitoring wells, each of which is monitored for five constituents (i.e., $k = 15$). Since n is fixed at 7 and k is fixed at 15, we must identify an optimum number of resamplings such that $\gamma \geq .95$. Inspection of Table 2.1 for $k = 15$ and $n = 7$ reveals that the maximum value of the seven background measurements will only have .655 probability of exceeding at least one of the next two measurements at each of the 15 monitoring wells; therefore the required confidence level is insufficient. With two resamplings (i.e., pass at least one out of three samples at each of the 15 wells), the confidence level is .882 (see Table 2.2 $k = 15$ and $n = 7$), which also falls short of the desired confidence level of .950. The confidence level for three resamplings is .955, which is the smallest number of resamplings that satisfies $\gamma \geq .95$. We could have saved ourselves a bit of trouble by going directly to Table 2.3, where entering with $k = 15$ and $n = 7$ reveals that three resamplings are required to achieve at least 95% confidence that at least one of the four measurements will be below the maximum of seven background measurements for each of the 15 comparisons.

This example is intended to illustrate the folly of basing a decision rule on a small number of background observations, to be applied to a large number of future measurements. As illustrated in the following example, a far more judicious choice is to use the method presented here to determine the number of background measurements required to ensure 95% confidence that the maximum background measurement exceeds at least one of two new monitoring measurements (i.e., a single resample) for each of k future comparisons.

In some cases owner/operators will obtain split verification resamples to minimize the cost of repeated verification resampling. These resamples are not independent (i.e., same sampling team, purging, day of analysis, and any temporal effects) and will lead to lower confidence levels (i.e., higher false positive rates) than predicted by the model under the assumption that they are independent.

Example 2.2

A facility with 20 monitoring wells and a single monitoring constituent is only permitted a single resampling by state regulation. How many background measurements must be obtained such that the maximum will have 95% proability of exceeding at least one of the two measurements at each of the 20 monitoring wells? The answer to this question can be found by inspecting Table 2.1 under the column $k = 20$. A background sample size of $n = 25$ yields $\gamma = .945$ and $n = 30$ yields $\gamma = .960$. Linear interpolation yields $n = 27$ ($\gamma = .951$). Alternatively, inspection of Table 2.3 reveals that for $k = 20$, $n = 30$ is the smallest number of background measurements that achieves 95% confidence with a single resampling. Assume that the list of monitoring constituents is increased to 10. Since the tables do not cover $k = 200$, we may obtain the same result by selecting $k = 20$ and $\gamma = .95/10 = .995$. Inspection of Table 2.1 reveals that $n = 90$ background measurements are now required. If 90 background measurements are not available, then two verification resamples, (i.e., $m = 3$) may be required, which would require a background sample size of $n = 28$ (see Table 2.2).

Example 2.3

A new landfill has one year to obtain four quarterly measurements in a single well before operation. State regulation requires a single resampling. How many monitoring wells should the owner/operator install in order to obtain 95% confidence that at least one of the two measurements at each monitoring well will not exceed the largest of the four background measurements? Inspection of Table 2.1 reveals that it is hopeless; even with a single monitoring well, we can at most have .933 confidence. Inspection of Table 2.3 reveals that, even with as few as three monitoring wells, three resamplings are required to achieve a 95% confidence level. Furthermore, from a hydrogeological perspective, characterizing background water quality with a single background well is not a reasonable choice. From a statistical perspective, a single background well confounds contamination and spatial variability in a way that cannot be resolved without additional background wells.

Example 2.4

Consider a facility with 5 upgradient wells monitored quarterly for one year ($n = 20$) and 12 downgradient monitoring wells ($k = 12$). In the event of a statistical failure, the owner/operator is permitted to resample the well in question, and if the resample does not exceed the statistical limit, the owner/operator may continue with normal detection monitoring the following quarter. Chromium is of particular interest at this facility due to high levels in the leachate. Of the 20 background measurements, there were 17 nondetects and 3 detected values of 10, 12, and 16 ppb, respectively. The upper prediction limit for chromium is therefore 16 ppb. Inspection of Table 2.1 reveals that, with $n = 20$, $k = 12$, and $m = 2$, we can have 94.9% confidence that either the initial measurement or the resample will not exceed 16 ppb in each of the 12 downgradient monitoring wells. For example, if 11 of the 12 downgradient wells yielded nondetects for chromium and one well yielded a value of 20 ppb, then we would only resample that well. If the resample exceeded 16 ppb, a statistical failure would be recorded; otherwise the owner/operator would continue with normal detection monitoring on the next quarterly sampling event. It is important to realize that we do not have to resample the 11 wells that did not exceed the limit on the first sample. The prediction limit is for *at least* one of two samples; therefore the requirement is met if the first sample does not exceed the limit.

In the examples considered here, we have only dealt with the case of a single variable (i.e., indicator parameter). This is rarely, if ever, the case in practice. When multiple variables are evaluated, the number of future measurements (k) is equal to the number of monitoring wells multiplied by the number of variables, assuming that the variables are independent. To the extent that the variables provide redundant information (i.e., correlated), the value of k decreases to an intermediate value between the number of monitoring wells and the number of monitoring wells multiplied by the number of variables. There is no direct way of estimating this intermediate value of k. For this reason it is critically important to limit both the number of monitoring wells and the number of indicator parameters.

The tables provided here are also directly applicable to intrawell comparisons. Recall that the approximation is based on a correction to the $k = 1$ result. As such, in the intrawell case, the value of k corresponds exactly to the total number of comparisons (i.e., monitoring wells and constituents) made on each monitoring event. Unfortunately, the number of available background measurements in any well is typically small (e.g., $n = 8$) and as such the number of verification resamples required for sitewide confidence of 95% for $k < 3$ is one, $k = 3, \ldots, 8$ is two, $k = 9, \ldots, 25$ is three, $k = 26, \ldots, 65$ is four, and $k > 65$ is five or more. A moderately sized facility with 10 monitoring wells and 10 constituents would therefore require 5 verification resamples to achieve an overall sitewide false positive rate of 5% with $n = 8$ background measurements available in each well. Alternatively, however, if all historical data were pooled into a single background (perhaps following a site assessment that indicated no impact from the facility and no evidence of spatial variability), then $n = 80$ background measurements would be available and a single resample would be adequate (see Table 2.3).

2.3 PASS $m - 1$ OF m SAMPLES

The second resampling plan to be considered requires that, in the presence of a significant exceedance, $m - 1$ resamples are obtained and failure is indicated if *any* resamples exceed the limit. Again, the m measurements refer to the initial sample plus the $m - 1$ resamples. Nonparametric prediction limits for this case have been described by Gibbons (1991a). This resampling plan is more conservative in the sense that in order to pass all resamples must be "in bounds" instead of only one as in the previously described resampling plan. The drawback, however, is that a far greater number of background measurements is required to achieve the same level of confidence. As a result, in many cases these limits are of theoretical but not practical value.

The probability statement that can be used to describe this second resampling strategy is

$$\Pr\left(Y_{1(m-1,m)} \leq X_{(\max, n)}, Y_{2(m-1,m)} \leq X_{(\max, n)}, \ldots, Y_{k(m-1,m)} \leq X_{(\max, n)}\right) = \gamma$$
(2.9)

Since there are $m - 1$ resamples, $Y_{j(m-1, m)} \leq X_{(\max, n)}$ requires that we pass all resamples in monitoring well j in order to return to normal detection monitoring.

To help further clarify the differences between the two methods, consider the following example. If the maximum of $n = 20$ background samples of benzene was 7 μg/l, and the initial sample in a monitoring well was 100 μg/l, followed by three verification resamples yielding 50 μg/l, 80 μg/l, and 5 μg/l, respectively, the previous resampling plan would not consider this to represent a statistically significant result, since one resample passed. However, for this resampling plan it would be a significant exceedance, since at least one resample failed. Alternatively, if the three resamples yielded 5 μg/l, 6μg/l, and 8 μg/l, respectively, the previous method (i.e., Gibbons, 1990) would not consider this to represent a statistically significant result, since two resamples passed, but the method in this section would consider this to be a significant exceedance, since one resample failed. Statistically, this change in the model (i.e., pass all $m - 1$ resamples versus pass only one of $m - 1$ resamples) leads to quite different probability expressions. The two methods are identical for the case of a single verification resample; however, the computational method described in this section provides exact probabilities.

Following Gibbons (1991a), the probability in (2.9) can be evaluated using a variant of the hypergeometric distribution (see Chou and Owen, 1986; Hall, Prarie, and Motlagh, 1975) as

$$
\gamma = \frac{n}{km + n} \sum_{j_1 = m-1}^{m} \sum_{j_2 = m-1}^{m} \cdots \sum_{j_k = m-1}^{m} \frac{\binom{m}{j_1}\binom{m}{j_2}\cdots\binom{m}{j_k}}{\binom{km + n - 1}{\sum_{i=1}^{k} j_i + n - 1}} \quad (2.10)
$$

In this case an exact solution can be obtained that is computationally tractable for large k regardless of the values of n and m. To begin, note that each term $\binom{m}{j_i}$ in the numerator of (2.10) can take on only two possible values. When $j_i = m - 1$ its value is $\binom{m}{m-1}$, and when $j_i = m$ then $\binom{m}{m} = 1$. Now, for the first term in the summation in which all $j_i = m - 1$, the right-hand side of (2.10) is simply

$$
\frac{\binom{m}{m-1}^k}{\binom{km + n - 1}{km + n - 1 - k}} \quad (2.11)
$$

For the last term in the summation, in which all $j_i = m$, the right-hand side of (2.10) is equal to 1. For the other $2^k - 2$ terms in the summation, there are $\binom{k}{1}$ terms in which the right-hand side of (2.10) is

$$
\frac{\binom{m}{m-1}^{k-1}}{\binom{km + n - 1}{km + n - 1 - (k - 1)}} \quad (2.12)
$$

(i.e., one of the $j_i = m$), $\binom{k}{2}$ terms in which the probability is

$$\frac{\left(\dfrac{m}{m-1}\right)^{k-2}}{\left(\dfrac{km+n-1}{km+n-1-(k-2)}\right)} \tag{2.13}$$

(i.e., two of the $j_i = m$), and so on. Taken together,

$$\gamma = \frac{n}{km+n}\left\{ \frac{\left(\dfrac{m}{m-1}\right)^{k}}{\left(\dfrac{km+n-1}{km+n-1-k}\right)} + \binom{k}{1}\frac{\left(\dfrac{m}{m-1}\right)^{k-1}}{\left(\dfrac{km+n-1}{km+n-1-(k-1)}\right)} \right.$$

$$\left. + \binom{k}{2}\frac{\left(\dfrac{m}{m-1}\right)}{\left(\dfrac{km+n-1}{km+n-1-(k-2)}\right)} + \cdots + \binom{k}{k-1}\frac{\left(\dfrac{m}{m-1}\right)}{\left(\dfrac{km+n-1}{km+n-2}\right)} + 1 \right\} \tag{2.14}$$

For example, with $k = 2$ monitoring wells, $m = 3$ (i.e., two resamples and an initial sample), and $n = 30$ upgradient measurements, there are $2^2 = 4$ terms in the summation in (2.10) which take on the values

$$\gamma = \frac{30}{2(3)+30}\left\{ \frac{\left(\dfrac{3}{2}\right)^{2}}{\left(\dfrac{35}{33}\right)} + \binom{2}{1}\frac{\left(\dfrac{3}{2}\right)}{\left(\dfrac{35}{34}\right)} + 1 \right\} = .989 \tag{2.15}$$

Probability values for $n = 4$ to 100 and $k = 1$ to 100 and one resample are the same as those given in Table 2.1 (i.e., $m = 2$). For two resamples, however (i.e., $m = 3$), confidence levels are given in Table 2.4.

Example 2.5

An owner/operator is required to monitor for chromium in 10 monitoring wells at a hazardous waste disposal facility. State law requires that, in the event of an intial exceedance, two resamples must be taken at weekly intervals and both must pass in order to return to normal detection monitoring. If the owner/operator is going to use the maximum of n upgradient measurements as the prediction limit, how many upgradient samples must be obtained in order to have 95% confidence in the detection monitoring program? Inspection of Table 2.4 (i.e., two out of three samples less than the prediction limit) reveals that for $k = 10$ monitoring wells the owner/operator would require $n = 30$ upgradient measurements to provide 95% confidence.

Example 2.6

An owner/operator has a facility with 20 monitoring wells requiring benzene monitoring. Benzene has been detected 3 times out of 50 upgradient samples at concentrations of 3, 4, and 6 $\mu g/1$. How many verification resamples can be taken in the event that an initial monitoring value in a downgradient well exceeds 6 $\mu g/l$? Inspection of Tables 2.1 and 2.4 reveals that with one

TABLE 2.4   Probability That at Least Two of Three Samples Will Be below the Maximum of n Background Measurements at Each of k Monitoring Wells

Previous	Number of Monitoring Wells (k)														
n	1	2	3	4	5	6	7	8	9	10	11	12	13	14	15
4	.857	.767	.702	.653	.613	.581	.553	.530	.509	.491	.474	.460	.446	.434	.423
5	.893	.818	.762	.717	.680	.649	.622	.599	.578	.559	.543	.528	.514	.501	.489
6	.917	.855	.805	.765	.731	.702	.677	.654	.634	.616	.599	.584	.570	.558	.546
7	.933	.881	.838	.803	.772	.745	.721	.700	.680	.663	.647	.632	.618	.606	.594
8	.945	.901	.864	.832	.804	.779	.757	.737	.719	.702	.687	.672	.659	.647	.635
9	.955	.916	.884	.855	.830	.807	.787	.768	.751	.735	.720	.707	.694	.682	.671
10	.962	.929	.900	.874	.851	.831	.812	.794	.778	.763	.749	.736	.724	.713	.702
11	.967	.938	.913	.890	.869	.850	.833	.816	.801	.787	.774	.762	.750	.739	.729
12	.971	.946	.923	.903	.884	.866	.850	.835	.821	.808	.796	.784	.773	.763	.753
13	.975	.953	.932	.913	.896	.880	.865	.852	.838	.826	.815	.804	.793	.783	.774
14	.978	.958	.940	.923	.907	.892	.878	.866	.853	.842	.831	.821	.811	.801	.792
15	.980	.962	.946	.930	.916	.902	.890	.878	.866	.856	.845	.836	.826	.817	.809
16	.982	.966	.951	.937	.924	.911	.900	.888	.878	.868	.858	.849	.840	.831	.823
17	.984	.970	.956	.943	.931	.919	.908	.898	.888	.878	.869	.861	.852	.844	.836
18	.986	.972	.960	.948	.937	.926	.916	.906	.897	.888	.879	.871	.863	.856	.848
19	.987	.975	.963	.952	.942	.932	.922	.913	.905	.896	.888	.881	.873	.866	.859
20	.988	.977	.966	.956	.946	.937	.928	.920	.912	.904	.896	.889	.882	.875	.868
25	.992	.984	.977	.970	.963	.956	.950	.944	.938	.932	.926	.921	.915	.910	.905
30	.994	.989	.983	.978	.973	.968	.963	.959	.954	.949	.945	.941	.936	.932	.928
35	.996	.992	.987	.983	.980	.976	.972	.968	.965	.961	.958	.954	.951	.947	.944
40	.997	.993	.990	.987	.984	.981	.978	.975	.972	.969	.966	.963	.961	.958	.955
45	.997	.995	.992	.990	.987	.985	.982	.980	.977	.975	.973	.970	.968	.966	.964
50	.998	.996	.994	.991	.989	.987	.985	.983	.981	.979	.977	.975	.973	.972	.970
60	.998	.997	.995	.994	.992	.991	.990	.988	.987	.985	.984	.982	.981	.980	.978
70	.999	.998	.997	.995	.994	.993	.992	.991	.990	.989	.988	.987	.986	.985	.984
80	.999	.998	.997	.996	.996	.995	.994	.993	.992	.991	.991	.990	.989	.988	.987
90	.999	.999	.998	.997	.997	.996	.995	.994	.994	.993	.992	.992	.991	.990	.990
100	.999	.999	.998	.998	.997	.997	.996	.995	.995	.994	.994	.993	.993	.992	992

TABLE 2.4   (*Continued*)

Previous n	Number of Monitoring Wells (k)														

Previous n	20	25	30	35	40	45	50	55	60	65	70	75	80	90	100
4	.378	.346	.321	.300	.284	.270	.258	.247	.238	.229	.222	.215	.209	.198	.189
5	.442	.406	.379	.356	.338	.322	.308	.296	.285	.276	.267	.259	.252	.239	.229
6	.496	.459	.430	.406	.386	.369	.354	.341	.329	.318	.309	.300	.292	.278	.266
7	.544	.506	.476	.451	.430	.412	.396	.382	.369	.358	.347	.338	.329	.314	.301
8	.586	.548	.517	.492	.470	.451	.434	.419	.406	.394	.383	.373	.364	.348	.333
9	.623	.585	.554	.528	.506	.486	.469	.454	.440	.428	.417	.406	.396	.379	.364
10	.655	.618	.587	.561	.539	.519	.502	.486	.472	.459	.447	.437	.427	.409	.393
11	.684	.647	.617	.591	.569	.549	.531	.516	.501	.488	.476	.465	.455	.437	.420
12	.709	.674	.644	.618	.596	.576	.559	.543	.528	.515	.503	.492	.481	.463	.446
13	.732	.698	.668	.643	.621	.602	.584	.568	.554	.540	.528	.517	.506	.487	.470
14	.752	.719	.691	.666	.644	.625	.607	.591	.577	.564	.551	.540	.529	.510	.493
15	.770	.738	.711	.687	.665	.646	.629	.613	.599	.585	.573	.562	.551	.532	.514
16	.787	.756	.729	.706	.685	.666	.649	.633	.619	.606	.594	.582	.571	.552	.535
17	.802	.772	.746	.723	.703	.684	.667	.652	.638	.625	.613	.601	.591	.571	.554
18	.815	.786	.761	.739	.719	.701	.685	.670	.656	.643	.630	.619	.609	.589	.572
19	.827	.800	.776	.754	.735	.717	.701	.686	.672	.659	.647	.636	.626	.606	.589
20	.838	.812	.789	.768	.749	.731	.716	.701	.687	.675	.663	.652	.642	.622	.605
25	.881	.859	.840	.822	.806	.791	.777	.764	.751	.740	.729	.719	.709	.691	.674
30	.909	.891	.875	.860	.846	.833	.821	.810	.799	.788	.778	.769	.760	.743	.728
35	.928	.914	.900	.888	.876	.865	.854	.844	.834	.825	.816	.808	.800	.784	.770
40	.942	.930	.919	.908	.898	.888	.879	.870	.862	.853	.846	.838	.831	.817	.804
45	.953	.943	.933	.924	.915	.906	.898	.890	.883	.876	.869	.862	.855	.843	.831
50	.961	.952	.944	.936	.928	.921	.913	.907	.900	.893	.887	.881	.875	.864	.853
60	.971	.965	.959	.953	.947	.941	.935	.930	.925	.920	.915	.910	.905	.896	.887
70	.978	.973	.969	.964	.959	.955	.950	.946	.942	.937	.933	.929	.925	.918	.911
80	.983	.979	.975	.971	.968	.964	.960	.957	.953	.950	.943	.943	.940	.934	.928
90	.986	.983	.980	.977	.974	.971	.968	.965	.962	.959	.956	.954	.951	.946	.941
100	.989	.986	.984	.981	.978	.976	.973	.971	.969	.966	.964	.961	.959	.955	.950

resample the overall confidence level is 98.5%, with two resamples the confidence level is 96.1%, and with three resamples the confidence level is 93% (see Gibbons, 1991a, Table 3). Using a criterion of 95% confidence, the owner/operator should select two verification resamples.

2.4   PASS FIRST OR ALL $m - 1$ RESAMPLES

Davis and McNichols (1993) and Willits (1993) have noted that the nonparametric prediction limits suggested by Gibbons (1991a) do not exactly apply to those verification resampling plans in which all $m - 1$ resamples are required to be "in bounds." They point out that if the first sample passes, all verification resamples could potentially exceed the limit but since they are never collected the well would nevertheless pass. They point out that the rule is in fact pass the first or all of $m - 1$ resamples, and they have provided exact formulas for computing the confidence of this decision rule for varying values of n, k, and m. In addition, Davis and McNichols (1993) have also presented exact results for one of m samples in bounds where the limit is the second highest measurement. They have kindly provided tables of these confidence levels which should be used in place of previously published tables by Gibbons (1991a) for $m > 2$, which are slightly conservative. Tables 2.5 to 2.8 present confidence levels for one of m samples in bounds for $m = 1$ to 4, respectively, where the limit is the highest background value from a sample of size n. Tables 2.9 to 2.12 present confidence levels for one of m samples in bounds for $m = 1$ to 4, respectively, where the limit is the second highest background value from a sample of size n. Tables 2.13 and 2.14 present confidence levels for the first or all of $m - 1$ resamples in bounds for $m = 3$ and 4, respectively, where the limit is the highest background value from a sample of size n. Tables 2.15 and 2.16 present confidence levels for the first or all of $m - 1$ resamples in bounds for $m = 3$ and 4, respectively, where the limit is the second highest background value from a sample of size n. Details of the derivation are presented in Davis and McNichols (1993).

2.5 SUMMARY

Application of nonparametric prediction limits in the previous examples illustrates an important point about groundwater detection monitoring programs often overlooked; namely, increasing the number of background (e.g., upgradient) measurements increases our confidence in our overall detection monitoring program while reducing the false negative rate (e.g., multiple verification resamples). Of course, this fact also applies in the parametric case in which we pay a huge price in our false negative rate for having uncertainty in our estimates of the background mean and variance. As n increases (i.e., the number of available upgradient measurements), our uncertainty in the sample-based estimates of the population mean and variance decreases, as does the prediction limit; hence the false negative rate also decreases. In contrast to the parametric case, however, confidence levels for the nonparametric prediction limits described here can only be increased by increasing the background sample size. As illustrated for the parametric case for fixed sample size, false positive rates are constant, but power can be increased by requiring only one of m measurements to be in bounds, since the limit itself decreases. For nonparametric prediction limits with fixed n, however, false positive rates are decreased by requiring only one of m measurements to be in bounds, but power is also decreased since the limit itself is constant (see Figure 2.1).

(*text continues on page* 76)

TABLE 2.5 Probability That at Least One Sample Will Be below the Maximum of n Background Measurements at Each of k Monitoring Wells

Previous n	\multicolumn{15}{c}{Number of Monitoring Wells (k)}														
	1	2	3	4	5	6	7	8	9	10	11	12	13	14	15
4	.8000	.6667	.5714	.5000	.4444	.4000	.3636	.3333	.3077	.2857	.2667	.2500	.2353	.2222	.2105
5	.8333	.7143	.6250	.5556	.5000	.4545	.4167	.3846	.3571	.3333	.3125	.2941	.2778	.2632	.2500
6	.8571	.7500	.6667	.6000	.5455	.5000	.4615	.4286	.4000	.3750	.3529	.3333	.3158	.3000	.2857
7	.8750	.7778	.7000	.6364	.5833	.5385	.5000	.4667	.4375	.4118	.3889	.3684	.3500	.3333	.3182
8	.8889	.8000	.7273	.6667	.6154	.5714	.5333	.5000	.4706	.4444	.4211	.4000	.3810	.3636	.3478
9	.9000	.8182	.7500	.6923	.6429	.6000	.5625	.5294	.5000	.4737	.4500	.4286	.4091	.3913	.3750
10	.9091	.8333	.7692	.7143	.6667	.6250	.5882	.5556	.5263	.5000	.4762	.4545	.4348	.4167	.4000
11	.9167	.8462	.7857	.7333	.6875	.6471	.6111	.5789	.5500	.5238	.5000	.4783	.4583	.4400	.4231
12	.9231	.8571	.8000	.7500	.7059	.6667	.6316	.6000	.5714	.5455	.5217	.5000	.4800	.4615	.4444
13	.9286	.8667	.8125	.7647	.7222	.6842	.6500	.6190	.5909	.5652	.5417	.5200	.5000	.4815	.4643
14	.9333	.8750	.8235	.7778	.7368	.7000	.6667	.6364	.6087	.5833	.5600	.5385	.5185	.5000	.4828
15	.9375	.8824	.8333	.7895	.7500	.7143	.6818	.6522	.6250	.6000	.5769	.5556	.5357	.5172	.5000
16	.9412	.8889	.8421	.8000	.7619	.7273	.6957	.6667	.6400	.6154	.5926	.5714	.5517	.5333	.5161
17	.9444	.8947	.8500	.8095	.7727	.7391	.7083	.6800	.6538	.6296	.6071	.5862	.5667	.5484	.5313
18	.9474	.9000	.8571	.8182	.7826	.7500	.7200	.6923	.6667	.6429	.6207	.6000	.5806	.5625	.5455
19	.9500	.9048	.8636	.8261	.7917	.7600	.7308	.7037	.6786	.6552	.6333	.6129	.5938	.5758	.5588
20	.9524	.9091	.8696	.8333	.8000	.7692	.7407	.7143	.6897	.6667	.6452	.6250	.6061	.5882	.5714
25	.9615	.9259	.8929	.8621	.8333	.8065	.7813	.7576	.7353	.7143	.6944	.6757	.6579	.6410	.6250
30	.9677	.9375	.9091	.8824	.8571	.8333	.8108	.7895	.7692	.7500	.7317	.7143	.6977	.6818	.6667
35	.9722	.9459	.9211	.8974	.8750	.8537	.8333	.8140	.7955	.7778	.7609	.7447	.7292	.7143	.7000
40	.9756	.9524	.9302	.9091	.8889	.8696	.8511	.8333	.8163	.8000	.7843	.7692	.7547	.7407	.7273
45	.9783	.9574	.9375	.9184	.9000	.8824	.8654	.8491	.8333	.8182	.8036	.7895	.7759	.7627	.7500
50	.9804	.9615	.9434	.9259	.9091	.8929	.8772	.8621	.8475	.8333	.8197	.8065	.7937	.7813	.7692
60	.9836	.9677	.9524	.9375	.9231	.9091	.8955	.8824	.8696	.8571	.8451	.8333	.8219	.8108	.8000
70	.9859	.9722	.9589	.9459	.9333	.9211	.9091	.8974	.8861	.8750	.8642	.8537	.8434	.8333	.8235
80	.9877	.9756	.9639	.9524	.9412	.9302	.9195	.9091	.8989	.8889	.8791	.8696	.8602	.8511	.8421
90	.9890	.9783	.9677	.9574	.9474	.9375	.9278	.9184	.9091	.9000	.8911	.8824	.8738	.8654	.8571
100	.9901	.9804	.9709	.9615	.9524	.9434	.9346	.9259	.9174	.9091	.9009	.8929	.8850	.8772	.8696

Previous

Number of Monitoring Wells (k)

n	20	25	30	35	40	45	50	55	60	65	70	75	80	90	100
4	.1667	.1379	.1176	.1026	.0909	.0816	.0741	.0678	.0625	.0580	.0541	.0506	.0476	.0426	.0385
5	.2000	.1667	.1429	.1250	.1111	.1000	.0909	.0833	.0769	.0714	.0667	.0625	.0588	.0526	.0476
6	.2308	.1935	.1667	.1463	.1304	.1176	.1071	.0984	.0909	.0845	.0789	.0741	.0698	.0625	.0566
7	.2593	.2188	.1892	.1667	.1489	.1346	.1228	.1129	.1045	.0972	.0909	.0854	.0805	.0722	.0654
8	.2857	.2424	.2105	.1860	.1667	.1509	.1379	.1270	.1176	.1096	.1026	.0964	.0909	.0816	.0741
9	.3103	.2647	.2308	.2045	.1837	.1667	.1525	.1406	.1304	.1216	.1139	.1071	.1011	.0909	.0826
10	.3333	.2857	.2500	.2222	.2000	.1818	.1667	.1538	.1429	.1333	.1250	.1176	.1111	.1000	.0909
11	.3548	.3056	.2683	.2391	.2157	.1964	.1803	.1667	.1549	.1447	.1358	.1279	.1209	.1089	.0991
12	.3750	.3243	.2857	.2553	.2308	.2105	.1935	.1791	.1667	.1558	.1463	.1379	.1304	.1176	.1071
13	.3939	.3421	.3023	.2708	.2453	.2241	.2063	.1912	.1781	.1667	.1566	.1477	.1398	.1262	.1150
14	.4118	.3590	.3182	.2857	.2593	.2373	.2188	.2029	.1892	.1772	.1667	.1573	.1489	.1346	.1228
15	.4286	.3750	.3333	.3000	.2727	.2500	.2308	.2143	.2000	.1875	.1765	.1667	.1579	.1429	.1304
16	.4444	.3902	.3478	.3137	.2857	.2623	.2424	.2254	.2105	.1975	.1860	.1758	.1667	.1509	.1379
17	.4595	.4048	.3617	.3269	.2982	.2742	.2537	.2361	.2208	.2073	.1954	.1848	.1753	.1589	.1453
18	.4737	.4186	.3750	.3396	.3103	.2857	.2647	.2466	.2308	.2169	.2045	.1935	.1837	.1667	.1525
19	.4872	.4318	.3878	.3519	.3220	.2969	.2754	.2568	.2405	.2262	.2135	.2021	.1919	.1743	.1597
20	.5000	.4444	.4000	.3636	.3333	.3077	.2857	.2667	.2500	.2353	.2222	.2105	.2000	.1818	.1667
25	.5556	.5000	.4545	.4167	.3846	.3571	.3333	.3125	.2941	.2778	.2632	.2500	.2381	.2174	.2000
30	.6000	.5455	.5000	.4615	.4286	.4000	.3750	.3529	.3333	.3158	.3000	.2857	.2727	.2500	.2308
35	.6364	.5833	.5385	.5000	.4667	.4375	.4118	.3889	.3684	.3500	.3333	.3182	.3043	.2800	.2593
40	.6667	.6154	.5714	.5333	.5000	.4706	.4444	.4211	.4000	.3810	.3636	.3478	.3333	.3077	.2857
45	.6923	.6429	.6000	.5625	.5294	.5000	.4737	.4500	.4286	.4091	.3913	.3750	.3600	.3333	.3103
50	.7143	.6667	.6250	.5882	.5556	.5263	.5000	.4762	.4545	.4348	.4167	.4000	.3846	.3571	.3333
60	.7500	.7059	.6667	.6316	.6000	.5714	.5455	.5217	.5000	.4800	.4615	.4444	.4286	.4000	.3750
70	.7778	.7368	.7000	.6667	.6364	.6087	.5833	.5600	.5385	.5185	.5000	.4828	.4667	.4375	.4118
80	.8000	.7619	.7273	.6957	.6667	.6400	.6154	.5926	.5714	.5517	.5333	.5161	.5000	.4706	.4444
90	.8182	.7826	.7500	.7200	.6923	.6667	.6429	.6207	.6000	.5806	.5625	.5455	.5294	.5000	.4737
100	.8333	.8000	.7692	.7407	.7143	.6897	.6667	.6452	.6250	.6061	.5882	.5714	.5556	.5263	.5000

Prepared by Charles Davis based on results in Davis and McNichols (1993).

TABLE 2.6 Probability That at Least One of Two Samples Will Be below the Maximum of *n* Background Measurements at Each of *k* Monitoring Wells

Previous *n*	Number of Monitoring Wells (*k*)														
	1	2	3	4	5	6	7	8	9	10	11	12	13	14	15
4	.9333	.8810	.8381	.8020	.7710	.7439	.7199	.6984	.6791	.6614	.6453	.6304	.6167	.6039	.5920
5	.9524	.9127	.8788	.8493	.8232	.7999	.7788	.7597	.7422	.7260	.7111	.6972	.6842	.6720	.6606
6	.9643	.9333	.9061	.8817	.8598	.8398	.8215	.8046	.7890	.7745	.7609	.7481	.7362	.7248	.7141
7	.9722	.9475	.9252	.9049	.8863	.8692	.8533	.8385	.8246	.8116	.7994	.7878	.7769	.7665	.7566
8	.9778	.9576	.9391	.9220	.9061	.8913	.8775	.8645	.8522	.8406	.8296	.8192	.8092	.7997	.7906
9	.9818	.9650	.9495	.9349	.9213	.9084	.8963	.8849	.8740	.8636	.8538	.8443	.8353	.8266	.8183
10	.9848	.9707	.9574	.9449	.9331	.9219	.9112	.9011	.8914	.8822	.8733	.8648	.8566	.8487	.8411
11	.9872	.9751	.9637	.9528	.9425	.9326	.9232	.9142	.9056	.8973	.8893	.8816	.8741	.8670	.8600
12	.9890	.9786	.9686	.9591	.9500	.9413	.9330	.9249	.9172	.9097	.9025	.8955	.8888	.8822	.8759
13	.9905	.9814	.9727	.9643	.9562	.9485	.9410	.9338	.9268	.9201	.9136	.9072	.9011	.8951	.8893
14	.9917	.9837	.9760	.9685	.9614	.9544	.9477	.9412	.9349	.9288	.9229	.9171	.9115	.9060	.9007
15	.9926	.9856	.9787	.9721	.9656	.9594	.9534	.9475	.9418	.9362	.9308	.9255	.9204	.9154	.9105
16	.9935	.9871	.9810	.9750	.9693	.9636	.9582	.9528	.9476	.9426	.9376	.9328	.9281	.9235	.9190
17	.9942	.9885	.9829	.9776	.9723	.9672	.9623	.9574	.9527	.9480	.9435	.9391	.9347	.9305	.9263
18	.9947	.9896	.9846	.9797	.9750	.9703	.9658	.9613	.9570	.9527	.9486	.9445	.9405	.9366	.9327
19	.9952	.9906	.9860	.9816	.9773	.9730	.9689	.9648	.9608	.9569	.9530	.9493	.9456	.9419	.9384
20	.9957	.9914	.9873	.9832	.9793	.9754	.9715	.9678	.9641	.9605	.9569	.9534	.9500	.9466	.9433
25	.9972	.9943	.9916	.9889	.9862	.9835	.9809	.9783	.9758	.9733	.9708	.9683	.9659	.9635	.9612
30	.9980	.9960	.9940	.9921	.9901	.9882	.9863	.9844	.9826	.9808	.9789	.9771	.9753	.9736	.9718
35	.9985	.9970	.9955	.9941	.9926	.9912	.9897	.9883	.9869	.9855	.9841	.9827	.9814	.9800	.9787
40	.9988	.9977	.9965	.9954	.9943	.9931	.9920	.9909	.9898	.9887	.9876	.9865	.9854	.9844	.9833
45	.9991	.9982	.9972	.9963	.9954	.9945	.9936	.9927	.9918	.9910	.9901	.9892	.9883	.9875	.9866
50	.9992	.9985	.9977	.9970	.9963	.9955	.9948	.9941	.9933	.9926	.9919	.9912	.9904	.9897	.9890
60	.9995	.9989	.9984	.9979	.9974	.9969	.9963	.9958	.9953	.9948	.9943	.9938	.9932	.9927	.9922
70	.9996	.9992	.9988	.9984	.9981	.9977	.9973	.9969	.9965	.9961	.9957	.9954	.9950	.9946	.9942
80	.9997	.9994	.9991	.9988	.9985	.9982	.9979	.9976	.9973	.9970	.9967	.9964	.9961	.9958	.9955
90	.9998	.9995	.9993	.9990	.9988	.9986	.9983	.9981	.9979	.9976	.9974	.9972	.9969	.9967	.9965
100	.9998	.9996	.9994	.9992	.9990	.9988	.9986	.9985	.9983	.9981	.9979	.9977	.9975	.9973	.9971

Number of Monitoring Wells (k)

Previous n	20	25	30	35	40	45	50	55	60	65	70	75	80	90	100
4	.5424	.5045	.4741	.4491	.4279	.4096	.3937	.3796	.3670	.3557	.3454	.3360	.3274	.3120	.2988
5	.6121	.5741	.5431	.5171	.4949	.4756	.4586	.4434	.4298	.4174	.4062	.3959	.3864	.3694	.3546
6	.6680	.6310	.6003	.5743	.5517	.5320	.5144	.4987	.4845	.4715	.4597	.4488	.4387	.4205	.4046
7	.7133	.6779	.6481	.6226	.6002	.5805	.5628	.5469	.5324	.5191	.5069	.4956	.4852	.4663	.4496
8	.7504	.7170	.6884	.6637	.6418	.6224	.6049	.5890	.5744	.5611	.5488	.5373	.5267	.5073	.4902
9	.7812	.7497	.7226	.6989	.6777	.6588	.6416	.6259	.6115	.5983	.5860	.5745	.5638	.5443	.5269
10	.8068	.7775	.7518	.7292	.7089	.6905	.6738	.6585	.6444	.6313	.6191	.6077	.5971	.5776	.5601
11	.8284	.8011	.7769	.7554	.7360	.7184	.7022	.6873	.6736	.6607	.6488	.6376	.6270	.6077	.5903
12	.8468	.8213	.7986	.7782	.7597	.7428	.7273	.7129	.6995	.6871	.6754	.6644	.6540	.6349	.6177
13	.8625	.8387	.8174	.7982	.7806	.7645	.7496	.7357	.7228	.7107	.6993	.6886	.6784	.6597	.6427
14	.8759	.8538	.8339	.8157	.7991	.7837	.7694	.7561	.7436	.7319	.7209	.7104	.7005	.6822	.6656
15	.8876	.8670	.8483	.8312	.8154	.8007	.7871	.7743	.7623	.7510	.7403	.7302	.7206	.7028	.6865
16	.8978	.8786	.8610	.8449	.8299	.8160	.8030	.7907	.7792	.7683	.7580	.7482	.7389	.7215	.7056
17	.9067	.8888	.8723	.8571	.8429	.8296	.8172	.8055	.7945	.7840	.7741	.7646	.7556	.7387	.7232
18	.9145	.8978	.8823	.8679	.8545	.8419	.8301	.8189	.8083	.7982	.7887	.7795	.7708	.7545	.7394
19	.9214	.9057	.8912	.8777	.8649	.8530	.8417	.8310	.8209	.8112	.8020	.7932	.7848	.7689	.7543
20	.9275	.9129	.8992	.8864	.8743	.8630	.8522	.8420	.8323	.8230	.8142	.8057	.7976	.7822	.7680
25	.9498	.9390	.9288	.9190	.9098	.9009	.8924	.8843	.8765	.8689	.8617	.8547	.8479	.8350	.8229
30	.9633	.9551	.9472	.9397	.9324	.9254	.9187	.9121	.9058	.8996	.8937	.8879	.8823	.8715	.8613
35	.9721	.9657	.9595	.9535	.9477	.9421	.9366	.9313	.9261	.9210	.9161	.9113	.9066	.8975	.8889
40	.9781	.9730	.9680	.9631	.9584	.9538	.9493	.9449	.9406	.9364	.9323	.9283	.9243	.9167	.9093
45	.9823	.9782	.9741	.9701	.9662	.9624	.9586	.9550	.9514	.9478	.9444	.9409	.9376	.9310	.9247
50	.9855	.9820	.9786	.9753	.9720	.9688	.9657	.9626	.9595	.9565	.9535	.9506	.9477	.9421	.9367
60	.9897	.9872	.9848	.9824	.9800	.9776	.9753	.9730	.9707	.9685	.9663	.9641	.9619	.9577	.9535
70	.9923	.9905	.9886	.9868	.9850	.9832	.9814	.9797	.9779	.9762	.9745	.9728	.9711	.9678	.9646
80	.9941	.9926	.9912	.9898	.9883	.9869	.9855	.9842	.9828	.9814	.9801	.9787	.9774	.9748	.9722
90	.9953	.9941	.9930	.9918	.9907	.9896	.9884	.9873	.9862	.9851	.9840	.9829	.9818	.9797	.9776
100	.9962	.9952	.9943	.9933	.9924	.9915	.9905	.9896	.9887	.9878	.9869	.9860	.9851	.9833	.9816

Prepared by Charles Davis based on results in Davis and McNichols (1993).

TABLE 2.7 Probability That at Least One of Three Samples Will Be below the Maximum of n Background Measurements at Each of k Monitoring Wells

Previous	Number of Monitoring Wells (k)														
n	1	2	3	4	5	6	7	8	9	10	11	12	13	14	15
4	.9714	.9476	.9272	.9092	.8933	.8789	.8657	.8537	.8426	.8322	.8226	.8135	.8050	.7970	.7893
5	.9821	.9665	.9524	.9397	.9281	.9174	.9075	.8982	.8895	.8813	.8736	.8663	.8593	.8527	.8463
6	.9881	.9773	.9673	.9581	.9496	.9415	.9339	.9268	.9200	.9136	.9074	.9015	.8959	.8905	.8853
7	.9917	.9839	.9767	.9698	.9634	.9573	.9514	.9459	.9405	.9354	.9305	.9258	.9213	.9169	.9126
8	.9939	.9882	.9828	.9776	.9727	.9679	.9634	.9590	.9548	.9507	.9468	.9430	.9393	.9357	.9322
9	.9955	.9911	.9869	.9829	.9791	.9754	.9718	.9683	.9649	.9616	.9584	.9553	.9523	.9494	.9465
10	.9965	.9931	.9899	.9867	.9837	.9807	.9778	.9750	.9723	.9696	.9670	.9645	.9620	.9596	.9572
11	.9973	.9946	.9920	.9895	.9870	.9846	.9823	.9800	.9778	.9756	.9734	.9713	.9693	.9673	.9653
12	.9978	.9957	.9936	.9915	.9895	.9876	.9856	.9838	.9819	.9801	.9783	.9766	.9749	.9732	.9715
13	.9982	.9965	.9948	.9931	.9914	.9898	.9882	.9866	.9851	.9836	.9821	.9806	.9792	.9778	.9764
14	.9985	.9971	.9957	.9943	.9929	.9915	.9902	.9889	.9876	.9863	.9851	.9838	.9826	.9814	.9802
15	.9988	.9976	.9964	.9952	.9940	.9929	.9918	.9907	.9896	.9885	.9874	.9864	.9853	.9843	.9833
16	.9990	.9979	.9969	.9960	.9950	.9940	.9930	.9921	.9912	.9902	.9893	.9884	.9875	.9866	.9857
17	.9991	.9983	.9974	.9965	.9957	.9949	.9941	.9932	.9924	.9916	.9908	.9901	.9893	.9885	.9878
18	.9992	.9985	.9978	.9970	.9963	.9956	.9949	.9942	.9935	.9928	.9921	.9914	.9908	.9901	.9894
19	.9994	.9987	.9981	.9974	.9968	.9962	.9956	.9950	.9943	.9937	.9931	.9926	.9920	.9914	.9908
20	.9994	.9989	.9983	.9978	.9972	.9967	.9961	.9956	.9951	.9945	.9940	.9935	.9930	.9925	.9919
25	.9997	.9994	.9991	.9988	.9985	.9982	.9979	.9976	.9973	.9970	.9967	.9964	.9961	.9958	.9956
30	.9998	.9996	.9995	.9993	.9991	.9989	.9987	.9985	.9984	.9982	.9980	.9978	.9977	.9975	.9973
35	.9999	.9998	.9996	.9995	.9994	.9993	.9992	.9991	.9989	.9988	.9987	.9986	.9985	.9984	.9982
40	.9999	.9998	.9998	.9997	.9996	.9995	.9994	.9994	.9993	.9992	.9991	.9990	.9990	.9989	.9988
45	.9999	.9999	.9998	.9998	.9997	.9997	.9996	.9995	.9995	.9994	.9994	.9993	.9993	.9992	.9991
50	.9999	.9999	.9999	.9998	.9998	.9997	.9997	.9997	.9996	.9996	.9995	.9995	.9994	.9994	.9994
60	.9999	.9999	.9999	.9999	.9999	.9998	.9998	.9998	.9998	.9997	.9997	.9997	.9997	.9996	.9996
70	.9999	.9999	.9999	.9999	.9999	.9999	.9999	.9999	.9999	.9998	.9998	.9998	.9998	.9998	.9998
80	.9999	.9999	.9999	.9999	.9999	.9999	.9999	.9999	.9999	.9999	.9999	.9998	.9998	.9998	.9998
90	.9999	.9999	.9999	.9999	.9999	.9999	.9999	.9999	.9999	.9999	.9999	.9999	.9999	.9999	.9999
100	.9999	.9999	.9999	.9999	.9999	.9999	.9999	.9999	.9999	.9999	.9999	.9999	.9999	.9999	.9999

Number of Monitoring Wells (k)

Previous n	20	25	30	35	40	45	50	55	60	65	70	75	80	90	100
4	.7563	.7296	.7072	.6879	.6711	.6562	.6427	.6306	.6195	.6093	.5999	.5911	.5829	.5681	.5549
5	.8184	.7950	.7751	.7576	.7421	.7282	.7156	.7041	.6935	.6837	.6745	.6660	.6580	.6433	.6302
6	.8620	.8420	.8246	.8092	.7954	.7828	.7713	.7607	.7508	.7416	.7331	.7250	.7174	.7034	.6907
7	.8932	.8764	.8614	.8480	.8358	.8246	.8142	.8046	.7957	.7873	.7794	.7719	.7648	.7517	.7398
8	.9161	.9019	.8891	.8775	.8668	.8569	.8477	.8391	.8311	.8235	.8163	.8095	.8030	.7909	.7798
9	.9332	.9211	.9102	.9002	.8909	.8822	.8741	.8664	.8592	.8524	.8459	.8398	.8339	.8228	.8126
10	.9460	.9359	.9265	.9179	.9098	.9022	.8950	.8883	.8819	.8758	.8699	.8644	.8590	.8490	.8397
11	.9559	.9473	.9393	.9318	.9247	.9181	.9118	.9058	.9002	.8947	.8895	.8845	.8797	.8706	.8621
12	.9636	.9562	.9493	.9429	.9367	.9309	.9254	.9201	.9151	.9102	.9056	.9011	.8967	.8885	.8808
13	.9696	.9633	.9574	.9518	.9464	.9413	.9364	.9318	.9273	.9230	.9188	.9148	.9109	.9035	.8965
14	.9745	.9690	.9639	.9590	.9543	.9498	.9455	.9414	.9374	.9335	.9298	.9262	.9227	.9160	.9097
15	.9783	.9736	.9692	.9649	.9608	.9568	.9530	.9493	.9458	.9424	.9390	.9358	.9327	.9266	.9209
16	.9815	.9774	.9735	.9697	.9661	.9626	.9593	.9560	.9528	.9498	.9468	.9439	.9410	.9356	.9304
17	.9840	.9805	.9771	.9738	.9706	.9675	.9645	.9616	.9587	.9560	.9533	.9507	.9482	.9433	.9386
18	.9862	.9831	.9800	.9771	.9743	.9715	.9689	.9663	.9637	.9613	.9589	.9565	.9542	.9498	.9455
19	.9879	.9852	.9825	.9799	.9774	.9750	.9726	.9703	.9680	.9658	.9636	.9615	.9594	.9554	.9515
20	.9894	.9870	.9846	.9823	.9801	.9779	.9758	.9737	.9716	.9696	.9677	.9658	.9639	.9602	.9567
25	.9941	.9927	.9914	.9900	.9887	.9874	.9861	.9849	.9836	.9824	.9812	.9800	.9788	.9765	.9743
30	.9964	.9956	.9947	.9939	.9930	.9922	.9914	.9906	.9898	.9890	.9882	.9874	.9867	.9852	.9837
35	.9977	.9971	.9965	.9960	.9954	.9949	.9943	.9938	.9932	.9927	.9922	.9916	.9911	.9901	.9891
40	.9984	.9980	.9976	.9972	.9968	.9965	.9961	.9957	.9953	.9949	.9946	.9942	.9938	.9931	.9924
45	.9989	.9986	.9983	.9980	.9977	.9975	.9972	.9969	.9966	.9963	.9961	.9958	.9955	.9950	.9945
50	.9992	.9989	.9987	.9985	.9983	.9981	.9979	.9977	.9975	.9973	.9971	.9969	.9967	.9963	.9959
60	.9995	.9994	.9992	.9991	.9990	.9989	.9988	.9986	.9985	.9984	.9983	.9981	.9980	.9978	.9975
70	.9997	.9996	.9995	.9994	.9994	.9993	.9992	.9991	.9990	.9990	.9989	.9988	.9987	.9986	.9984
80	.9998	.9997	.9997	.9996	.9996	.9995	.9995	.9994	.9994	.9993	.9992	.9992	.9991	.9990	.9989
90	.9998	.9998	.9998	.9997	.9997	.9997	.9996	.9996	.9995	.9995	.9995	.9994	.9994	.9993	.9992
100	.9999	.9999	.9998	.9998	.9998	.9997	.9997	.9997	.9997	.9996	.9996	.9996	.9995	.9995	.9994

Prepared by Charles Davis based on results in Davis and McNichols (1993).

TABLE 2.8 Probability That at Least One of Four Samples Will Be below the Maximum of *n* Background Measurements at Each of *k* Monitoring Wells

Previous *n*	Number of Monitoring Wells (*k*)														
	1	2	3	4	5	6	7	8	9	10	11	12	13	14	15
4	.9857	.9734	.9627	.9530	.9442	.9362	.9288	.9219	.9154	.9093	.9036	.8982	.8930	.8881	.8835
5	.9921	.9849	.9784	.9723	.9667	.9614	.9565	.9518	.9474	.9431	.9391	.9352	.9315	.9280	.9246
6	.9952	.9908	.9867	.9827	.9790	.9755	.9722	.9689	.9659	.9629	.9600	.9573	.9546	.9520	.9495
7	.9970	.9941	.9914	.9887	.9862	.9838	.9815	.9792	.9771	.9749	.9729	.9709	.9690	.9671	.9653
8	.9980	.9960	.9942	.9924	.9906	.9889	.9873	.9856	.9841	.9826	.9811	.9796	.9782	.9768	.9755
9	.9986	.9972	.9959	.9946	.9934	.9922	.9910	.9898	.9887	.9875	.9864	.9854	.9843	.9833	.9823
10	.9990	.9980	.9971	.9961	.9952	.9943	.9934	.9926	.9917	.9909	.9901	.9893	.9885	.9877	.9869
11	.9993	.9985	.9978	.9971	.9965	.9958	.9951	.9945	.9938	.9932	.9926	.9919	.9913	.9907	.9901
12	.9995	.9989	.9984	.9978	.9973	.9968	.9963	.9958	.9953	.9948	.9943	.9939	.9934	.9929	.9925
13	.9996	.9992	.9988	.9983	.9979	.9975	.9972	.9968	.9964	.9960	.9956	.9952	.9949	.9945	.9941
14	.9997	.9993	.9990	.9987	.9984	.9981	.9978	.9975	.9972	.9969	.9966	.9963	.9960	.9957	.9954
15	.9997	.9995	.9992	.9990	.9987	.9985	.9982	.9980	.9977	.9975	.9973	.9970	.9968	.9966	.9963
16	.9998	.9996	.9994	.9992	.9990	.9988	.9986	.9984	.9982	.9980	.9978	.9976	.9974	.9972	.9970
17	.9998	.9997	.9995	.9993	.9992	.9990	.9988	.9987	.9985	.9984	.9982	.9981	.9979	.9977	.9976
18	.9999	.9997	.9996	.9995	.9993	.9992	.9991	.9989	.9988	.9987	.9985	.9984	.9983	.9981	.9980
19	.9999	.9998	.9997	.9996	.9994	.9993	.9992	.9991	.9990	.9989	.9988	.9987	.9986	.9985	.9984
20	.9999	.9998	.9997	.9996	.9995	.9994	.9993	.9993	.9992	.9991	.9990	.9989	.9988	.9987	.9986
25	.9999	.9999	.9999	.9998	.9998	.9997	.9997	.9997	.9996	.9996	.9995	.9995	.9995	.9994	.9994
30	.9999	.9999	.9999	.9999	.9999	.9999	.9998	.9998	.9998	.9998	.9998	.9997	.9997	.9997	.9997
35	.9999	.9999	.9999	.9999	.9999	.9999	.9999	.9998	.9999	.9998	.9998	.9999	.9998	.9998	.9998
40	.9999	.9999	.9999	.9999	.9999	.9999	.9999	.9999	.9999	.9999	.9999	.9999	.9999	.9999	.9999
45	.9999	.9999	.9999	.9999	.9999	.9999	.9999	.9999	.9999	.9999	.9999	.9999	.9999	.9999	.9999
50	.9999	.9999	.9999	.9999	.9999	.9999	.9999	.9999	.9999	.9999	.9999	.9999	.9999	.9999	.9999
60	.9999	.9999	.9999	.9999	.9999	.9999	.9999	.9999	.9999	.9999	.9999	.9999	.9999	.9999	.9999
70	.9999	.9999	.9999	.9999	.9999	.9999	.9999	.9999	.9999	.9999	.9999	.9999	.9999	.9999	.9999
80	.9999	.9999	.9999	.9999	.9999	.9999	.9999	.9999	.9999	.9999	.9999	.9999	.9999	.9999	.9999
90	.9999	.9999	.9999	.9999	.9999	.9999	.9999	.9999	.9999	.9999	.9999	.9999	.9999	.9999	.9999
100	.9999	.9999	.9999	.9999	.9999	.9999	.9999	.9999	.9999	.9999	.9999	.9999	.9999	.9999	.9999

Number of Monitoring Wells (k)

Previous n	20	25	30	35	40	45	50	55	60	65	70	75	80	90	100
4	.8628	.8455	.8307	.8177	.8061	.7957	.7862	.7775	.7694	.7619	.7549	.7484	.7422	.7308	.7206
5	.9091	.8958	.8841	.8737	.8643	.8557	.8477	.8404	.8335	.8271	.8210	.8153	.8099	.7999	.7908
6	.9381	.9280	.9189	.9107	.9032	.8962	.8897	.8837	.8780	.8726	.8675	.8627	.8581	.8495	.8416
7	.9567	.9490	.9421	.9356	.9297	.9241	.9189	.9139	.9093	.9048	.9006	.8966	.8927	.8855	.8788
8	.9691	.9632	.9578	.9528	.9480	.9436	.9394	.9354	.9316	.9280	.9245	.9212	.9180	.9119	.9063
9	.9774	.9729	.9687	.9648	.9610	.9575	.9541	.9509	.9478	.9448	.9420	.9393	.9366	.9315	.9268
10	.9832	.9797	.9764	.9733	.9703	.9675	.9648	.9622	.9597	.9573	.9549	.9527	.9505	.9463	.9423
11	.9873	.9845	.9820	.9795	.9771	.9748	.9727	.9705	.9685	.9665	.9646	.9627	.9609	.9574	.9541
12	.9902	.9881	.9860	.9840	.9821	.9803	.9785	.9768	.9751	.9735	.9719	.9704	.9689	.9659	.9632
13	.9924	.9907	.9890	.9874	.9859	.9844	.9829	.9815	.9801	.9788	.9775	.9762	.9750	.9725	.9702
14	.9940	.9926	.9913	.9900	.9887	.9875	.9863	.9851	.9840	.9829	.9818	.9807	.9797	.9777	.9757
15	.9952	.9941	.9930	.9919	.9909	.9899	.9889	.9879	.9870	.9861	.9852	.9843	.9834	.9817	.9800
16	.9961	.9952	.9943	.9934	.9926	.9917	.9909	.9901	.9893	.9886	.9878	.9871	.9863	.9849	.9835
17	.9968	.9961	.9953	.9946	.9939	.9932	.9925	.9919	.9912	.9905	.9899	.9893	.9886	.9874	.9862
18	.9974	.9968	.9961	.9955	.9949	.9944	.9938	.9932	.9927	.9921	.9916	.9910	.9905	.9895	.9885
19	.9978	.9973	.9968	.9963	.9958	.9953	.9948	.9943	.9939	.9934	.9929	.9925	.9920	.9911	.9903
20	.9982	.9977	.9973	.9969	.9965	.9960	.9956	.9952	.9948	.9944	.9940	.9936	.9932	.9925	.9917
25	.9992	.9990	.9988	.9986	.9984	.9982	.9980	.9978	.9976	.9974	.9972	.9970	.9968	.9965	.9961
30	.9996	.9995	.9994	.9993	.9992	.9990	.9989	.9988	.9987	.9986	.9985	.9984	.9983	.9981	.9979
35	.9998	.9997	.9996	.9996	.9995	.9995	.9994	.9993	.9993	.9992	.9992	.9991	.9990	.9989	.9988
40	.9999	.9998	.9998	.9997	.9997	.9997	.9996	.9996	.9996	.9995	.9995	.9995	.9994	.9993	.9993
45	.9999	.9999	.9999	.9998	.9998	.9998	.9998	.9997	.9997	.9997	.9997	.9996	.9996	.9996	.9995
50	.9999	.9999	.9999	.9999	.9998	.9999	.9998	.9999	.9998	.9998	.9998	.9998	.9999	.9997	.9997
60	.9999	.9999	.9999	.9999	.9999	.9999	.9999	.9999	.9999	.9999	.9999	.9999	.9999	.9999	.9998
70	.9999	.9999	.9999	.9999	.9999	.9999	.9999	.9999	.9999	.9999	.9999	.9999	.9999	.9999	.9999
80	.9999	.9999	.9999	.9999	.9999	.9999	.9999	.9999	.9999	.9999	.9999	.9999	.9999	.9999	.9999
90	.9999	.9999	.9999	.9999	.9999	.9999	.9999	.9999	.9999	.9999	.9999	.9999	.9999	.9999	.9999
100	.9999	.9999	.9999	.9999	.9999	.9999	.9999	.9999	.9999	.9999	.9999	.9999	.9999	.9999	.9999

Prepared by Charles Davis based on results in Davis and McNichols (1993).

TABLE 2.9 Probability That at Least One Sample Will Be below the Second Largest of _n_ Background Measurements at Each of _k_ Monitoring Wells

Previous _n_	Number of Monitoring Wells (_k_)														
	1	2	3	4	5	6	7	8	9	10	11	12	13	14	15
4	.6000	.4000	.2857	.2143	.1667	.1333	.1091	.0909	.0769	.0659	.0571	.0500	.0441	.0392	.0351
5	.6667	.4762	.3571	.2778	.2222	.1818	.1515	.1282	.1099	.0952	.0833	.0735	.0654	.0585	.0526
6	.7143	.5357	.4167	.3333	.2727	.2273	.1923	.1648	.1429	.1250	.1103	.0980	.0877	.0789	.0714
7	.7500	.5833	.4667	.3818	.3182	.2692	.2308	.2000	.1750	.1544	.1373	.1228	.1105	.1000	.0909
8	.7778	.6222	.5091	.4242	.3590	.3077	.2667	.2333	.2059	.1830	.1637	.1474	.1333	.1212	.1107
9	.8000	.6545	.5455	.4615	.3956	.3429	.3000	.2647	.2353	.2105	.1895	.1714	.1558	.1423	.1304
10	.8182	.6818	.5769	.4945	.4286	.3750	.3309	.2941	.2632	.2368	.2143	.1948	.1779	.1630	.1500
11	.8333	.7051	.6044	.5238	.4583	.4044	.3595	.3216	.2895	.2619	.2381	.2174	.1993	.1833	.1692
12	.8462	.7253	.6286	.5500	.4853	.4314	.3860	.3474	.3143	.2857	.2609	.2391	.2200	.2031	.1880
13	.8571	.7429	.6500	.5735	.5098	.4561	.4105	.3714	.3377	.3083	.2826	.2600	.2400	.2222	.2063
14	.8667	.7583	.6691	.5948	.5322	.4789	.4333	.3939	.3597	.3297	.3033	.2800	.2593	.2407	.2241
15	.8750	.7721	.6863	.6140	.5526	.5000	.4545	.4150	.3804	.3500	.3231	.2991	.2778	.2586	.2414
16	.8824	.7843	.7018	.6316	.5714	.5195	.4743	.4348	.4000	.3692	.3419	.3175	.2956	.2759	.2581
17	.8889	.7953	.7158	.6476	.5887	.5375	.4928	.4533	.4185	.3875	.3598	.3350	.3126	.2925	.2742
18	.8947	.8053	.7286	.6623	.6047	.5543	.5100	.4708	.4359	.4048	.3768	.3517	.3290	.3085	.2898
19	.9000	.8143	.7403	.6759	.6196	.5700	.5262	.4872	.4524	.4212	.3931	.3677	.3448	.3239	.3048
20	.9048	.8225	.7510	.6884	.6333	.5846	.5413	.5026	.4680	.4368	.4086	.3831	.3598	.3387	.3193
25	.9231	.8547	.7937	.7389	.6897	.6452	.6048	.5682	.5348	.5042	.4762	.4505	.4267	.4049	.3846
30	.9355	.8770	.8239	.7754	.7311	.6905	.6532	.6188	.5870	.5577	.5305	.5052	.4817	.4598	.4394
35	.9444	.8934	.8464	.8030	.7628	.7256	.6911	.6589	.6290	.6010	.5749	.5504	.5275	.5060	.4857
40	.9512	.9059	.8638	.8245	.7879	.7536	.7216	.6915	.6633	.6367	.6118	.5882	.5660	.5451	.5253
45	.9565	.9158	.8777	.8418	.8082	.7765	.7466	.7184	.6918	.6667	.6429	.6203	.5989	.5786	.5593
50	.9608	.9238	.8890	.8560	.8249	.7955	.7675	.7411	.7160	.6921	.6694	.6478	.6272	.6076	.5889
60	.9672	.9360	.9063	.8780	.8510	.8252	.8005	.7770	.7545	.7329	.7123	.6925	.6735	.6553	.6378
70	.9718	.9448	.9189	.8941	.8703	.8474	.8254	.8042	.7838	.7642	.7454	.7272	.7097	.6928	.6765
80	.9753	.9515	.9286	.9065	.8852	.8646	.8447	.8255	.8069	.7890	.7717	.7549	.7387	.7229	.7077
90	.9780	.9568	.9362	.9163	.8970	.8783	.8602	.8426	.8256	.8091	.7931	.7775	.7624	.7478	.7335
100	.9802	.9610	.9423	.9242	.9066	.8895	.8729	.8567	.8410	.8257	.8108	.7963	.7822	.7685	.7551

Number of Monitoring Wells (k)

Previous n	20	25	30	35	40	45	50	55	60	65	70	75	80	90	100
4	.0217	.0148	.0107	.0081	.0063	.0051	.0042	.0035	.0030	.0026	.0022	.0019	.0017	.0014	.0011
5	.0333	.0230	.0168	.0128	.0101	.0082	.0067	.0056	.0048	.0041	.0036	.0032	.0028	.0022	.0018
6	.0462	.0323	.0238	.0183	.0145	.0118	.0097	.0082	.0070	.0060	.0053	.0046	.0041	.0033	.0027
7	.0598	.0423	.0315	.0244	.0194	.0158	.0132	.0111	.0095	.0082	.0072	.0063	.0056	.0045	.0037
8	.0741	.0530	.0398	.0310	.0248	.0203	.0169	.0143	.0123	.0107	.0093	.0082	.0073	.0059	.0048
9	.0887	.0642	.0486	.0381	.0306	.0252	.0210	.0179	.0153	.0133	.0117	.0103	.0092	.0074	.0061
10	.1034	.0756	.0577	.0455	.0367	.0303	.0254	.0216	.0186	.0162	.0142	.0126	.0112	.0091	.0075
11	.1183	.0873	.0671	.0531	.0431	.0357	.0301	.0256	.0221	.0193	.0170	.0150	.0134	.0109	.0090
12	.1331	.0991	.0767	.0611	.0498	.0414	.0349	.0299	.0258	.0226	.0199	.0176	.0158	.0128	.0106
13	.1477	.1110	.0864	.0691	.0566	.0472	.0399	.0342	.0297	.0260	.0229	.0204	.0182	.0148	.0123
14	.1622	.1228	.0962	.0774	.0636	.0532	.0451	.0388	.0337	.0295	.0261	.0232	.0208	.0170	.0141
15	.1765	.1346	.1061	.0857	.0707	.0593	.0505	.0435	.0378	.0332	.0294	.0262	.0235	.0192	.0160
16	.1905	.1463	.1159	.0941	.0779	.0656	.0559	.0483	.0421	.0370	.0328	.0293	.0263	.0216	.0180
17	.2042	.1580	.1258	.1026	.0852	.0719	.0615	.0532	.0465	.0410	.0364	.0325	.0292	.0240	.0200
18	.2176	.1694	.1356	.1110	.0926	.0783	.0672	.0582	.0509	.0450	.0400	.0358	.0322	.0265	.0222
19	.2308	.1808	.1454	.1195	.0999	.0848	.0729	.0633	.0555	.0491	.0437	.0391	.0353	.0291	.0244
20	.2436	.1919	.1551	.1279	.1073	.0913	.0787	.0685	.0601	.0532	.0474	.0426	.0384	.0317	.0266
25	.3030	.2449	.2020	.1695	.1442	.1242	.1081	.0949	.0840	.0749	.0672	.0606	.0549	.0458	.0387
30	.3551	.2929	.2458	.2091	.1801	.1568	.1377	.1218	.1086	.0974	.0879	.0797	.0726	.0609	.0519
35	.4007	.3362	.2861	.2464	.2144	.1883	.1667	.1486	.1333	.1202	.1090	.0992	.0908	.0768	.0658
40	.4407	.3750	.3230	.2811	.2468	.2185	.1948	.1747	.1576	.1429	.1301	.1190	.1092	.0930	.0802
45	.4760	.4099	.3568	.3133	.2773	.2472	.2217	.2000	.1813	.1651	.1510	.1387	.1277	.1095	.0948
50	.5072	.4414	.3877	.3431	.3059	.2744	.2475	.2244	.2043	.1869	.1716	.1581	.1461	.1259	.1096
60	.5601	.4958	.4419	.3964	.3576	.3242	.2952	.2700	.2479	.2284	.2111	.1957	.1819	.1584	.1392
70	.6030	.5409	.4879	.4423	.4028	.3684	.3382	.3116	.2880	.2670	.2482	.2313	.2161	.1899	.1681
80	.6384	.5788	.5271	.4821	.4426	.4077	.3769	.3494	.3248	.3027	.2828	.2648	.2484	.2200	.1962
90	.6681	.6110	.5609	.5168	.4776	.4428	.4116	.3836	.3584	.3356	.3149	.2960	.2788	.2486	.2231
100	.6933	.6387	.5903	.5473	.5087	.4741	.4430	.4147	.3892	.3659	.3446	.3251	.3073	.2757	.2487

Prepared by Charles Davis based on results in Davis and McNichols (1993).

TABLE 2.10 Probability That at Least One of Two Samples Will Be below the Second Largest of n Background Measurements at Each of k Monitoring Wells

Previous n	Number of Monitoring Wells (k)														
	1	2	3	4	5	6	7	8	9	10	11	12	13	14	15
4	.8000	.6714	.5810	.5134	.4609	.4187	.3840	.3550	.3302	.3089	.2903	.2739	.2594	.2464	.2347
5	.8571	.7540	.6753	.6131	.5623	.5201	.4843	.4535	.4267	.4031	.3822	.3636	.3468	.3315	.3177
6	.8929	.8095	.7424	.6870	.6402	.6002	.5654	.5349	.5078	.4837	.4619	.4423	.4244	.4080	.3929
7	.9167	.8485	.7914	.7427	.7005	.6635	.6308	.6016	.5754	.5516	.5300	.5102	.4920	.4752	.4596
8	.9333	.8768	.8280	.7853	.7476	.7140	.6837	.6564	.6314	.6086	.5877	.5683	.5503	.5336	.5180
9	.9455	.8979	.8559	.8186	.7849	.7545	.7268	.7015	.6781	.6566	.6366	.6179	.6005	.5843	.5690
10	.9545	.9141	.8777	.8449	.8149	.7875	.7622	.7388	.7172	.6970	.6781	.6604	.6438	.6281	.6133
11	.9615	.9267	.8950	.8660	.8392	.8145	.7915	.7700	.7500	.7312	.7135	.6968	.6810	.6661	.6519
12	.9670	.9368	.9090	.8832	.8592	.8368	.8159	.7962	.7777	.7603	.7438	.7281	.7132	.6991	.6856
13	.9714	.9450	.9203	.8973	.8758	.8555	.8365	.8184	.8014	.7852	.7698	.7551	.7412	.7278	.7151
14	.9750	.9516	.9297	.9091	.8897	.8713	.8539	.8373	.8216	.8066	.7923	.7786	.7655	.7529	.7409
15	.9779	.9572	.9376	.9190	.9014	.8847	.8687	.8535	.8390	.8251	.8118	.7990	.7868	.7750	.7636
16	.9804	.9618	.9442	.9274	.9114	.8961	.8815	.8675	.8541	.8412	.8288	.8169	.8054	.7944	.7837
17	.9825	.9657	.9498	.9346	.9200	.9060	.8926	.8797	.8673	.8553	.8438	.8326	.8219	.8115	.8014
18	.9842	.9691	.9546	.9407	.9274	.9146	.9022	.8903	.8788	.8676	.8569	.8465	.8364	.8267	.8172
19	.9857	.9720	.9588	.9461	.9338	.9220	.9106	.8996	.8889	.8785	.8685	.8588	.8493	.8402	.8313
20	.9870	.9745	.9624	.9508	.9395	.9286	.9180	.9078	.8978	.8882	.8788	.8697	.8609	.8522	.8439
25	.9915	.9831	.9750	.9670	.9593	.9517	.9443	.9370	.9299	.9230	.9162	.9095	.9030	.8966	.8903
30	.9940	.9880	.9822	.9764	.9708	.9653	.9598	.9544	.9492	.9440	.9389	.9338	.9289	.9240	.9192
35	.9955	.9911	.9867	.9823	.9781	.9739	.9697	.9656	.9615	.9575	.9536	.9496	.9458	.9420	.9382
40	.9965	.9931	.9897	.9863	.9829	.9796	.9764	.9731	.9699	.9667	.9636	.9605	.9574	.9543	.9513
45	.9972	.9945	.9917	.9890	.9864	.9837	.9811	.9784	.9758	.9733	.9707	.9682	.9657	.9632	.9607
50	.9977	.9955	.9933	.9910	.9888	.9867	.9845	.9823	.9802	.9781	.9759	.9738	.9718	.9697	.9676
60	.9984	.9968	.9953	.9937	.9921	.9906	.9891	.9875	.9860	.9845	.9830	.9815	.9800	.9785	.9770
70	.9988	.9977	.9965	.9953	.9942	.9930	.9919	.9907	.9896	.9885	.9873	.9862	.9851	.9840	.9828
80	.9991	.9982	.9973	.9964	.9955	.9946	.9937	.9928	.9920	.9911	.9902	.9893	.9885	.9876	.9867
90	.9993	.9986	.9979	.9971	.9964	.9957	.9950	.9943	.9936	.9929	.9922	.9915	.9908	.9901	.9894
100	.9994	.9988	.9983	.9977	.9971	.9965	.9959	.9954	.9948	.9942	.9937	.9931	.9925	.9919	.9914

Number of Monitoring Wells (k)

Previous n	20	25	30	35	40	45	50	55	60	65	70	75	80	90	100
4	.1901	.1602	.1387	.1224	.1097	.0994	.0909	.0838	.0778	.0725	.0680	.0640	.0605	.0545	.0495
5	.2636	.2260	.1983	.1769	.1598	.1458	.1342	.1244	.1160	.1086	.1022	.0965	.0915	.0828	.0757
6	.3328	.2897	.2571	.2314	.2107	.1936	.1792	.1669	.1562	.1469	.1387	.1314	.1249	.1136	.1043
7	.3962	.3494	.3133	.2844	.2608	.2410	.2242	.2098	.1972	.1861	.1763	.1675	.1596	.1459	.1345
8	.4534	.4045	.3660	.3348	.3090	.2871	.2684	.2521	.2379	.2252	.2140	.2039	.1947	.1788	.1655
9	.5046	.4548	.4149	.3821	.3547	.3312	.3110	.2932	.2776	.2637	.2512	.2399	.2297	.2118	.1967
10	.5501	.5003	.4598	.4261	.3976	.3730	.3516	.3327	.3160	.3010	.2875	.2753	.2641	.2445	.2278
11	.5907	.5415	.5010	.4668	.4376	.4123	.3900	.3703	.3527	.3369	.3226	.3095	.2976	.2765	.2585
12	.6267	.5786	.5384	.5043	.4748	.4490	.4262	.4059	.3877	.3712	.3562	.3425	.3300	.3077	.2885
13	.6587	.6120	.5726	.5387	.5092	.4832	.4601	.4395	.4208	.4038	.3884	.3742	.3611	.3378	.3177
14	.6872	.6421	.6036	.5703	.5410	.5151	.4919	.4710	.4520	.4348	.4190	.4044	.3909	.3669	.3459
15	.7125	.6692	.6319	.5992	.5704	.5446	.5215	.5005	.4815	.4640	.4480	.4331	.4194	.3947	.3731
16	.7352	.6937	.6576	.6257	.5974	.5720	.5491	.5282	.5091	.4916	.4754	.4604	.4465	.4213	.3992
17	.7555	.7158	.6810	.6501	.6224	.5974	.5748	.5541	.5351	.5175	.5013	.4862	.4722	.4468	.4243
18	.7738	.7358	.7023	.6724	.6454	.6210	.5987	.5782	.5594	.5420	.5258	.5107	.4966	.4710	.4482
19	.7902	.7540	.7218	.6928	.6666	.6428	.6209	.6008	.5822	.5649	.5488	.5338	.5197	.4940	.4712
20	.8050	.7705	.7395	.7116	.6862	.6630	.6416	.6218	.6035	.5865	.5705	.5556	.5416	.5159	.4930
25	.8606	.8335	.8086	.7855	.7642	.7442	.7256	.7081	.6916	.6761	.6615	.6476	.6344	.6099	.5876
30	.8961	.8747	.8546	.8357	.8179	.8011	.7851	.7700	.7557	.7420	.7289	.7165	.7045	.6821	.6614
35	.9199	.9027	.8863	.8707	.8559	.8418	.8293	.8153	.8029	.7910	.7795	.7685	.7579	.7378	.7190
40	.9366	.9225	.9090	.8961	.8836	.8717	.8602	.8491	.8384	.8280	.8180	.8083	.7990	.7811	.7642
45	.9486	.9369	.9257	.9148	.9043	.8941	.8843	.8747	.8655	.8565	.8478	.8393	.8310	.8151	.8001
50	.9575	.9477	.9382	.9290	.9201	.9113	.9029	.8946	.8866	.8787	.8711	.8636	.8563	.8423	.8288
60	.9696	.9625	.9555	.9487	.9420	.9355	.9290	.9227	.9166	.9105	.9046	.8988	.8931	.8819	.8712
70	.9773	.9719	.9666	.9613	.9562	.9511	.9461	.9412	.9363	.9316	.9269	.9222	.9177	.9088	.9001
80	.9824	.9782	.9740	.9698	.9658	.9617	.9577	.9538	.9499	.9461	.9423	.9386	.9349	.9276	.9205
90	.9860	.9826	.9792	.9758	.9725	.9693	.9660	.9628	.9597	.9565	.9534	.9503	.9473	.9413	.9354
100	.9886	.9858	.9830	.9802	.9775	.9748	.9721	.9695	.9669	.9642	.9617	.9591	.9565	.9515	.9466

Prepared by Charles Davis based on results in Davis and McNichols (1993).

TABLE 2.11 Probability That at Least One of Three Samples Will Be below the Second Largest of *n* Background Measurements at Each of *k* Monitoring Wells

Previous *n*	Number of Monitoring Wells														
	1	2	3	4	5	6	7	8	9	10	11	12	13	14	15
4	.8857	.8048	.7432	.6941	.6536	.6195	.5902	.5647	.5422	.5220	.5040	.4876	.4727	.4590	.4464
5	.9286	.8723	.8262	.7873	.7539	.7247	.6988	.6757	.6548	.6359	.6185	.6025	.5878	.5741	.5613
6	.9524	.9123	.8779	.8477	.8209	.7968	.7751	.7552	.7369	.7201	.7045	.6899	.6763	.6635	.6515
7	.9667	.9374	.9114	.8879	.8666	.8471	.8291	.8124	.7968	.7823	.7687	.7559	.7438	.7323	.7214
8	.9758	.9538	.9339	.9155	.8985	.8826	.8679	.8540	.8409	.8286	.8169	.8058	.7952	.7852	.7756
9	.9818	.9650	.9494	.9348	.9212	.9084	.8962	.8847	.8738	.8633	.8534	.8439	.8348	.8260	.8176
10	.9860	.9729	.9606	.9489	.9378	.9273	.9173	.9077	.8985	.8897	.8813	.8731	.8653	.8578	.8504
11	.9890	.9786	.9687	.9592	.9502	.9415	.9332	.9252	.9175	.9100	.9028	.8959	.8892	.8826	.8763
12	.9912	.9828	.9747	.9670	.9595	.9523	.9453	.9386	.9321	.9258	.9197	.9137	.9079	.9023	.8968
13	.9929	.9860	.9793	.9729	.9667	.9606	.9548	.9491	.9436	.9382	.9330	.9279	.9229	.9180	.9133
14	.9941	.9884	.9829	.9775	.9723	.9672	.9622	.9574	.9527	.9481	.9436	.9392	.9349	.9307	.9265
15	.9951	.9903	.9857	.9811	.9767	.9724	.9682	.9640	.9600	.9560	.9521	.9483	.9446	.9409	.9373
16	.9959	.9918	.9879	.9840	.9803	.9765	.9729	.9694	.9659	.9624	.9591	.9558	.9525	.9493	.9462
17	.9965	.9931	.9897	.9864	.9831	.9799	.9768	.9737	.9707	.9677	.9648	.9619	.9590	.9562	.9535
18	.9970	.9940	.9911	.9883	.9855	.9827	.9800	.9773	.9746	.9720	.9695	.9669	.9644	.9620	.9595
19	.9974	.9948	.9923	.9898	.9874	.9850	.9826	.9803	.9779	.9756	.9734	.9712	.9690	.9668	.9646
20	.9977	.9955	.9933	.9911	.9890	.9869	.9848	.9827	.9807	.9787	.9767	.9747	.9728	.9708	.9689
25	.9988	.9976	.9964	.9952	.9940	.9928	.9916	.9905	.9893	.9882	.9871	.9859	.9848	.9837	.9826
30	.9993	.9985	.9978	.9971	.9964	.9957	.9949	.9942	.9935	.9928	.9921	.9914	.9907	.9900	.9894
35	.9995	.9991	.9986	.9981	.9976	.9972	.9967	.9962	.9958	.9953	.9949	.9944	.9940	.9935	.9930
40	.9997	.9994	.9990	.9987	.9984	.9981	.9977	.9974	.9971	.9968	.9965	.9962	.9958	.0055	.9952
45	.9998	.9995	.9993	.9991	.9988	.9986	.9984	.9982	.9979	.9977	.9975	.9973	.9970	.9968	.9966
50	.9998	.9997	.9995	.9993	.9991	.9990	.9988	.9986	.9985	.9983	.9981	.9980	.9978	.9976	.9975
60	.9999	.9998	.9997	.9996	.9995	.9994	.9993	.9992	.9991	.9990	.9989	.9988	.9987	.9986	.9985
70	.9999	.9999	.9998	.9997	.9997	.9996	.9996	.9995	.9994	.9994	.9993	.9992	.9992	.9991	.9990
80	.9999	.9999	.9999	.9998	.9998	.9997	.9997	.9997	.9996	.9996	.9995	.9995	.9994	.9994	.9993
90	.9999	.9999	.9999	.9999	.9998	.9998	.9998	.9998	.9997	.9997	.9997	.9996	.9996	.9996	.9995
100	.9999	.9999	.9999	.9999	.9999	.9999	.9998	.9998	.9998	.9998	.9998	.9997	.9997	.9997	.9997

Number of Monitoring Wells (k)

Previous n	20	25	30	35	40	45	50	55	60	65	70	75	80	90	100
4	.3954	.3581	.3292	.3060	.2868	.2706	.2567	.2445	.2339	.2243	.2158	.2081	.2011	.1888	.1784
5	.5083	.4679	.4357	.4092	.3870	.3679	.3512	.3366	.3236	.3119	.3013	.2917	.2828	.2673	.2538
6	.6003	.5601	.5272	.4996	.4760	.4555	.4374	.4213	.4068	.3937	.3818	.3709	.3608	.3429	.3273
7	.6742	.6359	.6040	.5767	.5530	.5321	.5136	.4969	.4817	.4679	.4553	.4436	.4328	.4133	.3963
8	.7330	.6977	.6677	.6416	.6186	.5982	.5798	.5631	.5479	.5339	.5210	.5090	.4979	.4777	.4598
9	.7799	.7479	.7202	.6958	.6741	.6545	.6368	.6206	.6056	.5918	.5790	.5671	.5559	.5356	.5174
10	.8173	.7886	.7634	.7409	.7207	.7023	.6855	.6700	.6557	.6424	.6299	.6183	.6073	.5872	.5691
11	.8473	.8218	.7990	.7785	.7599	.7428	.7271	.7125	.6989	.6862	.6743	.6631	.6525	.6330	.6153
12	.8714	.8488	.8284	.8099	.7928	.7771	.7625	.7489	.7362	.7242	.7129	.7023	.6921	.6734	.6563
13	.8911	.8710	.8528	.8361	.8206	.8062	.7927	.7801	.7683	.7571	.7465	.7364	.7268	.7090	.6926
14	.9071	.8894	.8731	.8580	.8440	.8308	.8185	.8069	.7959	.7855	.7756	.7662	.7572	.7403	.7247
15	.9203	.9046	.8900	.8765	.8638	.8518	.8406	.8299	.8198	.8101	.8009	.7921	.7837	.7678	.7531
16	.9312	.9173	.9043	.8921	.8806	.8698	.8595	.8497	.8404	.8314	.8229	.8147	.8069	.7920	.7782
17	.9402	.9279	.9163	.9053	.8949	.8851	.8757	.8668	.8582	.8500	.8421	.8345	.8272	.8134	.8004
18	.9478	.9368	.9264	.9166	.9072	.8983	.8897	.8815	.8737	.8661	.8588	.8518	.8450	.8321	.8200
19	.9543	.9444	.9351	.9262	.9177	.9096	.9018	.8943	.8871	.8802	.8735	.8670	.8607	.8487	.8374
20	.9597	.9509	.9425	.0345	.9268	.9194	.9123	.9055	.8989	.8925	.8863	.8803	.8745	.8633	.8528
25	.9772	.9720	.9669	.9619	.9572	.9525	.9479	.9435	.9392	.9350	.9308	.9268	.9229	.9152	.9079
30	.9860	.9827	.9794	.9762	.9731	.9700	.9670	.9641	.9612	.9583	.9555	.9527	.9500	.9447	.9396
35	.9908	.9886	.9864	.9842	.9821	.9800	.9780	.9759	.9739	.9719	.9700	.9680	.9661	.9623	.9578
40	.9937	.9921	.9906	.9891	.9876	.9861	.9846	.9832	.9817	.9803	.9789	.9775	.9761	.9734	.9707
45	.9954	.9943	.9932	.9921	.9910	.9900	.9889	.9878	.9868	.9857	.9847	.9836	.9826	.9806	.9785
50	.9966	.9958	.9950	.9941	.9933	.9925	.9917	.9909	.9901	.9893	.9885	.9877	.9870	.9854	.9839
60	.9980	.9975	.9970	.9965	.9960	.9955	.9951	.9946	.9941	.9936	.9931	.9926	.9922	.9912	.9903
70	.9987	.9984	.9981	.9978	.9975	.9971	.9968	.9965	.9962	.9959	.9956	.9953	.9950	.9943	.9937
80	.9991	.9989	.9987	.9985	.9983	.9981	.9978	.9976	.9974	.9972	.9970	.9968	.9966	.9961	.9957
90	.9994	.9992	.9991	.9989	.9988	.9986	.9985	.9983	.9982	.9980	.9979	.9977	.9976	.9973	.9970
100	.9995	.9994	.9993	.9992	.9991	.9990	.9989	.9988	.9987	.9985	.9984	.9983	.9982	.9980	.9978

Prepared by Charles Davis based on results in Davis and McNichols (1993).

TABLE 2.12 Probability That at Least One of Four Samples Will Be below the Second Largest of n Background Measurements at Each of k Monitoring Wells

Previous n	Number of Monitoring Wells (k)														
	1	2	3	4	5	6	7	8	9	10	11	12	13	14	15
4	.9286	.8753	.8331	.7983	.7688	.7433	.7209	.7009	.6830	.6667	.6518	.6381	.6255	.6138	.6028
5	.9603	.9276	.8998	.8757	.8543	.8352	.8178	.8020	.7875	.7741	.7616	.7500	.7390	.7288	.7191
6	.9762	.9554	.9369	.9202	.9050	.8910	.8781	.8661	.8549	.8443	.8344	.8250	.8161	.8077	.7996
7	.9848	.9711	.9585	.9468	.9360	.9258	.9163	.9073	.8987	.8906	.8829	.8755	.8685	.8617	.8552
8	.9899	.9805	.9717	.9634	.9556	.9481	.9410	.9343	.9278	.9216	.9157	.9099	.9044	.8991	.8939
9	.9930	.9864	.9801	.9741	.9683	.9628	.9575	.9524	.9475	.9428	.9382	.9337	.9294	.9252	.9211
10	.9950	.9902	.9856	.9812	.9769	.9728	.9689	.9649	.9611	.9574	.9539	.9504	.9470	.9437	.9405
11	.9963	.9928	.9894	.9860	.9828	.9796	.9766	.9736	.9706	.9678	.9650	.9623	.9596	.9570	.9544
12	.9973	.9946	.9920	.9894	.9869	.9845	.9821	.9798	.9775	.9753	.9731	.9709	.9688	.9667	.9647
13	.9979	.9958	.9938	.9919	.9899	.9880	.9861	.9843	.9825	.9807	.9790	.9773	.9756	.9739	.9723
14	.9984	.9968	.9952	.9936	.9921	.9906	.9891	.9876	.9862	.9848	.9834	.9820	.9806	.9793	.9780
15	.9987	.9974	.9962	.9949	.9937	.9925	.9913	.9902	.9890	.9878	.9867	.9856	.9845	.9834	.9823
16	.9990	.9979	.9969	.9959	.9950	.9940	.9930	.9921	.9911	.9902	.9893	.9883	.9874	.9865	.9856
17	.9992	.9983	.9975	.9967	.9959	.9951	.9943	.9935	.9928	.9920	.9912	.9905	.9897	.9890	.9882
18	.9993	.9986	.9980	.9973	.9966	.9960	.9953	.9947	.9940	.9934	.9928	.9921	.9915	.9909	.9903
19	.9994	.9989	.9983	.9978	.9972	.9967	.9961	.9956	.9951	.9945	.9940	.9935	.9930	.9924	.9919
20	.9995	.9991	.9986	.9981	.9977	.9972	.9968	.9963	.9959	.9954	.9950	.9945	.9941	.9937	.9932
25	.9998	.9996	.9994	.9992	.9990	.9987	.9985	.9983	.9981	.9979	.9977	.9975	.9973	.9971	.9969
30	.9999	.9998	.9997	.9996	.9995	.9994	.9992	.9991	.9990	.9989	.9988	.9987	.9986	.9985	.9984
35	.9999	.9999	.9998	.9998	.9997	.9996	.9996	.9995	.9995	.9994	.9993	.9993	.9992	.9992	.9991
40	.9999	.9999	.9999	.9999	.9998	.9998	.9997	.9997	.9997	.9996	.9996	.9996	.9995	.9995	.9994
45	.9999	.9999	.9999	.9999	.9999	.9999	.9998	.9998	.9998	.9998	.9997	.9997	.9997	.9997	.9996
50	.9999	.9999	.9999	.9999	.9999	.9999	.9999	.9999	.9999	.9998	.9998	.9998	.9998	.9998	.9998
60	.9999	.9999	.9999	.9999	.9999	.9999	.9999	.9999	.9999	.9999	.9999	.9999	.9999	.9999	.9999
70	.9999	.9999	.9999	.9999	.9999	.9999	.9999	.9999	.9999	.9999	.9999	.9999	.9999	.9999	.9999
80	.9999	.9999	.9999	.9999	.9999	.9999	.9999	.9999	.9999	.9999	.9999	.9999	.9999	.9999	.9999
90	.9999	.9999	.9999	.9999	.9999	.9999	.9999	.9999	.9999	.9999	.9999	.9999	.9999	.9999	.9999
100	.9999	.9999	.9999	.9999	.9999	.9999	.9999	.9999	.9999	.9999	.9999	.9999	.9999	.9999	.9999

Number of Monitoring Wells (k)

Previous n	20	25	30	35	40	45	50	55	60	65	70	75	80	90	100
4	.5573	.5224	.4943	.4711	.4513	.4342	.4193	.4060	.3940	.3832	.3734	.3644	.3561	.3413	.3285
5	.6776	.6444	.6170	.5937	.5736	.5558	.5401	.5259	.5130	.5012	.4904	.4805	.4712	.4545	.4398
6	.7643	.7351	.7103	.6888	.6699	.6530	.6377	.6239	.6112	.5995	.5886	.5785	.5691	.5520	.5367
7	.8262	.8015	.7800	.7611	.7442	.7288	.7148	.7020	.6901	.6791	.6688	.6592	.6501	.6334	.6185
8	.8704	.8500	.8319	.8157	.8009	.7875	.7750	.7635	.7528	.7427	.7333	.7244	.7160	.7004	.6863
9	.9023	.8855	.8705	.8568	.8442	.8326	.8218	.8117	.8022	.7932	.7848	.7768	.7691	.7549	.7419
10	.9254	.9118	.8993	.8879	.8772	.8673	.8580	.8493	.8410	.8332	.8257	.8186	.8118	.7991	.7873
11	.9423	.9313	.9210	.9115	.9026	.8942	.8862	.8787	.8716	.8648	.8583	.8520	.8460	.8348	.8243
12	.9549	.9459	.9374	.9295	.9220	.9150	.9082	.9018	.8957	.8898	.8842	.8787	.8735	.8636	.8543
13	.9644	.9570	.9500	.9434	.9372	.9312	.9255	.9200	.9148	.9097	.9048	.9001	.8956	.8869	.8788
14	.9715	.9655	.9597	.9542	.9490	.9439	.9391	.9344	.9299	.9256	.9214	.9173	.9134	.9059	.8987
15	.9770	.9720	.9672	.9626	.9582	.9540	.9499	.9459	.9421	.9384	.9347	.9312	.9278	.9212	.9150
16	.9813	.9771	.9731	.9693	.9656	.9620	.9585	.9551	.9518	.9486	.9455	.9425	.9395	.9338	.9284
17	.9846	.9812	.9778	.9746	.9714	.9684	.9654	.9625	.9597	.9569	.9543	.9516	.9491	.9441	.9394
18	.9873	.9844	.9815	.9788	.9761	.9735	.9710	.9685	.9661	.9637	.9614	.9591	.9569	.9526	.9484
19	.9894	.9869	.9845	.9822	.9799	.9777	.9755	.9734	.9713	.9693	.9673	.9653	.9634	.9596	.9560
20	.9911	.9890	.9870	.9850	.9830	.9811	.9792	.9774	.9756	.9738	.9721	.9704	.9687	.9654	.9622
25	.9959	.9949	.9939	.9930	.9920	.9911	.9902	.9892	.9883	.9874	.9865	.9857	.9848	.9831	.9814
30	.9979	.9974	.9968	.9963	.9958	.9953	.9948	.9943	.9938	.9933	.9928	.9924	.9919	.9909	.9900
35	.9988	.9985	.9982	.9979	.9976	.9973	.9970	.9967	.9965	.9962	.9959	.9956	.9953	.9948	.9942
40	.9993	.9991	.9989	.9987	.9985	.9984	.9982	.9980	.9978	.9977	.9975	.9973	.9971	.9968	.9964
45	.9995	.9994	.9993	.9992	.9991	.9989	.9988	.9987	.9986	.9985	.9984	.9983	.9981	.9979	.9977
50	.9997	.9996	.9995	.9994	.9994	.9993	.9992	.9991	.9991	.9990	.9989	.9988	.9987	.9986	.9984
60	.9998	.9998	.9998	.9997	.9997	.9996	.9996	.9996	.9995	.9995	.9995	.9994	.9994	.9993	.9992
70	.9999	.9999	.9999	.9998	.9998	.9998	.9998	.9998	.9997	.9997	.9997	.9997	.9997	.9996	.9996
80	.9999	.9999	.9999	.9999	.9999	.9999	.9999	.9999	.9998	.9998	.9998	.9998	.9998	.9998	.9997
90	.9999	.9999	.9999	.9999	.9999	.9999	.9999	.9999	.9999	.9999	.9999	.9999	.9999	.9999	.9998
100	.9999	.9999	.9999	.9999	.9999	.9999	.9999	.9999	.9999	.9999	.9999	.9999	.9999	.9999	.9999

Prepared by Charles Davis based on results in Davis and McNichols (1993).

TABLE 2.13 Probability That the First or Both of Two Resamples Will Be below the Maximum of n Background Measurements at Each of k Monitoring Wells

Previous n	\multicolumn{15}{c}{Number of Monitoring Wells (k)}
	1	2	3	4	5	6	7	8	9	10	11	12	13	14	15
4	.8952	.8206	.7637	.7183	.6809	.6493	.6221	.5983	.5773	.5585	.5416	.5263	.5123	.4994	.4875
5	.9226	.8633	.8156	.7762	.7427	.7137	.6883	.6657	.6455	.6272	.6106	.5954	.5813	.5683	.5563
6	.9405	.8924	.8524	.8182	.7886	.7624	.7391	.7182	.6992	.6818	.6659	.6512	.6375	.6248	.6129
7	.9528	.9132	.8793	.8497	.8235	.8001	.7789	.7597	.7420	.7258	.7107	.6967	.6837	.6714	.6599
8	.9616	.9285	.8995	.8738	.8506	.8297	.8105	.7929	.7767	.7616	.7475	.7344	.7220	.7104	.6994
9	.9682	.9402	.9152	.8926	.8721	.8533	.8360	.8200	.8051	.7911	.7781	.7657	.7541	.7431	.7327
10	.9732	.9492	.9274	.9076	.8894	.8725	.8569	.8423	.8286	.8157	.8036	.7921	.7812	.7709	.7610
11	.9771	.9563	.9373	.9197	.9034	.8883	.8741	.8608	.8482	.8363	.8251	.8144	.8043	.7946	.7853
12	.9802	.9620	.9452	.9296	.9150	.9013	.8884	.8763	.8648	.8538	.8434	.8335	.8240	.8150	.8063
13	.9827	.9667	.9518	.9378	.9247	.9123	.9005	.8894	.8788	.8687	.8591	.8499	.8411	.8326	.8244
14	.9848	.9706	.9573	.9447	.9328	.9215	.9108	.9006	.8909	.8816	.8726	.8641	.8558	.8479	.8402
15	.9865	.9738	.9618	.9505	.9397	.9294	.9196	.9103	.9013	.8926	.8844	.8764	.8687	.8613	.8541
16	.9880	.9766	.9657	.9554	.9456	.9362	.9272	.9186	.9103	.9023	.8946	.8872	.8800	.8730	.8663
17	.9892	.9789	.9691	.9597	.9507	.9421	.9338	.9258	.9181	.9107	.9036	.8966	.8899	.8834	.8771
18	.9902	.9809	.9719	.9634	.9551	.9472	.9395	.9322	.9250	.9181	.9115	.9050	.8987	.8926	.8867
19	.9911	.9826	.9744	.9666	.9590	.9517	.9446	.9377	.9311	.9247	.9185	.9124	.9065	.9008	.8952
20	.9919	.9841	.9766	.9694	.9624	.9556	.9490	.9427	.9365	.9305	.9247	.9190	.9135	.9081	.9029
25	.9946	.9894	.9842	.9792	.9744	.9696	.9650	.9604	.9560	.9516	.9474	.9432	.9391	.9351	.9312
30	.9962	.9924	.9887	.9850	.9815	.9780	.9745	.9711	.9678	.9645	.9613	.9581	.9550	.9519	.9489
35	.9971	.9943	.9915	.9887	.9860	.9833	.9807	.9780	.9755	.9729	.9704	.9679	.9655	.9631	.9607
40	.9978	.9955	.9934	.9912	.9890	.9869	.9848	.9828	.9807	.9787	.9767	.9747	.9727	.9708	.9689
45	.9982	.9964	.9947	.9929	.9912	.9895	.9878	.9861	.9845	.9828	.9812	.9795	.9779	.9763	.9748
50	.9985	.9971	.9956	.9942	.9928	.9914	.9900	.9886	.9872	.9858	.9845	.9831	.9818	.9805	.9791
60	.9990	.9979	.9969	.9959	.9949	.9939	.9929	.9919	.9909	.9899	.9890	.9880	.9870	.9860	.9851
70	.9992	.9985	.9977	.9970	.9962	.9954	.9947	.9940	.9932	.9925	.9917	.9910	.9903	.9896	.9888
80	.9994	.9988	.9982	.9976	.9971	.9965	.9959	.9953	.9947	.9942	.9936	.9930	.9925	.9919	.9913
90	.9995	.9991	.9986	.9981	.9977	.9972	.9967	.9963	.9958	.9954	.9949	.9944	.9940	.9935	.9931
100	.9996	.9992	.9989	.9985	.9981	.9977	.9973	.9970	.9966	.9962	.9958	.9955	.9951	.9947	.9943

Number of Monitoring Wells (k)

Previous n	20	25	30	35	40	45	50	55	60	65	70	75	80	90	100
4	.4393	.4037	.3759	.3534	.3347	.3187	.3050	.2929	.2822	.2727	.2641	.2563	.2491	.2365	.2257
5	.5065	.4689	.4391	.4147	.3942	.3766	.3613	.3479	.3359	.3251	.3153	.3065	.2983	.2839	.2715
6	.5631	.5248	.4940	.4684	.4468	.4281	.4117	.3972	.3843	.3726	.3620	.3523	.3433	.3275	.3137
7	.6112	.5729	.5418	.5157	.4934	.4740	.4569	.4417	.4281	.4157	.4044	.3941	.3846	.3676	.3528
8	.6522	.6146	.5836	.5574	.5348	.5151	.4976	.4819	.4678	.4550	.4432	.4324	.4225	.4046	.3890
9	.6875	.6509	.6204	.5944	.5718	.5519	.5342	.5183	.5039	.4908	.4787	.4676	.4573	.4388	.4225
10	.7180	.6826	.6529	.6273	.6049	.5851	.5674	.5514	.5368	.5235	.5112	.4999	.4894	.4704	.4536
11	.7444	.7105	.6816	.6566	.6346	.6150	.5974	.5814	.5668	.5534	.5411	.5296	.5189	.4996	.4825
12	.7676	.7351	.7072	.6828	.6613	.6420	.6246	.6088	.5942	.5809	.5685	.5570	.5463	.5268	.5095
13	.7879	.7568	.7300	.7064	.6854	.6665	.6494	.6337	.6194	.6061	.5938	.5823	.5715	.5520	.5345
14	.8057	.7762	.7504	.7276	.7072	.6888	.6720	.6566	.6424	.6293	.6170	.6056	.5949	.5754	.5579
15	.8215	.7934	.7687	.7467	.7270	.7090	.6926	.6775	.6636	.6506	.6386	.6273	.6166	.5972	.5797
16	.8356	.8089	.7852	.7640	.7449	.7275	.7115	.6968	.6831	.6704	.6585	.6473	.6368	.6175	.6001
17	.8481	.8227	.8001	.7797	.7613	.7444	.7288	.7145	.7011	.6886	.6769	.6659	.6555	.6364	.6192
18	.8593	.8352	.8136	.7940	.7762	.7599	.7448	.7307	.7177	.7055	.6940	.6832	.6730	.6541	.6370
19	.8694	.8465	.8258	.8070	.7899	.7741	.7594	.7458	.7331	.7211	.7099	.6993	.6892	.6706	.6538
20	.8785	.8567	.8369	.8189	.8024	.7871	.7729	.7597	.7473	.7356	.7246	.7142	.7044	.6861	.6695
25	.9126	.8956	.8798	.8652	.8516	.8388	.8268	.8154	.8047	.7945	.7848	.7756	.7668	.7502	.7350
30	.9344	.9209	.9082	.8963	.8851	.8744	.8643	.8546	.8454	.8366	.8282	.8201	.8123	.7976	.7839
35	.9492	.9383	.9279	.9181	.9087	.8998	.8912	.8830	.8751	.8675	.8602	.8531	.8463	.8333	.8211
40	.9595	.9506	.9420	.9338	.9260	.9184	.9111	.9041	.8973	.8908	.8844	.8782	.8723	.8608	.8500
45	.9670	.9596	.9525	.9456	.9389	.9324	.9262	.9202	.9143	.9086	.9030	.8977	.8924	.8823	.8727
50	.9727	.9664	.9604	.9545	.9488	.9432	.9379	.9326	.9275	.9225	.9177	.9129	.9083	.8994	.8908
60	.9804	.9758	.9713	.9669	.9626	.9584	.9543	.9503	.9464	.9425	.9388	.9350	.9314	.9243	.9175
70	.9853	.9817	.9783	.9749	.9716	.9683	.9651	.9620	.9589	.9558	.9528	.9499	.9470	.9413	.9357
80	.9885	.9858	.9831	.9804	.9777	.9751	.9726	.9700	.9675	.9651	.9626	.9602	.9579	.9532	.9487
90	.9908	.9886	.9864	.9842	.9821	.9800	.9779	.9758	.9738	.9717	.9697	.9678	.9658	.9619	.9582
100	.9925	.9907	.9889	.9871	.9853	.9835	.9818	.9801	.9784	.9767	.9750	.9734	.9717	.9685	.9653

Prepared by Charles Davis based on results in Davis and McNichols (1993).

TABLE 2.14 Probability That the First or All of Three Resamples Will Be below the Maximum of n Background Measurements at Each of k Monitoring Wells

Previous	Number of Monitoring Wells (k)														
n	1	2	3	4	5	6	7	8	9	10	11	12	13	14	15
4	.8714	.7838	.7193	.6692	.6287	.5952	.5668	.5423	.5209	.5019	.4850	.4697	.4559	.4433	.4317
5	.9028	.8312	.7756	.7307	.6934	.6618	.6344	.6105	.5892	.5702	.5531	.5375	.5233	.5102	.4981
6	.9238	.8646	.8167	.7769	.7430	.7136	.6879	.6651	.6446	.6261	.6092	.5938	.5796	.5664	.5542
7	.9386	.8890	.8476	.8123	.7817	.7548	.7308	.7093	.6899	.6721	.6559	.6409	.6270	.6140	.6020
8	.9495	.9073	.8713	.8400	.8125	.7879	.7658	.7457	.7274	.7106	.6950	.6806	.6672	.6546	.6428
9	.9577	.9215	.8900	.8621	.8373	.8149	.7945	.7759	.7588	.7429	.7282	.7144	.7016	.6895	.6781
10	.9648	.9327	.9049	.8800	.8576	.8372	.8185	.8012	.7852	.7703	.7564	.7434	.7311	.7196	.7086
11	.9690	.9416	.9170	.8947	.8744	.8558	.8385	.8226	.8077	.7937	.7807	.7683	.7567	.7457	.7352
12	.9731	.9489	.9270	.9069	.8885	.8714	.8556	.8408	.8269	.8139	.8016	.7900	.7790	.7685	.7585
13	.9764	.9549	.9352	.9171	.9003	.8847	.8701	.8564	.8435	.8313	.8198	.8089	.7984	.7885	.7790
14	.9791	.9599	.9422	.9258	.9104	.8961	.8826	.8699	.8579	.8465	.8357	.8254	.8156	.8062	.7972
15	.9814	.9641	.9481	.9331	.9191	.9059	.8934	.8816	.8704	.8598	.8496	.8399	.8307	.8218	.8132
16	.9833	.9677	.9532	.9395	.9266	.9144	.9029	.8919	.8815	.8715	.8620	.8528	.8441	.8357	.8276
17	.9849	.9708	.9575	.9450	.9331	.9218	.9111	.9009	.8912	.8818	.8729	.8643	.8560	.8480	.8404
18	.9863	.9735	.9613	.9498	.9388	.9284	.9184	.9089	.8997	.8910	.8826	.8745	.8667	.8591	.8519
19	.9875	.9758	.9646	.9539	.9438	.9341	.9248	.9159	.9074	.8992	.8912	.8836	.8762	.8691	.8622
20	.9886	.9778	.9675	.9576	.9482	.9392	.9305	.9222	.9142	.9065	.8990	.8918	.8848	.8781	.8715
25	.9923	.9849	.9778	.9709	.9642	.9577	.9514	.9453	.9393	.9336	.9279	.9225	.9171	.9119	.9068
30	.9945	.9891	.9839	.9788	.9738	.9689	.9642	.9595	.9550	.9505	.9462	.9419	.9378	.9337	.9296
35	.9958	.9918	.9878	.9839	.9800	.9762	.9725	.9689	.9653	.9618	.9584	.9550	.9517	.9484	.9452
40	.9968	.9936	.9904	.9873	.9843	.9813	.9783	.9754	.9725	.9697	.9669	.9642	.9614	.9588	.9561
45	.9974	.9948	.9923	.9898	.9873	.9849	.9825	.9801	.9777	.9754	.9731	.9708	.9686	.9664	.9642
50	.9979	.9957	.9937	.9916	.9895	.9875	.9855	.9835	.9816	.9796	.9777	.9758	.9739	.9720	.9702
60	.9985	.9970	.9955	.9940	.9926	.9911	.9897	.9882	.9868	.9854	.9840	.9826	.9812	.9799	.9785
70	.9989	.9978	.9966	.9955	.9944	.9933	.9923	.9912	.9901	.9890	.9880	.9869	.9859	.9848	.9838
80	.9991	.9983	.9974	.9965	.9957	.9948	.9940	.9931	.9923	.9915	.9906	.9898	.9890	.9882	.9873
90	.9993	.9986	.9979	.9972	.9966	.9959	.9952	.9945	.9938	.9932	.9925	.9918	.9912	.9905	.9899
100	.9994	.9989	.9983	.9977	.9972	.9966	.9961	.9955	.9950	.9944	.9939	.9933	.9928	.9922	.9917

Number of Monitoring Wells (k)

Previous n	20	25	30	35	40	45	50	55	60	65	70	75	80	90	100
4	.3854	.3517	.3258	.3051	.2879	.2735	.2610	.2502	.2406	.2321	.2245	.2175	.2112	.2001	.1907
5	.4490	.4127	.3843	.3613	.3421	.3258	.3118	.2994	.2885	.2788	.2700	.2620	.2547	.2418	.2307
6	.5039	.4660	.4360	.4115	.3909	.3733	.3580	.3445	.3325	.3217	.3120	.3031	.2950	.2806	.2682
7	.5516	.5130	.4821	.4565	.4349	.4163	.4001	.3857	.3729	.3613	.3508	.3412	.3324	.3168	.3033
8	.5932	.5545	.5231	.4970	.4748	.4555	.4386	.4235	.4101	.3979	.3868	.3766	.3673	.3507	.3362
9	.6296	.5912	.5599	.5335	.5109	.4912	.4738	.4583	.4443	.4317	.4202	.4096	.3998	.3823	.3671
10	.6616	.6239	.5928	.5665	.5437	.5238	.5061	.4903	.4760	.4630	.4511	.4402	.4301	.4120	.3961
11	.6899	.6531	.6225	.5963	.5736	.5536	.5358	.5198	.5053	.4921	.4799	.4687	.4584	.4397	.4234
12	.7150	.6793	.6492	.6234	.6009	.5809	.5631	.5470	.5324	.5190	.5067	.4953	.4848	.4658	.4490
13	.7373	.7027	.6734	.6481	.6258	.6060	.5882	.5722	.5575	.5441	.5317	.5202	.5095	.4902	.4731
14	.7572	.7239	.6954	.6706	.6486	.6291	.6115	.5955	.5808	.5674	.5550	.5434	.5327	.5132	.4959
15	.7751	.7430	.7153	.6911	.6696	.6504	.6329	.6171	.6025	.5891	.5767	.5652	.5544	.5348	.5173
16	.7912	.7603	.7335	.7099	.6889	.6700	.6528	.6371	.6227	.6094	.5970	.5855	.5747	.5551	.5376
17	.8057	.7760	.7501	.7272	.7067	.6881	.6713	.6558	.6415	.6283	.6161	.6046	.5938	.5742	.5567
18	.8188	.7903	.7653	.7430	.7230	.7049	.6884	.6732	.6591	.6460	.6339	.6225	.6118	.5922	.5747
19	.8307	.8033	.7792	.7576	.7382	.7205	.7043	.6893	.6755	.6626	.6506	.6393	.6287	.6093	.5918
20	.8415	.8153	.7920	.7711	.7522	.7349	.7191	.7044	.6908	.6781	.6663	.6551	.6446	.6253	.6080
25	.8831	.8618	.8425	.8249	.8086	.7936	.7796	.7665	.7542	.7427	.7318	.7215	.7117	.6935	.6770
30	.9107	.8933	.8773	.8624	.8485	.8355	.8233	.8118	.8009	.7906	.7808	.7714	.7625	.7458	.7304
35	.9297	.9154	.9020	.8894	.8776	.8664	.8558	.8457	.8361	.8269	.8181	.8097	.8017	.7865	.7724
40	.9434	.9314	.9202	.9095	.8993	.8896	.8804	.8716	.8631	.8550	.8472	.8396	.8324	.8187	.8058
45	.9535	.9434	.9338	.9247	.9159	.9075	.8994	.8917	.8842	.8770	.8701	.8633	.8568	.8444	.8328
50	.9612	.9526	.9443	.9364	.9288	.9215	.9144	.9076	.9010	.8946	.8884	.8824	.8765	.8653	.8547
60	.9718	.9654	.9592	.9531	.9473	.9416	.9361	.9307	.9255	.9204	.9154	.9106	.9058	.8967	.8880
70	.9787	.9737	.9689	.9641	.9595	.9550	.9506	.9463	.9421	.9380	.9339	.9300	.9261	.9186	.9113
80	.9833	.9794	.9755	.9717	.9680	.9644	.9608	.9573	.9538	.9505	.9471	.9439	.9406	.9344	.9283
90	.9866	.9834	.9803	.9772	.9741	.9711	.9682	.9653	.9624	.9596	.9568	.9541	.9514	.9461	.9410
100	.9890	.9864	.9838	.9812	.9786	.9761	.9737	.9712	.9688	.9665	.9641	.9618	.9595	.9550	.9506

Prepared by Charles Davis based on results in Davis and McNichols (1993).

TABLE 2.15 Probability That the First or Both of Two Resamples Will Be below the Second Largest of n Background Measurements at Each of k Monitoring Wells

Previous	Number of Monitoring Wells (k)														
n	1	2	3	4	5	6	7	8	9	10	11	12	13	14	15
4	.7143	.5571	.4577	.3890	.3387	.3001	.2696	.2449	.2244	.2072	.1924	.1797	.1686	.1588	.1501
5	.7857	.6501	.5561	.4869	.4338	.3915	.3571	.3285	.3043	.2836	.2656	.2499	.2360	.2236	.2125
6	.8333	.7175	.6320	.5659	.5132	.4700	.4340	.4034	.3772	.3543	.3342	.3163	.3004	.2861	.2732
7	.8667	.7677	.6910	.6295	.5789	.5366	.5005	.4693	.4421	.4181	.3967	.3776	.3604	.3448	.3306
8	.8909	.8059	.7375	.6810	.6335	.5928	.5575	.5267	.4994	.4750	.4531	.4333	.4154	.3989	.3839
9	.9091	.8356	.7746	.7231	.6788	.6403	.6065	.5765	.5496	.5254	.5035	.4835	.4652	.4484	.4329
10	.9231	.8590	.8046	.7578	.7168	.6807	.6486	.6197	.5937	.5700	.5483	.5285	.5102	.4933	.4776
11	.9341	.8779	.8292	.7866	.7488	.7151	.6848	.6573	.6322	.6093	.5882	.5687	.5507	.5339	.5182
12	.9429	.8932	.8495	.8107	.7760	.7446	.7161	.6900	.6661	.6441	.6237	.6047	.5870	.5705	.5551
13	.9500	.9059	.8665	.8311	.7991	.7699	.7432	.7186	.6959	.6748	.6552	.6369	.6197	.6036	.5885
14	.9559	.9164	.8808	.8485	.8190	.7919	.7669	.7437	.7222	.7021	.6833	.6657	.6491	.6335	.6187
15	.9608	.9253	.8930	.8634	.8362	.8110	.7876	.7658	.7455	.7264	.7084	.6915	.6756	.6605	.6461
16	.9649	.9329	.9034	.8763	.8511	.8277	.8058	.7853	.7661	.7480	.7309	.7147	.6994	.6848	.6710
17	.9684	.9393	.9124	.8874	.8641	.8423	.8219	.8026	.7845	.7673	.7510	.7356	.7209	.7069	.6936
18	.9714	.9449	.9203	.8972	.8756	.8553	.8361	.8180	.8009	.7846	.7691	.7544	.7404	.7270	.7142
19	.9740	.9498	.9271	.9058	.8857	.8667	.8488	.8317	.8156	.8002	.7855	.7714	.7580	.7452	.7329
20	.9763	.9540	.9331	.9133	.8946	.8769	.8601	.8440	.8288	.8142	.8002	.7869	.7741	.7618	.7500
25	.9841	.9689	.9544	.9404	.9270	.9141	.9016	.8896	.8780	.8669	.8560	.8456	.8354	.8256	.8161
30	.9886	.9776	.9670	.9566	.9466	.9368	.9274	.9182	.9092	.9005	.8920	.8837	.8756	.8677	.8600
35	.9915	.9831	.9750	.9671	.9593	.9517	.9443	.9371	.9300	.9230	.9162	.9096	.9030	.8966	.8903
40	.9934	.9868	.9804	.9742	.9680	.9620	.9560	.9502	.9445	.9388	.9333	.9278	.9225	.9172	.9120
45	.9947	.9894	.9843	.9792	.9742	.9693	.9644	.9597	.9549	.9503	.9457	.9412	.9367	.9323	.9280
50	.9956	.9913	.9871	.9829	.9788	.9747	.9707	.9667	.9627	.9588	.9550	.9512	.9474	.9437	.9401
60	.9969	.9939	.9909	.9879	.9849	.9820	.9791	.9762	.9733	.9705	.9677	.9649	.9622	.9594	.9567
70	.9977	.9954	.9932	.9910	.9887	.9865	.9843	.9822	.9800	.9779	.9757	.9736	.9715	.9694	.9673
80	.9982	.9965	.9947	.9930	.9913	.9896	.9879	.9862	.9845	.9828	.9811	.9795	.9778	.9762	.9745
90	.9986	.9972	.9958	.9944	.9930	.9917	.9903	.9889	.9876	.9862	.9849	.9836	.9822	.9808	.9796
100	.9989	.9977	.9966	.9955	.9943	.9932	.9921	.9910	.9899	.9888	.9877	.9866	.9855	.9844	.9833

Number of Monitoring Wells (k)

Previous n	20	25	30	35	40	45	50	55	60	65	70	75	80	90	100
4	.1180	.0973	.0829	.0723	.0641	.0576	.0523	.0479	.0442	.0410	.0383	.0359	.0338	.0302	.0274
5	.1705	.1428	.1230	.1081	.0965	.0872	.0796	.0732	.0678	.0631	.0591	.0555	.0524	.0471	.0428
6	.2234	.1895	.1648	.1461	.1312	.1192	.1093	.1009	.0938	.0876	.0822	.0775	.0733	.0661	.0603
7	.2748	.2359	.2071	.1849	.1671	.1526	.1405	.1302	.1214	.1138	.1070	.1011	.0958	.0868	.0793
8	.3239	.2811	.2488	.2236	.2032	.1865	.1724	.1603	.1500	.1409	.1329	.1258	.1195	.1086	.0996
9	.3701	.3243	.2894	.2616	.2390	.2203	.2044	.1908	.1789	.1686	.1594	.1512	.1439	.1312	.1207
10	.4132	.3654	.3283	.2985	.2741	.2536	.2362	.2211	.2080	.1964	.1861	.1769	.1687	.1543	.1424
11	.4532	.4040	.3653	.3340	.3081	.2861	.2673	.2511	.2368	.2241	.2128	.2027	.1936	.1777	.1644
12	.4901	.4402	.4005	.3679	.3408	.3177	.2978	.2804	.2651	.2515	.2393	.2284	.2184	.2011	.1865
13	.5241	.4740	.4336	.4002	.3721	.3481	.3272	.3089	.2928	.2784	.2654	.2537	.2431	.2244	.2086
14	.5555	.5055	.4648	.4308	.4020	.3773	.3557	.3366	.3197	.3046	.2909	.2786	.2673	.2475	.2306
15	.5843	.5347	.4940	.4598	.4305	.4052	.3830	.3633	.3458	.3301	.3159	.3029	.2911	.2702	.2524
16	.6107	.5619	.5214	.4871	.4575	.4318	.4092	.3891	.3710	.3548	.3401	.3266	.3143	.2926	.2738
17	.6351	.5872	.5470	.5128	.4831	.4572	.4342	.4138	.3954	.3787	.3635	.3497	.3370	.3144	.2949
18	.6575	.6106	.5710	.5370	.5074	.4813	.4582	.4374	.4187	.4017	.3863	.3721	.3590	.3357	.3155
19	.6781	.6324	.5934	.5598	.5303	.5042	.4810	.4601	.4411	.4239	.4082	.3937	.3803	.3564	.3357
20	.6971	.6526	.6144	.5812	.5519	.5260	.5027	.4817	.4626	.4452	.4293	.4146	.4010	.3766	.3553
25	.7724	.7342	.7005	.6704	.6433	.6187	.5964	.5758	.5569	.5395	.5232	.5081	.4940	.4683	.4456
30	.8241	.7919	.7627	.7362	.7119	.6896	.6689	.6498	.6319	.6152	.5995	.5848	.5709	.5454	.5224
35	.8606	.8335	.8085	.7854	.7640	.7440	.7253	.7078	.6913	.6758	.6611	.6472	.6340	.6094	.5871
40	.8873	.8643	.8428	.8228	.8040	.7863	.7696	.7537	.7387	.7245	.7109	.6980	.6857	.6626	.6413
45	.9071	.8875	.8691	.8516	.8351	.8195	.8046	.7904	.7768	.7639	.7515	.7396	.7282	.7067	.6867
50	.9223	.9055	.8895	.8743	.8597	.8459	.8326	.8199	.8077	.7959	.7847	.7738	.7633	.7434	.7249
60	.9435	.9308	.9185	.9068	.8954	.8845	.8739	.8636	.8537	.8441	.8348	.8258	.8170	.8002	.7843
70	.9572	.9473	.9377	.9284	.9194	.9106	.9020	.8937	.8856	.8777	.8700	.8625	.8552	.8410	.8275
80	.9665	.9586	.9509	.9434	.9361	.9290	.9220	.9151	.9084	.9019	.8955	.8892	.8830	.8710	.8595
90	.9731	.9667	.9604	.9543	.9482	.9423	.9365	.9308	.9252	.9197	.9143	.9090	.9038	.8936	.8837
100	.9779	.9726	.9674	.9623	.9573	.9523	.9474	.9426	.9379	.9332	.9286	.9241	.9196	.9109	.9024

Prepared by Charles Davis based on results in Davis and McNichols (1993).

TABLE 2.16 Probability That the First or All of Three Resamples Will Be below the Second Largest of n Background Measurements at Each of k Monitoring Wells

Previous n	Number of Monitoring Wells (k)														
	1	2	3	4	5	6	7	8	9	10	11	12	13	14	15
4	.6714	.5013	.3989	.3308	.2825	.2465	.2187	.1965	.1784	.1633	.1506	.1398	.1304	.1222	.1150
5	.7460	.5943	.4940	.4229	.3699	.3289	.2962	.2695	.2473	.2286	.2125	.1986	.1864	.1757	.1661
6	.7976	.6644	.5702	.5000	.4457	.4024	.3670	.3375	.3125	.2911	.2726	.2563	.2419	.2291	.2176
7	.8348	.7183	.6315	.5644	.5107	.4669	.4303	.3993	.3726	.3495	.3292	.3112	.2952	.2808	.2678
8	.8626	.7605	.6813	.6181	.5663	.5230	.4863	.4548	.4273	.4031	.3817	.3626	.3454	.3299	.3157
9	.8839	.7940	.7221	.6631	.6138	.5719	.5357	.5043	.4765	.4519	.4299	.4101	.3922	.3759	.3609
10	.9006	.8210	.7557	.7010	.6545	.6143	.5792	.5483	.5208	.4962	.4740	.4538	.4355	.4187	.4032
11	.9139	.8432	.7838	.7332	.6895	.6513	.6175	.5874	.5605	.5361	.5140	.4938	.4753	.4582	.4425
12	.9247	.8615	.8075	.7607	.7198	.6836	.6513	.6222	.5960	.5721	.5503	.5303	.5118	.4947	.4789
13	.9336	.8768	.8276	.7844	.7461	.7119	.6811	.6532	.6279	.6046	.5833	.5636	.5453	.5283	.5125
14	.9410	.8898	.8447	.8048	.7690	.7368	.7076	.6809	.6565	.6340	.6132	.5939	.5759	.5592	.5435
15	.9472	.9008	.8595	.8225	.7891	.7588	.7311	.7056	.6822	.6605	.6403	.6215	.6040	.5875	.5721
16	.9525	.9103	.8723	.8380	.8068	.7782	.7520	.7278	.7053	.6845	.6650	.6468	.6297	.6136	.5984
17	.9571	.9184	.8835	.8516	.8224	.7956	.7707	.7477	.7262	.7062	.6874	.6698	.6532	.6375	.6227
18	.9610	.9256	.8933	.8636	.8363	.8110	.7875	.7656	.7451	.7259	.7079	.6908	.6748	.6596	.6451
19	.9644	.9318	.9019	.8743	.8487	.8249	.8026	.7818	.7623	.7439	.7265	.7101	.6946	.6798	.6658
20	.9673	.9373	.9095	.8837	.8597	.8373	.8162	.7965	.7778	.7602	.7436	.7278	.7128	.6985	.6849
25	.9778	.9569	.9371	.9184	.9006	.8837	.8675	.8522	.8375	.8234	.8099	.7970	.7845	.7726	.7611
30	.9839	.9686	.9538	.9397	.9261	.9130	.9005	.8883	.8766	.8653	.8543	.8437	.8335	.8235	.8139
35	.9878	.9761	.9647	.9537	.9431	.9327	.9227	.9129	.9034	.8942	.8852	.8765	.8680	.8597	.8516
40	.9905	.9812	.9722	.9634	.9548	.9465	.9383	.9303	.9226	.9149	.9075	.9002	.8931	.8861	.8793
45	.9923	.9849	.9775	.9704	.9633	.9564	.9497	.9431	.9366	.9302	.9240	.9179	.9119	.9059	.9001
50	.9937	.9875	.9815	.9755	.9697	.9639	.9582	.9527	.9472	.9418	.9365	.9313	.9262	.9211	.9161
60	.9955	.9911	.9868	.9825	.9782	.9741	.9699	.9658	.9618	.9578	.9539	.9500	.9462	.9424	.9386
70	.9967	.9934	.9901	.9869	.9837	.9805	.9774	.9742	.9712	.9681	.9651	.9621	.9591	.9562	.9532
80	.9974	.9949	.9923	.9898	.9873	.9848	.9823	.9799	.9775	.9751	.9727	.9703	.9679	.9656	.9633
90	.9979	.9959	.9939	.9918	.9898	.9878	.9859	.9839	.9819	.9800	.9780	.9761	.9742	.9723	.9704
100	.9983	.9966	.9950	.9933	.9917	.9900	.9884	.9868	.9852	.9836	.9820	.9804	.9788	.9772	.9757

Numbering of Monitoring Wells (k)

Previous n	20	25	30	35	40	45	50	55	60	65	70	75	80	90	100
4	.0888	.0723	.0610	.0528	.0465	.0416	.0376	.0343	.0316	.0292	.0272	.0255	.0239	.0213	.0193
5	.1308	.1080	.0921	.0803	.0712	.0640	.0581	.0532	.0491	.0456	.0425	.0399	.0375	.0336	.0304
6	.1744	.1458	.1254	.1101	.0982	.0887	.0809	.0743	.0688	.0640	.0599	.0563	.0531	.0477	.0433
7	.2180	.1843	.1599	.1413	.1267	.1150	.1052	.0971	.0901	.0841	.0789	.0742	.0702	.0632	.0575
8	.2607	.2227	.1947	.1732	.1561	.1422	.1307	.1209	.1125	.1053	.0990	.0934	.0884	.0799	.0729
9	.3020	.2604	.2293	.2052	.1858	.1700	.1567	.1455	.1358	.1273	.1199	.1133	.1075	.0974	.0892
10	.3413	.2969	.2633	.2369	.2155	.1979	.1830	.1704	.1594	.1498	.1414	.1339	.1272	.1157	.1061
11	.3787	.3320	.2963	.2679	.2448	.2256	.2093	.1954	.1833	.1726	.1632	.1548	.1473	.1343	.1235
12	.4139	.3656	.3282	.2982	.2736	.2529	.2354	.2203	.2071	.1955	.1852	.1760	.1677	.1533	.1413
13	.4469	.3975	.3588	.3275	.3016	.2798	.2611	.2449	.2308	.2183	.2071	.1971	.1881	.1725	.1594
14	.4779	.4278	.3881	.3557	.3288	.3059	.2863	.2692	.2542	.2408	.2289	.2182	.2086	.1917	.1775
15	.5069	.4565	.4161	.3829	.3551	.3313	.3109	.2930	.2772	.2631	.2505	.2392	.2289	.2109	.1957
16	.5340	.4835	.4427	.4089	.3804	.3560	.3348	.3162	.2997	.2850	.2718	.2598	.2490	.2300	.2138
17	.5592	.5090	.4680	.4338	.4047	.3797	.3580	.3388	.3217	.3064	.2927	.2802	.2688	.2489	.2318
18	.5828	.5329	.4919	.4575	.4281	.4027	.3804	.3607	.3431	.3274	.3131	.3002	.2883	.2675	.2497
19	.6048	.5555	.5147	.4801	.4505	.4247	.4021	.3820	.3640	.3478	.3331	.3197	.3075	.2859	.2673
20	.6253	.5767	.5362	.5017	.4719	.4459	.4230	.4025	.3842	.3676	.3526	.3389	.3263	.3039	.2847
25	.7093	.6654	.6276	.5946	.5655	.5395	.5162	.4950	.4759	.4583	.4422	.4273	.4135	.3887	.3671
30	.7697	.7311	.6971	.6667	.6394	.6146	.5921	.5714	.5524	.5349	.5186	.5034	.4892	.4635	.4407
35	.8140	.7805	.7502	.7228	.6978	.6748	.6537	.6341	.6159	.5989	.5830	.5680	.5540	.5283	.5052
40	.8472	.8181	.7914	.7669	.7443	.7233	.7037	.6854	.6683	.6522	.6370	.6227	.6091	.5840	.5613
45	.8726	.8472	.8237	.8019	.7815	.7624	.7445	.7276	.7117	.6967	.6824	.6688	.6559	.6318	.6099
50	.8923	.8701	.8494	.8299	.8116	.7944	.7781	.7626	.7479	.7339	.7206	.7078	.6957	.6729	.6518
60	.9205	.9033	.8869	.8714	.8566	.8425	.8290	.8160	.8036	.7917	.7803	.7692	.7586	.7385	.7197
70	.9390	.9254	.9124	.8998	.8878	.8761	.8649	.8541	.8436	.8335	.8238	.8143	.8051	.7875	.7709
80	.9519	.9409	.9303	.9200	.9100	.9004	.8910	.8819	.8731	.8645	.8561	.8479	.8400	.8247	.8102
90	.9611	.9521	.9433	.9348	.9264	.9183	.9104	.9027	.8952	.8878	.8806	.8736	.8667	.8535	.8407
100	.9680	.9604	.9531	.9459	.9388	.9319	.9252	.9186	.9121	.9058	.8996	.8935	.8876	.8760	.8648

Prepared by Charles Davis based on results in Davis and McNichols (1993).

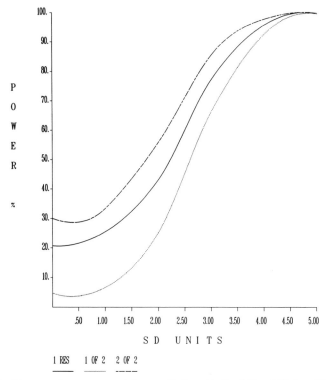

Figure 2.1 Power of nonparametric prediction limits ($N = 16$) for 10 wells and 5 constituents and 3 resampling plans.

Potential users of nonparametric methods should be aware that nonparametric methods are generally less powerful (i.e., higher false negative rates) than their parametric counterparts. However, this is only necessarily true when the distribution assumed by the parametric method is correct. This is rarely the case for groundwater monitoring data which are plagued by nondetects and irregularly shaped distributions. As such, the nonparametric alternatives may in fact be more powerful than their parametric counterparts in this context. Even if groundwater monitoring data were to behave exactly as assumed by the statistic, the loss in power of the nonparametric method would be generally no more than 10% (see Gibbons, 1991a).

However, when taking the number of constituents into account, the required background sample size for nonparametric procedures is often too large to be practical even in the absence of spatial variability. As noted previously, if intrawell comparisons are needed to control for inherent spatial variability, nonparametric procedures are often simply not available.

3 Prediction Intervals for Other Distributions

3.1 OVERVIEW

As discussed in Chapter 2, it is often the case that the distribution of a particular constituent in background groundwater quality samples is not normal nor can the data be suitably transformed to approximate a normal distribution. In many cases the previously described nonparametric approach is viable; however, it often requires large background sample sizes. When samples of the requisite size are unavailable, there may be incentive to pursue prediction limits for alternative parametric distributions so that the attractive power characteristics and ability to control sitewide false positive rates at nominal levels can be realized. Also, for those constituents that are only rarely detected, the detection frequency itself may be the outcome of interest, and prediction limits for qualitative or rare event distributions such as the Poisson may be of interest. In this chapter some useful parametric prediction limits for nonnormal distributions are examined.

3.2 LOGNORMAL DISTRIBUTION

When data exhibit a few elevated values such that the frequency distribution is "skewed" with a long right tail, estimates based on the assumption of normality do not apply. In practice, transformations of data are used to stabilize variance and bring about normality (see Box and Cox, 1964). Perhaps the most commonly used transformation in this situation is the natural log transformation (Aroian, 1941), where x is a lognormal random variable, such that $y = \log_e x \sim N(\mu, \sigma^2)$. When μ and σ^2 are unknown, we may substitute the sample estimates $\bar{y}$ and s_y^2 where

$$\bar{y} = \sum_{i=1}^{n} \frac{\log_e(x_i)}{n}$$

and

$$s_y^2 = \sum_{i=1}^{n} \frac{\left(\log_e(x_i) - \bar{y}\right)^2}{n-1}$$

TABLE 3.1   Eight Quarterly TOX Measurements

Upgradient Well	Quarter	Year	TOX	$\log_e(\text{TOX})$
MW01	1	1985	48	3.87
MW01	2	1985	5	1.61
MW01	3	1985	44	3.78
MW01	4	1985	24	3.18
MW01	1	1986	17	2.83
MW01	2	1986	6	1.79
MW01	3	1986	12	2.48
MW01	4	1986	60	4.09
			$\bar{x} = 27.00$	$\bar{y} = 2.96$
			$s = 20.97$	$s_y = 0.95$
			$n = 8$	$n = 8$

Based on a result obtained by Dahiya and Guttman (1982), the $(1 - \alpha)100\%$ upper prediction limit has the familiar form

$$\exp\left\{ \bar{y} + \sqrt{1 + \frac{1}{n}} \, t_{[n-1, \alpha]} s_y \right\}$$

where $\bar{y}$ and s_y are the mean and standard deviation computed on natural-log-transformed data. In light of this the tables in Chapter 1 may be used to obtain the one-sided 95% prediction limit for the next k measurements.

Gibbons (1987a) suggests that caution must be raised regarding the use of log transformation in calculating prediction limits. This point is best illustrated by an example. Consider the data in Table 3.1 for total organic halogen (TOX), and, for simplicity, assume that these measurements were collected from a single upgradient well, quarterly for 2 years so that s_y^2 is an unbiased estimate of σ^2.

Inspection of the original TOX values reveals a distribution that is skewed to the right; therefore log transformation appears to be a natural choice for better approximating the assumed normality of the statistical procedure. Indeed, the transformation appears to have somewhat normalized the distribution of these measurements.

Assuming that we are interested in evaluating a single new downgradient measurement, the 99% prediction limit based on the original data is

$$27.00 + 20.97(3.16) = 93.27 \text{ ppb}$$

whereas the 99% prediction limit for the log-transformed data is

$$\exp[2.96 + .95(3.16)] = 388.39 \text{ ppb}$$

The interval estimate obtained following log transformation of the data, a seemingly standard and benign practice, is extremely high and is likely to lead to elevated false negative rates even for liberal effect magnitudes. In fact, the limit value for the log-transformed data is almost an entire order of magnitude higher than the maximum value in the background sample. It is for this reason that log transformation must be used cautiously when constructing prediction intervals. It is interesting to note that in recent guidance from the EPA (1992), lognormality is taken as the default distribution.

3.3   POISSON DISTRIBUTION

In 1837, Poisson published the derivation of a distribution that bears his name. Poisson's approach was to derive a distribution for a series of independent events in which the number of trials was large, the probability of the occurrence of the outcome was small, and the probability remained constant over trials. In the classic illustration of the Poisson distribution, Bortkiewicz (1898) considered the number of deaths from being kicked by mules, per annum, in the Prussian Army Corps. In this case the probability of death from this cause was small and the number of soldiers exposed to risk was large. In 1907, "Student" (W. S. Gosset), used the Poisson distribution to represent the number of particles falling in a small area when a large number of particles are spread at random over a surface. Similarly, Rutherford and Geiger (1910) used the Poisson distribution to model variations in the number of particles emitted by a radioactive source per unit time.

In more recent years, the Poisson distribution has been used to characterize rare events such as suicide rates and number of mutated cells on plates containing 10^6 cells.

In the context of groundwater monitoring, two applications of the Poisson distribution are suggested. First, the number of "hits" out of a large number of VOC measurements could be modeled by the Poisson distribution. Second, we might consider the molecule the unit of observation and postulate that the number of molecules of a particular compound of a much larger number of molecules of water is the result of a Poisson process. For example, we might consider 12 parts per billion (ppb) of benzene to represent a count of 12 units of benzene for every billion units examined. In this context Poisson's approach is justified in that the number of units (i.e., molecules) examined is large and the probability of the occurrence (i.e., a molecule being classified as benzene) is small.

To begin, let us examine the question of how many detected compounds on average can be expected to occur per scan of 32 compounds. The probability of a scan with exactly x detected compounds is

$$f(x,\mu) = \frac{\mu^x}{x!}e^{-\mu} \qquad x = 0, 1, 2, \ldots \qquad e = 2.718 \qquad (3.1)$$

The term μ is the mean of the Poisson distribution, that is,

$$\mu = \sum_{x=0}^{\infty} xf(x) \qquad (3.2)$$

for which the sample mean is the unbiased estimator, that is,

$$\hat{\mu} = \bar{x} = \sum_{i=1}^{N} \frac{x_i}{n} \qquad (3.3)$$

For example, the probability of a sample with no detected compounds is

$$f(0, \bar{x}) = e^{-\bar{x}} \qquad (3.4)$$

and the probability of a sample with three detected compounds is

$$f(3, \bar{x}) = \frac{\bar{x}^3 e^{-\bar{x}}}{(2)(3)} \qquad (3.5)$$

The expected number of samples with three detected compounds is therefore $nf(3, \bar{x})$, where n is the number of samples.

Example 3.1

Assume that in 100 historical VOC scans, each consisting of 32 constituents, 10 detections were observed; therefore $\bar{x} = 10/100$ or 0.10 detections per scan. The probability of a sample with one detected compound is therefore

$$f(1, 0.10) = \frac{(0.10)^1 e^{-0.10}}{1!} = .09$$

and we would expect $100(.09) = 9$ out of 100 samples to have at least one detected compound.

Example 3.2

As an alternative view of the Poisson process, Gibbons (1987b) has applied the Poisson distribution to the distribution of low-level VOC concentrations themselves. Using the molecule as the unit of observation (i.e., ppb), he presents data for field blanks in which a total of 2120 ppb were detected in 61 samples; therefore $\bar{x} = 2120/61$ or 34.75 ppb per sample. The probability of a sample with exactly 40 ppb is therefore

$$f(40, 34.75) = \frac{(34.75)^{40} e^{-34.75}}{40!} = .04$$

The principal advantage of this approach is that it jointly considers all 32 VOCs on the priority pollutant list (or any other set of constituents infrequently detected), dramatically reducing an otherwise enormous multiple comparison problem. The drawback, of course, is that no distinction is made between constituents on the priority pollutant list. It has been suggested (EPA, 1992) that this limits application since we never know with certainty which constituent is responsible for the exceedance, and, in fact, the method is sensitive to small elevations in several constituents as would be expected with an actual release. This criticism is without force because the objective of detection monitoring is to identify groundwater monitoring results inconsistent with chance expectations. Identification of the specific constituents that may be implicated in a release from the facility is the role of assessment monitoring, where much larger lists of chemical analyses are often performed to identify specific pollutants in groundwater.

3.3.1 Poisson Prediction Limits

In the previous two examples, the probability of various outcomes has been considered, but this does not answer the question of what can be expected in the next k samples from that particular background distribution. Cox and Hinkley (1974) consider the case in which y has a Poisson distribution with mean μ. Having observed y, their goal is to predict y^*, which has a Poisson distribution with mean $c\mu$, where c is a known constant. In the context of groundwater monitoring, y is the number of events observed in n previous samples, and y^* is the number of events observed in a single future sample; therefore $c = 1/n$. Following Cox and Hinkley (1974), Gibbons (1987b) derived a prediction limit using as an approximation the fact that the left-hand side of

$$\left(y^* - \frac{c(y + y^*)}{(1 + c)} \right)^2 \bigg/ \left(\frac{c(y + y^*)}{(1 + c)^2} \right) < t^2_{[n-1, \alpha]} \tag{3.6}$$

is approximately a standard normal deviate; therefore the $(1 - \alpha)100\%$ prediction limit for y^* is formed from Student's t-distribution. Upon solving for y^* the upper limit value is found as the positive root of the quadratic equation:

$$y^* = \frac{2cy + t^2c + \sqrt{\left(-2cy - t^2c\right)^2 - 4\left[\left(cy\right)^2 - t^2cy\right]}}{2} \tag{3.7}$$

which after a bit of algebra simplifies somewhat to

$$y^* = cy + \frac{t^2c}{2} + tc\sqrt{y\left(1 + \frac{1}{c}\right) + \frac{t^2}{4}} \tag{3.8}$$

Example 3.3

To illustrate the computation, consider a facility with 20 monitoring wells. In $n = 16$ upgradient samples for which a 32-constituent VOC scan was conducted, there were $y = 5$ detections. Using Table 1.2 and $k = 20$, the value of Student's t-distribution required to provide an overall sitewide false positive rate of 5% is 3.28. The Poisson prediction limit is therefore

$$y^* = \frac{5}{16} + \frac{(3.28)^2}{2(16)} + \frac{3.28}{16}\sqrt{5(1 + 16) + \frac{(3.28)^2}{4}}$$

$$= 2.57 \text{ detections per scan}$$

Example 3.4

To illustrate the molecular approach, consider the same facility with 20 monitoring wells. In $n = 16$ upgradient samples for which a 32-constituent VOC scan was conducted, the 5 detections totaled 20 ppb. The Poisson prediction limit for the total detected concentration per scan is therefore

$$y^* = \frac{20}{16} + \frac{(3.28)^2}{2(16)} + \frac{3.28}{16}\sqrt{20(1 + 16) + \frac{(3.28)^2}{4}}$$

$$= 5.38 \text{ ppb detected per scan}$$

3.3.2 Discussion

The method developed for computing Poisson prediction limits is approximate due to the previously described normality assumption which leads to use of Student's t-distribution. These estimates should be adequate for practical purposes. Hahn and Meeker (1991) discuss two methods of computing Poisson prediction limits, one similar to the approach described here, for which some computational factors are displayed graphically for $\alpha = .025$ and

.005, and another based on large-sample assumptions originally described by Nelson (1982). The large-sample result is obtained by noting that the Poisson mean and variance are identical; therefore the large-sample prediction limit is simply

$$\bar{x} + z_{1-\alpha} \sqrt{\bar{x}\left(1 + \frac{1}{n}\right)} \tag{3.9}$$

For example, use of the large-sample limit in Example 3.3. yields

$$\frac{5}{16} + 2.81\sqrt{\frac{5}{16}\left(1 + \frac{1}{16}\right)} = 1.93$$

where 2.81 is the normal deviate corresponding to $\alpha = .05/20 = .0025$. This result is similar to the small-sample result of 2.57 detections per scan, although it is smaller because it ignores uncertainty in $\bar{x}$. Had we substituted $t = 3.28$ for $z = 2.81$, which is a reasonable alternative, the appropriate limit would be 2.20 detections per scan.

Exact $(1 - \alpha)100\%$ prediction limits for the Poisson distribution do not appear to have been considered in the statistical literature, although they may be obtained using Bayesian methods for interval estimation (Guttman, 1970). Similarly, no work has been done on the problem of computing Poisson prediction limits for r of m samples in each of k future events (e.g., monitoring wells). Nevertheless, Poisson prediction limits may prove enormously useful as long lists of rarely detected constituents are required for detection monitoring programs, as is the case for recent regulations governing disposal of hazardous (EPA, 1988) and municipal (EPA, 1991) waste.

3.4 SUMMARY

The methods described in this chapter complement those in the first two chapters by providing two general parametric alternatives that may be useful when the assumption of normality is untenable or when the number of background measurements is insufficient to provide a reasonable overall level of confidence. The use of Poisson prediction limits may be particularly useful for applications involving VOCs since it not only provides a reasonable method of constructing a limit but also reduces a large number of potential comparisons into a single statistical comparison, thereby dramatically reducing the overall sitewide false positive rate. There are, of course, prediction limits associated with a great many other distributions that may also have relevance to certain environmental applications. The interested reader is referred to the book by Hahn and Meeker (1991).

4 Tolerance Intervals

4.1 OVERVIEW

In the previous chapters focus has been on prediction of a potentially small and finite number of future observations drawn from the same distribution from which n previous samples are available. In some cases, however, the number of future measurements is either large or unknown such that it may be unreasonable to attempt to include *all* of these measurements with $(1 - \alpha)100\%$ confidence. In these cases we may relax the content of the interval from including 100% of the next k measurements to $P(100)\%$ coverage of *all* future measurements from the distribution with $(1 - \alpha)100\%$ confidence. This type of interval is known as a tolerance interval. Guttman (1970) introduced the distinction between β-expectation tolerance limits and β-content tolerance limits, where the former refer to prediction limits and the latter refer to tolerance limits in the nomenclature of this book. Considerable confusion between prediction and tolerance intervals exists in regulatory guidance for groundwater detection monitoring ranging from suggestions that the distinction is merely a matter of preference (EPA, 1986) to interchangeable use of expectation and coverage tolerance limits (i.e., prediction and tolerance limits) depending on the type of application. These two types of limits are *not* the same.

Under current regulatory guidance, tolerance limits are useless since tolerance limits have a built-in failure rate of $1 - P$. Since failure is indicated if any constituent in any monitoring well exceeds the statistical limit, the chance failure rate associated with tolerance limits is enormous (i.e., they are not designed to include all future measurements). If regulatory guidance permitted a small exceedance rate (e.g., 1% to 5% of all measurements could exceed the statistical limit without triggering a site assessment), then savings in cost of verification resampling and corresponding increases in the power of detecting small differences could be realized (i.e., when the number of future comparisons is large, the tolerance limit is typically smaller than the corresponding prediction limit needed to contain all k future measurements). A possible exception is a two-stage procedure suggested by Gibbons (1991b) in which tolerance limits are used as an initial screening tool, and prediction limits for the next $(1 - P)k$ measurements (i.e., the expected chance failure rate) are used for verification of only those measurements that exceeded the initial tolerance limit. Again, if a small expected chance failure rate were allowable, the costly verification stage would not be required. This two-stage procedure is quite similar to the prediction limit approach described by Davis and McNichols (1987).

4.2 NORMAL TOLERANCE LIMITS

Assume that we have available estimates $\bar{x}$ and s of the mean and standard deviation based on n background observations with degrees of freedom $f = n - 1$ from a normal distribution. We require the factor K from the two-sided interval

$$\bar{x} \pm Ks \tag{4.1}$$

which leads to the statement, "At least a proportion P of the normal population is between $\bar{x} - Ks$ and $\bar{x} + Ks$ with confidence $1 - \alpha$." Wald and Wolfowitz (1946) showed that K can be approximated by

$$K \sim ru \tag{4.2}$$

where r is a function of n and P and is determined from the normal distribution

$$\frac{1}{\sqrt{2\pi}} \int_{(1/\sqrt{n})-r}^{(1/\sqrt{n})+r} \exp\left(\frac{-x^2}{2}\right) dx = P \tag{4.3}$$

and u is a function of f and α and is defined as the $(1 - \alpha)100\%$ of the chi-square distribution as

$$u = \sqrt{\frac{f}{\chi^2_{\alpha, f}}} \tag{4.4}$$

By selecting a coverage probability P, (4.3) may be solved for r (since n is known), and by selecting a confidence level P, (4.4) may be solved for u (since $f = n - 1$ is known). Two-sided values of K are provided in Table 4.1 for $n = 4$ to ∞, 95% confidence and 95% and 99% coverage.

For one-sided tolerance limits $\bar{x} + Ks$, we require the factor K which leads to the statement, "At least a proportion P of the normal population is less than $\bar{x} + Ks$ with confidence $1 - \alpha$." Owen (1962) determines K by

$$\Pr\{(\text{noncentral } t \text{ with } \delta = z\sqrt{n}) \leq K\sqrt{n}\} = 1 - \alpha \tag{4.5}$$

where δ is the noncentrality parameter of the noncentral t-distribution with

TABLE 4.1 Factors (K) for Constructing Two-Sided Normal Tolerance Limits ($\bar{x} \pm Ks$) for 95% Confidence and 95% and 99% Coverage

n	95% Coverage	99% Coverage
4	6.370	8.299
5	5.079	6.634
6	4.414	5.775
7	4.007	5.248
8	3.732	4.891
9	3.532	4.631
10	3.379	4.433
11	3.259	4.277
12	3.169	4.150
13	3.081	4.044
14	3.012	3.955
15	2.954	3.878
16	2.903	3.812
17	2.858	3.754
18	2.819	3.702
19	2.784	3.656
20	2.752	3.615
21	2.723	3.577
22	2.697	3.543
23	2.673	3.512
24	2.651	3.483
25	2.631	3.457
30	2.549	3.350
35	2.490	3.272
40	2.445	3.212
50	2.379	3.126
60	2.333	3.066
80	2.272	2.986
100	2.233	2.934
500	2.070	2.721
∞	1.960	2.576

$f = n - 1$ degrees of freedom, and z is defined by

$$\frac{1}{\sqrt{2\pi}} \int_{-\infty}^{z} \exp\left(\frac{-x^2}{2}\right) dx = P \qquad (4.6)$$

One-sided values of K are provided in Table 4.2 for $n = 4$ to ∞, 95% confidence and 95% and 99% coverage.

To illustrate the differences between tolerance and prediction limits, Figure 4.1 displays power curves for a 95% confidence normal prediction

TABLE 4.2   Factors (K) for Constructing One-Sided Normal Tolerance Limits ($\bar{x} + Ks$) for 95% Confidence and 95% and 99% Coverage

n	95% Coverage	99% Coverage
4	5.144	7.042
5	4.210	5.749
6	3.711	5.065
7	3.401	4.643
8	3.188	4.355
9	3.032	4.144
10	2.911	3.981
11	2.815	3.852
12	2.736	3.747
13	2.670	3.659
14	2.614	3.585
15	2.566	3.520
16	2.523	3.463
17	2.486	3.414
18	2.453	3.370
19	2.423	3.331
20	2.396	3.295
21	2.371	3.262
22	2.350	3.233
23	2.329	3.206
24	2.309	3.181
25	2.292	3.158
30	2.220	3.064
35	2.166	2.994
40	2.126	2.941
50	2.065	2.863
60	2.022	2.807
80	1.965	2.733
100	1.927	2.684
500	1.763	2.475
∞	1.645	2.326

limit for the next $k = 100$ measurements based on a previous sample of $n = 20$, and a corresponding 95% confidence 95% coverage normal tolerance limit and 95% confidence 99% coverage normal tolerance limit. Inspection of Figure 4.1 reveals that the probability of failing at least one of the 100 comparisons by chance alone is much greater for the tolerance limits which have expected failure rates of 1% and 5%, respectively, versus the prediction limit that is designed to include 100% of the next 100 measurements with 95% confidence. Use of these two alternative limits for groundwater detection monitoring is anything but a "matter of personal preference."

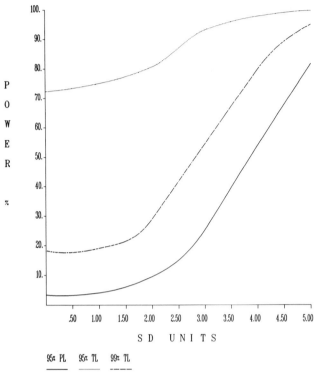

Figure 4.1   Power of prediction versus tolerance limits for 20 wells and 5 constituents ($N = 20$).

It should be noted that if we could be content with including 99% of all future measurements (e.g., 99 of the next 100 measurements), an extremely useful and general statistical monitoring program results. For example, with 20 background measurements, the tolerance limit is

$$\bar{x} + 3.295s$$

regardless of the number of future comparisons. Alternatively, the 95% confidence prediction limit for the next 50 comparisons is

$$\bar{x} + 3.668s$$

for the next 100 comparisons

$$\bar{x} + 3.929s$$

for the next 500 comparisons

$$\bar{x} + 4.696s$$

and for the next 1000 comparisons

$$\bar{x} + 5.00s$$

The tolerance limit will be more sensitive to smaller differences than the prediction limit in all of these cases. However, using the 95% confidence 99% coverage tolerance limit, we must allow one significant result per 100 comparisons in order to ensure a sitewide false positive rate of 5%. This nominal sitewide false positive rate is ensured regardless of the number of future comparisons (i.e., wells and constituents). A major advantage is that as long as no more than $(1 - P)k$ comparisons exceed the tolerance limit, there is no need for verification resampling. Since site impacts are rarely confined to a single monitoring constituent, this is quite a reasonable and cost-effective alternative to traditional approaches to groundwater detection monitoring.

As an alternative, Gibbons (1991b) has proposed a two-stage procedure in which on stage 1 a normal tolerance limit is used, and for those well and constituent combinations that exceed the tolerance limit, resampled values are compared to a prediction limit for the next $(1 - P)k$ measurements (i.e., stage 2).

Example 4.1

An owner/operator has a facility with 50 monitoring wells and is required to evaluate 20 inorganic constituents on a quarterly basis. The total number of quarterly comparisons is 1000. Application of a 95% confidence 99% coverage tolerance limit should include all but 10 of the new monitoring measurements. After a 2-year period, a total of 40 background measurements are available from 5 upgradient wells. The tolerance limit is therefore

$$\bar{x} + 2.941s$$

which is to be computed separately for each constituent. In the event that more than 10 exceedances are observed, those wells should be resampled only for those constituents that failed and compared to a 95% confidence prediction limit for the next $k = 10$ measurements (i.e., the expected failure rate), that is,

$$\bar{x} + 2.708s\sqrt{1 + \frac{1}{40}}$$

The conservative approach would be to obtain verification resamples for the tolerance limit exceedances on every quarter even if there are more than 10, and compare them to the prediction limit for the next 10 measurements. It is important to point out that the prediction limit on stage 2 is based on the expected number of failures [i.e., $(1 - P)k$] and not the number of failures that happen to be observed. Of course, it is expensive to use the more conservative approach. A good compromise is to apply the more restrictive prediction limit on the next regular monitoring event to wells and constituents which exceed the tolerance limit on the preceding monitoring event. In this way consistent increases will be quickly identified at no additional cost.

4.3   POISSON TOLERANCE LIMITS

As in the case of prediction limits, tolerance limits for a Poisson random variable can be derived. These limits may also be particularly useful in the analysis of nonnaturally occurring compounds rarely detected in background groundwater samples (e.g., VOCs).

The uniformly most accurate upper tolerance limit for the Poisson distribution is given by Zacks (1970) and was adapted to the problem of groundwater detection monitoring by Gibbons (1987b). In terms of future decisions, we can construct an upper tolerance limit for the Poisson distribution that will cover $P(100)\%$ of the population of future measurements with $(1 - \alpha)100\%$ confidence. The derivation begins by obtaining the cumulative probability that a or more occurrences will be observed in the next sample:

$$\Pr(a,\mu) = \sum_{x=a}^{\infty} f(x,\mu) \tag{4.7}$$

which can be approximated by

$$\Pr(a,\mu) = \Pr(\chi^2[2a + 2]) \geq 2\mu \tag{4.8}$$

where $\chi^2[f]$ designates a chi-square random variable with f degrees of freedom. This relationship between the Poisson and chi-square distribution was first described by Hartley and Pearson (1950). Given n independent and identically distributed Poisson random variables (i.e., one count per sample), the sum

$$T_n = \sum_{i=1}^{n} x_i \tag{4.9}$$

also has a Poisson distribution.

Substituting T_n for μ, we can find the value for which the cumulative probability is $1 - \alpha$, that is,

$$K_{1-\alpha}(T_n) = \frac{1}{2n}\chi^2_{1-\alpha}[2T_n + 2] \tag{4.10}$$

The $P(100)\%$ upper tolerance limit is therefore $\mathrm{Pr}^{-1}[P; K_{1-\alpha}(T_n)]$ = smallest j (≥ 0) such that

$$\chi_\gamma^2[2j + 2] > 2K_{1-\alpha}(T_n) \qquad (4.11)$$

The required probability points of the chi-square distribution can be most easily obtained using the Peizer and Pratt approximation described by Maindonald (1984), page 294.

Example 4.2

As an illustration, the 95% confidence 99% coverage upper Poisson tolerance limit is computed for the data from Example 3.3. Recall that in 16 VOC scans, each consisting of 32 constituents, 5 detections were observed; therefore $T_n = 5$ and $n = 16$. The cumulative 95% probability point is

$$K_{.95}(5) = \frac{1}{2(16)} \chi_{.95}^2[2(5) + 2] = .688$$

The 99% upper tolerance limit is obtained by finding the smallest nonnegative integer j such that

$$\chi_{.01}^2[2j + 2] > 2(.688) = 1.375$$

Inspection of the chi-square distribution (see Table 4.3, extracted as follows for $j = 2$ to 4) reveals that the value of j that satisfies this equation is 3:

j	$\chi^2[2j + 2]$
2	0.87
3	1.7
4	2.6

Therefore the $\mathrm{Pr}^{-1}[.99; K_{.95}(5)]$ upper tolerance limit is 3 detections per scan, which compares favorably with the Poisson prediction limit for the next 20 measurements of 2.57 detections per scan.

Example 4.3

As a second illustration, the 95% confidence 99% coverage upper Poisson tolerance limit is computed for the concentration data from Example 3.4. Recall that in 16 VOC scans, each consisting of 32 constituents, a total of 20 ppb were detected; therefore $T_n = 20$ and $n = 16$. The cumulative 95%

probability point is

$$K_{.95}(20) = \frac{1}{2(16)} \chi^2_{.95}[2(20) + 2] = 1.844$$

The 99% upper tolerance limit is obtained by finding the smallest nonnegative integer j such that

$$\chi^2_{.01}[2j + 2] > 2(1.844) = 3.688$$

Inspection of the chi-square distribution (see Table 4.3, extracted as follows

TABLE 4.3   1%, 5%, 95%, and 99% Points of the Chi-Square Distribution for Computing Poisson Tolerance Limits

df	$\chi^2_{.01}$	$\chi^2_{.05}$	$\chi^2_{.95}$	$\chi^2_{.99}$
1	—	—	3.8	6.6
2	0.02	0.10	6.0	9.2
3	0.11	0.35	7.8	11.3
4	0.30	0.71	9.5	13.3
5	0.55	1.1	11.1	15.1
6	0.87	1.6	12.6	16.8
7	1.2	2.2	14.1	18.5
8	1.7	2.7	15.5	20.1
9	2.1	3.3	16.9	21.7
10	2.6	3.9	18.3	23.2
11	3.1	4.6	19.7	24.7
12	3.6	5.2	21.0	26.2
13	4.1	5.9	22.4	27.7
14	4.7	6.6	23.7	29.1
15	5.2	7.3	25.0	30.6
16	5.8	8.0	26.3	32.0
17	6.4	8.7	27.6	33.4
18	7.0	9.4	28.9	34.8
19	7.6	10.1	30.1	36.2
20	8.3	10.9	31.4	37.6
21	8.9	11.6	32.7	38.9
22	9.5	12.3	33.9	40.3
23	10.2	13.1	35.2	41.6
24	10.9	13.8	36.4	43.0
25	11.5	14.6	37.7	44.3
26	12.2	15.4	38.9	45.6
27	12.9	16.2	40.1	47.0
28	13.6	16.9	41.3	48.3
29	14.3	17.7	42.6	49.6
30	15.0	18.5	43.8	50.9

for $j = 5$ to 7) reveals that the value of j that satisfies this equation is 6:

j	$\chi^2[2j + 2]$
5	3.6
6	4.7
7	5.8

Therefore the $\mathrm{Pr}^{-1}[.99; K_{.95}(20)]$ upper tolerance limit is 6 detections per scan, which compares favorably with the Poisson prediction limit for the next 20 measurements of 5.38 detections per scan.

4.4   NONPARAMETRIC TOLERANCE LIMITS

The general procedure for nonparametric tolerance limits was derived by Wilks (1941). The nonparametric tolerance limit is defined in terms of the number of measurements required to be $(1 - \alpha)100\%$ certain that at least $P(100)\%$ of any population with a continuous cumulative distribution function lies below the largest value in a random sample from that population. The required inequality is

$$P^n \leq \alpha \tag{4.12}$$

which when solved for n yields

$$n = \frac{\log_e(\alpha)}{\log_e(P)} \tag{4.13}$$

For 95% confidence and 95% coverage, $n = 59$ background measurements are required. For 95% confidence and 99% coverage, $n = 299$ background measurements are required. For 99% confidence and 99% coverage, $n = 459$ background measurements are required. Due to these large background sample size requirements, it would seem that nonparametric tolerance limits would rarely be applicable to problems in groundwater detection monitoring.

4.5   SUMMARY

In many ways statistical tolerance limits are ideal for groundwater monitoring applications in that they provide a very effective means of dealing with the multiple comparison problem, namely providing coverage for a proportion of all future measurements. If current regulations tolerated a small proportion of statistical exceedances and if this proportion were less than or equal to the coverage proportion of the statistical tolerance limit, the facility would pass the test and remain in a detection monitoring mode. Unfortunately, this has not been permitted under past and current regulation in the United States. Under current regulation, the best that can be attained is to verify the initial exceedances using a prediction limit for the expected number of initial exceedances based on the coverage of the tolerance limit (i.e., a two-stage procedure). Note that current municipal solid waste regulations call for semiannual statistical evaluation of groundwater quality. As such, on the first quarter, all wells and constituents could be evaluated using a tolerance limit, and on the second quarter, any statistical exceedances could be reevaluated using the corresponding prediction limit. The net result is a semiannual statistical decision rule that can be applied to large detection monitoring programs without excessive false positive rates or sampling costs.

5 Method Detection Limits

5.1 OVERVIEW

As concern over "low level" chemical health hazards has grown, the *method detection limit* (MDL) has become a key player in environmental monitoring programs. In monitoring hazardous waste facilities, for example, the detection of a single volatile organic compound in a groundwater sample is often taken as evidence that the facility has had an impact on environmental quality. As regulations begin to use detection decisions for determining environmental impact, the statistical properties of various detection limit estimators have undergone more careful scrutiny. Clayton et al. (1987) have pointed out the following list of flaws with traditional detection limit estimators:

1. Traditional techniques for determining detection limits have been concerned only with providing protection against Type I errors, or false positive conclusions (i.e., reporting an analyte present when it is not). They have not considered the corresponding need for similar protection against Type II errors, or false negative assertions (i.e., reporting an analyte as not present when it is).
2. The influence of the calibration has been ignored altogether.
3. The measurement error variance has been assumed known (or a large-sample, highly precise estimate of it is assumed available).
4. A single straight-line calibration with known model parameters (slope and intercept) has been assumed.
5. Mathematical or logical fallacies have occurred in the development of the detection limit estimators.

From the perspective of groundwater detection monitoring, the point at which a nonnaturally occurring substance is declared as being present in a groundwater sample is critically important. Constituents that are rarely detected have a greater probability of being detected in a downgradient point of compliance well than in a background well simply because there are invariably more downgradient wells than background wells. As estimated MDLs approach zero, the probability that false positive detection decisions will be incorrectly interpreted as site impacts dramatically increases. It must never be overlooked that the MDL is a statistical estimate of a true population value, often based on very limited data, with corresponding false positive and false negative rates. A proper detection monitoring program must account for the propagation of these errors. In the following sections various detection limit estimators are reviewed, compared, and contrasted.

In general, there are two major categories of MDL estimators: those based on a single concentration design and those based on calibration designs. The major disadvantage of single concentration designs is that we must assume that variability at concentration x is identical to variability at the true MDL. Calibration designs, in which multiple concentrations in the range of the MDL are used, provide a method of modeling the variance that may often be a function of concentration (e.g., absolute variability may be proportional to concentration). In this way the resulting MDL estimate is not simply a function of what concentration the analyst decided to spike the sample. Historically, however, single concentration designs have been used exclusively by the EPA largely because they are computationally less complex. As will be shown, single concentration designs and their associated detection limit estimators are rarely, if ever, justified.

5.2   SINGLE CONCENTRATION DESIGNS

The most common definition of the method detection limit is the minimum concentration of a substance that can be identified, measured, and reported with 99% confidence that the analytic concentration is greater than zero (Kaiser, 1965). In the following sections we describe several strategies that have been proposed for the estimation of MDLs from single or fixed concentration designs for either blank or spiked samples.

5.2.1   Kaiser–Currie Method

Based on developments due to Kaiser (1956, 1965, 1966), Currie (1968) described a two-stage procedure for calculating the MDL. At the first level of analysis, Currie defined the critical level L_C. The critical level is the concentration at which the binary decision of detection can be made with a specified level of confidence. Statistically, Currie defined the critical level as

$$L_C = z_{1-\alpha}\sigma_0 \tag{5.1}$$

where σ_0 is the population standard deviation of the response signal when the true concentration (C) is zero (i.e., the standard deviation of the net signal found in the population of blank samples), and $z_{1-\alpha}$ is a multiplication factor based on the $(1 - \alpha)100\%$ point of the standardized normal distribution.

For example, the one-sided 99% point of the normal distribution is 2.33; therefore the critical value is defined as

$$L_C = z_{1-\alpha}\sigma_0 = 2.33\sigma_0 \tag{5.2}$$

Although the critical level places a restriction on the Type I error rate (i.e., false positives), no such restriction is placed on Type II error rates (i.e., false negative rates). When σ_0 is known, the Type II error rate for the critical level is 50%. That is, we have a 50% chance of declaring that the analyte is not present when it, in fact, is present. In order to provide an acceptable Type II error rate, Currie defines the detection limit (L_D) as

$$L_D = L_C + z_{1-\beta}\sigma_D \tag{5.3}$$

where σ_D is the population standard deviation of the response signal at L_D (or net response signal after subtracting the background signal), and β is the acceptable Type II error rate (i.e., false negative rate).

Currie points out that if we make the simplifying assumption that $\sigma_0 = \sigma_D$ (i.e., the variability of the signal is constant in the range of L_C to L_D) and that the risk of false positive and false negative rates is equivalent (i.e., $z_{1-\alpha} = z_{1-\beta} = z$), then the MDL is simply

$$L_D = L_C + z_\beta\sigma_D = z(\sigma_0 + \sigma_D) = 2L_C \tag{5.4}$$

or twice the critical level. For $\alpha = \beta = .01$ the MDL is therefore $4.66\sigma_0$.

In reviewing Currie's method it is critically important to note that he only considers the case in which the population values σ_0 and σ_D are known. In practice, however, σ_0 and σ_D may be equal but they are rarely, if ever, known. In this case σ_0 must be replaced with its estimate s_0 obtained from a sample of n_1 blank measurements (i.e., s_0), and σ_D must be replaced with its estimate s_D obtained from a series of n_2 spiked samples in the region of the MDL (i.e., s_D). Alternatively, we may assume that $s_D = s_0$, if there is evidence to suggest that such an assumption is reasonable. In this case the general consensus (Currie, 1988, page 28) appears to be to replace $z_{1-\alpha}$ and $z_{1-\beta}$ with the corresponding values from Student's t-distribution yielding

$$L_D = 2t_{[1-\alpha, n-1]}s_0 \tag{5.5}$$

or, in the procedure adopted by the EPA (Glaser et al., 1981),

$$\text{MDL} = t_{[.01, 6]}s_D = 3.143s_D \tag{5.6}$$

where s_D is the standard deviation of seven samples in which the analyte was spiked at a concentration of two to five times the suspected MDL.

There are two fundamental problems with this approach. First, the interval proposed by Currie for the case in which σ is known is clearly a tolerance or coverage interval. When σ is unknown and replaced by its sample-based estimate s, we cannot simply substitute $t_{1-\alpha}$ for $z_{1-\alpha}$. In this case we must equate the critical level to the corresponding tolerance limit

$$L_C = K_{[1-\alpha, P, n]}s \tag{5.7}$$

where $K_{[1-\alpha, P, n]}$ is the appropriate multiplier extracted from Table 4.1 depending on the desired coverage proportion P, confidence level $1 - \alpha$, and number of background samples (possibly blanks) used to establish s. Alternatively, if we only intend to use the critical level and corresponding detection limit for determining whether the analyte in question is present in a single test sample, the critical limit becomes a prediction limit of the form

$$L_C = t_{[1-\alpha, n-1]}s \sqrt{1 + \frac{1}{n}} \tag{5.8}$$

Of course, if the L_C and L_D are to be used for the next k determinations, where, for example, k is the number of test samples to be analyzed on that particular day, the critical level becomes

$$L_C = t_{[1-\alpha/k, n-1]}s \sqrt{1 + \frac{1}{n}} \tag{5.9}$$

Finally, if we seek to determine if the analyte was detected based on the average signal obtained from m test samples, the appropriate value of L_C is given by

$$L_C = t_{[1-\alpha, n-1]}s \sqrt{\frac{1}{m} + \frac{1}{n}} \tag{5.10}$$

The choice of how to compute L_C when σ is unknown is clearly tied to the specific application. Simple substitution of $t_{1-\alpha}$ for $z_{1-\alpha}$ does not provide a statistically rigorous solution to any of these cases.

The second problem encountered when σ is unknown involves the conversion of the critical level (L_C) to the detection limit (L_D). For prediction limits the false positive rate is described by Student's t-distribution; however, the false negative rate is governed by the noncentral t-distribution. For this reason Clayton et al. (1987) have proposed that the detection limit be computed as

$$L_D = \phi s \sqrt{1 + \frac{1}{n}} \tag{5.11}$$

where ϕ is the noncentrality parameter of the noncentral t-distribution with $n - 1$ degrees of freedom and specified Type I and II error rates (e.g., $\alpha = \beta = .01$). This formulation has identical properties to the prediction limit strategy, with the additional advantage of simultaneously controlling both false positive and false negative rates. To date, it is the only MDL calculation that correctly specifies both false positive and false negative rates when σ is unknown.

Unlike prediction intervals for which the distribution under the alternative hypothesis can be specified (i.e., the noncentral t-distribution), no such alternative distribution is available for tolerance intervals. In this case the detection limit can be approximated as

$$L_D = K_{[1-\alpha, P]} s_0 + K_{[1-\beta, P]} s_D = L_C + K_{[1-\beta, P]} s_D \qquad (5.12)$$

Of course, if it is reasonable to assume that the variability is constant in the range of L_C to L_D (i.e., $s_0 = s_D = s$) and that the risk of false positive and false negative results is the same (i.e., $\alpha = \beta$), then the detection limit is simply

$$L_D = 2 K_{[1-\alpha, P]} s \qquad (5.13)$$

Fortunately, an exact solution for a detection limit based on tolerance limits is possible. Guttman (1970) has explored the relationship between prediction intervals (in his terms tolerance intervals of β-expectation) and tolerance intervals (tolerance intervals of β-content). For example, he demonstrates that for $n = 100$ the probability that a 99% prediction limit for the next single observation will actually cover 99% of the entire population of measurements is only .5861; that is, we can only have 59% confidence that the prediction limit will cover 99% of all future measurements. We can also determine the confidence level α^* for a prediction interval that corresponds to a tolerance interval for coverage P and confidence $1 - \alpha$. This relation is

$$t_{[1-\alpha^*, n-1]} = K \Big/ \sqrt{1 + \frac{1}{n}} \qquad (5.14)$$

(see Guttman, 1970, page 89, Equation 4.42). Substitution of α^* for α in the detection limit formulation based on the noncentral t-distribution due to Clayton et al. (1987) will provide a detection limit that can be applied to a large and potentially unknown number of future sample determinations.

To illustrate these three approaches, let us consider the cases in which (1) σ is known and $\sigma_0 = \sigma_D = \sigma$ and $\alpha = \beta = .01$, (2) σ is unknown and $s_0 = s_D = s$ is estimated from $n = 7$ blank samples or fixed concentration samples, $\alpha = \beta = .01$, and we wish to construct a detection limit for the purpose of deciding whether the analyte is present in the next single test sample, and (3) σ is unknown, $s_0 = s_D = s$, $\alpha = \beta = .01$, and we wish to construct our detection limit not to be exceeded by 99% of all future test samples in which the true concentration of the analyte is zero.

With respect to case 1, the detection limit is

$$L_D = 2(2.33)\sigma = 4.66\sigma$$

For case 2, the detection limit based on the noncentral t-distribution prediction limit is

$$L_D = 6.21s\sqrt{1 + \frac{1}{7}} = 6.64s$$

For case 3, the tolerance limit can be expressed as a prediction limit, and the detection limit can be estimated using the method described in case 2, substituting α^* for α. In the present example, the value of K for a one-sided interval, based on $n = 7$, $P = .99$, and $\alpha = .01$, is $K = 6.411$ (see Table 4.2). Solving for α^* yields

$$t_{[1-\alpha^*,6]} = \frac{6.411}{\sqrt{1 + 1/7}} = 5.997$$

therefore $\alpha^* = .0005$. In light of this result, the prediction limit is $t_{[.9995,6]}\sqrt{1 + 1/7}\,s$; this is a limit that will include the next single measurement with .9995 confidence and will also provide 99% coverage of all future measurements from the population with 99% confidence.

We can verify this result by computing the probability γ that the $(1 - \alpha^*)100\%$ expectation interval has coverage $c \geq 1 - \beta$. Guttman (1970), page 89, shows that this probability is given by

$$\gamma = \Pr(c \geq 1 - \beta) = \Pr\left[T_{n-1}^*\left(\sqrt{n}\,z_{1-\beta}\right) \leq \sqrt{n+1}\,t_{[1-\alpha^*,n-1]}\right] \quad (5.15)$$

This probability can be approximated (see Winer, 1971, page 35) as $\Phi(z_\gamma)$ where

$$z_\gamma = \frac{\sqrt{n+1}\,t_{[1-\alpha^*,n-1]} - \sqrt{n}\,z_{1-\beta}}{\sqrt{1 + \left[\left(\sqrt{n+1}\,t_{[1-\alpha,n-1]}\right)^2 / 2(n-1)\right]}} \quad (5.16)$$

In the present example,

$$\sqrt{n}\,z_{1-\beta} = \sqrt{7}\,(2.326) = 6.15$$

$$\sqrt{n+1}\,t_{[1-\alpha^*,n-1]} = \sqrt{8}\,(5.997) = 16.96$$

Therefore

$$\Phi(z_\gamma) = \Phi\left(\frac{16.96 - 6.15}{\sqrt{1 + (16.96)^2/12}}\right) = \Phi(2.16) = .985$$

This value falls slightly short of the required value of $\Phi(2.33) = .990$; however, it appears to be more than adequate for practical purposes.

Given this equivalence, we can specify a detection limit L_D that properly balances false positive and false negative rates by referring to the noncentral t-distribution with α^*, β, and $n - 1$ degrees of freedom. The noncentrality parameter of the noncentral t-distribution can also be approximated as

$$\phi = -z_\beta \sqrt{1 + \frac{t_{[1-\alpha^*, n-1]}^2}{2(n-1)}} + t_{[1-\alpha^*, n-1]} \qquad (5.17)$$

where z_β is the lower $\beta(100)\%$ point of the unit normal distribution, and $t_{[1-\alpha^*, n-1]}$ is the upper $(1 - \alpha^*)100\%$ point of Student's t-distribution. In the present example, $z_{.01} = -2.326$ and $t_{[.9995, 6]} = 5.997$; therefore the noncentrality parameter $\phi(\alpha^*, \beta)$ is approximately

$$\phi = -(-2.326)\sqrt{1 + \frac{(5.997)^2}{12}} + 5.997 = 10.647$$

Values of $\phi(\alpha^*, \beta)$ for $\alpha = \beta = .05$ and $\alpha = \beta = .01$ and 95% and 99% coverage are given in Table 5.1 for $df = n - 1$ ranging from 5 to 49.

The detection limit is therefore

$$L_D = 10.647s\sqrt{1 + \frac{1}{7}} = 11.382s$$

This interval will cover 99% of all future values (for $x = 0$) with corresponding false positive and false negative rates of 1%. If the same conditions applied to an experiment with $n = 30$, as opposed to $n = 7$, the detection limit for case 3 would be $5.936s\sqrt{1 + 1/30} = 6.034s$. The method described by Clayton et al. provides the same level of assurance, but only for the next single observation. In the present example of $\alpha = .01$, $\beta = .01$, and $n = 7$, the method of Clayton et al. will only provide .8264 confidence that 99% of the population of future measurements will be covered (see Guttman, 1970, Table 4.7). The Clayton et al. MDL for $n = 7$ is $6.642s$, and for $n = 30$ the MDL is $4.969s$.

The reader may wonder why we can have a confidence level of 83% for $n = 7$, $\alpha = .01$, and $\beta = .01$, and only 59% confidence for $n = 100$, since

TABLE 5.1   Values of $\phi(\alpha^*, \beta)$ for Computing MDLs Based on Normal Tolerance Limits for Fixed Concentration Designs (df $= n - 1$)

	$\alpha = \beta = .05$ Coverage		$\alpha = \beta = .01$ Coverage	
df	95%	99%	95%	99%
5	5.860	8.845	8.089	12.299
6	5.412	8.071	7.093	10.647
7	5.116	7.556	6.463	9.603
8	4.901	7.191	6.027	8.887
9	4.741	6.917	5.714	8.367
10	4.616	6.703	5.473	7.968
11	4.514	6.532	5.283	7.658
12	4.431	6.392	5.130	7.407
13	4.361	6.275	5.003	7.199
14	4.302	6.175	4.895	7.030
15	4.249	6.088	4.804	6.881
16	4.205	6.016	4.724	6.755
17	4.165	5.949	4.657	6.644
18	4.129	5.891	4.596	6.547
19	4.098	5.839	4.539	6.460
20	4.068	5.792	4.493	6.382
21	4.044	5.750	4.448	6.314
22	4.019	5.712	4.406	6.249
23	3.996	5.677	4.372	6.192
24	3.977	5.644	4.336	6.142
29	3.895	5.516	4.205	5.936
34	3.835	5.423	4.110	5.791
39	3.791	5.354	4.038	5.685
49	3.724	5.254	3.962	5.530

intuitively we would expect the opposite. The answer is twofold. First, values of the t-distribution are far more extreme for small degrees of freedom, which, of course, produces *greater coverage*. Second, the factor $\sqrt{1 + 1/n}$ is equal to 1.069 for $n = 7$, but is only equal to 1.005 for $n = 100$. The larger the multiplier, the larger the interval and, of course, the greater the coverage.

These results also shed light on the question of what constitutes a large number of samples. In the present example, we find that a prediction interval with $\alpha^* = .0005$ provides the required coverage of 99% with 99% confidence. Based on the Bonferroni inequality, the overall experimentwise Type I error rate will be $\alpha = .01$ given an individual comparison Type I error rate of $\alpha^* = .0005$, when the experiment consists of $j = 20$ test sample comparisons; that is,

$$\alpha = 1 - (1 - \alpha^*)^j = 1 - (1 - .0005)^{20} = .01$$

In light of this result, as the number of test samples increases beyond 20 and the number of background samples used in establishing the MDL is small, say $n \leq 10$, method detection limits should be based on tolerance limits. Exact values of j could, of course, be computed for varying levels of α, β, and n.

These illustrations should make it clear that we pay a very high price for replacing σ with its sample-based estimate s, particularly when n is small (e.g., $n = 7$ as suggested by the EPA). Furthermore, if MDLs are to be used as regulatory thresholds, then the size of the detection limit must increase even further in order to provide the same overall protection from false positive and false negative rates.

5.2.2   EPA–Glaser et al. Method

Glaser et al. (1981) constructed a model for the MDL by assuming that the variability is a linear function of concentration, such that for a limited number of analyses

$$s_C = b_0 + b_1 C + e_C \tag{5.18}$$

where s_C is the standard deviation of n replicate analyses at concentration C, b_0 and b_1 are the intercept and the slope of the linear regression, respectively, and e_C is the random error associated with concentration C distributed $N(0, 1)$ over concentrations.

To avoid a negative variance estimate at $C = 0$, Glaser et al. divided through by C and obtained a new regression equation:

$$\frac{s_C}{C} = \frac{b_0}{C} + b_1 \tag{5.19}$$

The slope of the new regression is now b_0 and the intercept is b_1, estimates for which can be obtained by regressing $1/C$ on s_C/C.

Let us now define

$$t_C = \frac{C}{s_C/\sqrt{n}} \tag{5.20}$$

which is the t-value for a test of significance of the ratio of the concentration to its standard error of measurement. The regression equation can now be written as

$$\frac{\sqrt{n}}{t_C} = \frac{s_C}{C} = \frac{b_0}{C} + b_1 \tag{5.21}$$

If we set t_C equal to its critical value of $t_{[.01, n-1]}$ and solve for C, we find that

$$\text{MDL} = C = \frac{t_{[.01, n-1]}b_0}{\sqrt{n} - b_1 t_{[.01, n-1]}} \tag{5.22}$$

At this point Glaser et al. made two simplifying assumptions. First, they set the intercept b_1 to zero; therefore

$$\text{MDL} = \frac{t_{[.01, n-1]}b_0}{\sqrt{n}} \tag{5.23}$$

Second, they assume that $s_C = b_0/\sqrt{n}$; therefore

$$\text{MDL} = t_{[.01, n-1]}s_C \tag{5.24}$$

where they define s_C as the standard deviation of n analytical replicates. However, this equation is not a confidence interval or any other usual statistical interval estimate (i.e., a prediction or tolerance interval). Returning to their original regression equation:

$$\frac{s_C}{C} = \frac{b_0}{C} + b_1 \tag{5.25}$$

if we follow the first assumption of Glaser et al. (1981) and set $b_1 = 0$, then their second assumption $s_C = b_0/\sqrt{n}$ is violated and $s_C = b_0$. In light of this, the correct equation should be

$$\text{MDL} = \frac{t_{[.01, n-1]}s_C}{\sqrt{n}} \tag{5.26}$$

which is, in fact, a 99% confidence limit estimate, assuming that the mean of the seven replicate samples is zero.

Glaser et al. appear to be interchanging the concepts of the standard error $s_C/\sqrt{n}$ with the standard deviation s_C of n replicate determinations of a fixed concentration. The only case in which these two estimators are the same is for $n = 1$; however, if $n = 1$ how can we possibly estimate s_C?

In addition, the Glaser et al. method will underestimate the MDL as defined by Currie and others, because they assume that $L_C = 0$; that is, they have 99% confidence that any signal greater than zero can be detected (i.e., present or absent). This assumption highlights a major distinction between this and the other methods reviewed here, and is, of course, demonstrably false in practice. Furthermore, it seems odd that the initial assumption of this model is that variability is proportional to concentration yet, in practice, the observed standard deviation for seven replicates is used irrespective of the concentration at which the sample is spiked. In light of this, if regulations require a laboratory to attain a low MDL, the laboratory simply needs to spike at a low concentration. In this way laboratories maintain the illusion created by the EPA.

5.3 CALIBRATION DESIGNS

An alternative method for estimating MDLs is to use a calibration design. In this case a series of samples are spiked at known concentrations in the range of the hypothesized MDL, and variability is determined by examining the deviations of the actual response signals from the fitted regression line of response signal on known concentrations. In these designs it is generally assumed that the distribution of these deviations from the fitted regression line is normal with constant variance across the range of concentrations used in the study. Again, the concepts of confidence limits, tolerance limits, and prediction limits apply, and again, there appears to be some confusion regarding the choice of the appropriate interval and in some cases discrepancies between what is computed and what it is called (Hubaux and Vos, 1970). In the following discussion a brief description of these three statistical interval estimates in the calibration setting is presented.

As preparation for the following discussion, we generally conceive of the relationship between response signal (Y) and spiking concentration (X) in the region of the MDL as a linear function of the form

$$Y = \alpha + \beta(X - \overline{X}) + \varepsilon = \alpha + \beta x + \varepsilon \tag{5.27}$$

where ε is a random variable that describes the deviations from the regression line, which is distributed with mean 0 and constant variance $\sigma^2_{Y \cdot X}$. The sample regression coefficient

$$b = \frac{\Sigma x_i y_i}{\Sigma x_i^2} \tag{5.28}$$

provides an estimate of the population parameter β, where x_i and y_i denote deviations from the mean concentration and response signal, respectively (i.e., $x_i = X_i - \overline{X}$ and $y_i = Y_i - \overline{Y}$). The sample intercept

$$a = \overline{Y} - b\overline{X} \tag{5.29}$$

provides an estimate of the population parameter α. An unbiased sample estimate of $\sigma^2_{Y \cdot X}$ (i.e., the variance of deviations from the population regression line) is given by

$$s_{Y \cdot X}^2 = \sum_{i=1}^{n} \frac{\left(Y_i - \hat{Y}_i\right)^2}{n-2} \tag{5.30}$$

where $\hat{Y}_i = \bar{Y} + b(X_i - \bar{X}) = \bar{Y} + bx_i$.

5.3.1 Confidence Intervals for Calibration Lines

A $(1 - \alpha)100\%$ confidence interval for the population calibration line is given by

$$\bar{Y} + bx \pm \sqrt{2F}\, s_{Y \cdot X} \tag{5.31}$$

where F is the "variance ratio" extracted from a table of the F-distribution with 2 and $n - 2$ degrees of freedom. As X covers the range of spiking concentrations, the confidence bands form smooth curves that are two branches of a hyperbola.

Confidence bands define a region within which the true population regression line may be found with a certain level of confidence (e.g., 95% confidence). However, the precision with which the sample regression line approximates the population regression line has little to do with the problem of determining MDLs. In the context of the calibration design, we are concerned with establishing concentration limits, in which we can have a certain level of confidence that the concentration of the analyte in the solution is not zero. This problem is only indirectly related to the confidence limit problem in that they both require an estimate of $\sigma_{Y \cdot X}^2$.

5.3.2 Tolerance Intervals for Calibration Lines

The notion of a tolerance interval for a random sample of measurements can also be extended to the regression setting in which the intervals are simultaneous in each possible value of the independent variable X (e.g., spiking concentration levels). Lieberman and Miller (1963) give four techniques for deriving such intervals; the simplest and most robust is based on the Bonferroni inequality.

For the predicted response signal $\hat{Y}$ at concentration X, the interval is

$$\hat{Y}_X \pm s_{Y \cdot X} \left\{ \left(2F_{2,n-2}^{1-\alpha/2}\right)^{1/2} \left[\frac{1}{n} + \frac{x_i^2}{\sum_{i=1}^{n} x_i^2}\right]^{1/2} + \Phi(P) \left(\frac{n-2}{\alpha/2 \chi_{n-2}^2}\right)^{1/2} \right\} \tag{5.32}$$

where $F_{2,n-2}^{1-\alpha/2}$ is the upper $(1 - \alpha/2)$ percentile point of the F-distribution on 2 and $n - 2$ degrees of freedom, $^{\alpha/2}\chi_{n-2}^2$ is the lower $\alpha/2$ percentile point of the χ^2-distribution with $n - 2$ degrees of freedom, and $\Phi(P)$ is the two-sided P percentile point of the unit normal distribution.

For the case of $X = 0$, the upper tolerance limit is

$$Y_C = \bar{Y} - b\bar{X} + s_{Y \cdot X}\left\{\left(2F_{2,n-2}^{1-\alpha/2}\right)^{1/2}\left[\frac{1}{n} + \frac{\bar{X}^2}{\sum_{i=1}^{n}x_i^2}\right]^{1/2} + \Phi(P)\left(\frac{n-2}{\alpha/2\chi_{n-2}^2}\right)^{1/2}\right\}$$

(5.33)

The value Y_C in the previous equation specifies a proportion P of the population of response signals that are possible when the true concentration is zero, given a $(1 - \alpha)100\%$ confidence level. This interval estimate corresponds to the concept of a "critical level" defined by Currie (1968), for the case in which the data arise from a calibration experiment, μ and σ at $X = 0$ are unknown, and we wish to provide coverage of a proportion of the population of possible test samples and not just the next single test sample. Again, this approach is well suited to the case of MDLs as regulatory thresholds, or when they are to be applied to a large and/or potentially unknown number of future sample determinations, but not for the case in which MDLs are *continuously* reevaluated in the laboratory.

5.3.3 Prediction Intervals for Calibration Lines

In the regression case prediction limits for a single new measurement parallel those for the case of a fixed concentration design. In this case the estimated standard error of the prediction $\hat{Y}_X$ for a new value of Y at point X is

$$s(\hat{Y}_X) = s_{Y \cdot X}\sqrt{1 + \frac{1}{n} + \frac{x^2}{\sum_{i=1}^{n}(x_i)^2}}$$

(5.34)

The prediction interval for a response signal obtained from a new test sample given a concentration X (e.g., $X = 0$) is therefore

$$\bar{Y} + bx \pm t_{[1-\alpha/2, n-2]}s(\hat{Y}_X)$$

(5.35)

For example, at a concentration of $X = 0$, $x = 0 - \bar{X} = -\bar{X}$, and the prediction limit is

$$\bar{Y} - b\bar{X} \pm t_{[1-\alpha/2, n-2]}s(\hat{Y}_X)$$

(5.36)

As in the previous examples, the one-sided limit can be obtained by substituting α for $\alpha/2$. Similarly, prediction limits for r future test samples can be obtained via the Bonferroni inequality by substituting $\alpha/2r$ for $\alpha/2$. Finally, prediction limits for the average of m test sample determinations may be obtained by replacing the previous standard error of the prediction with

$$s\left(\hat{Y}_X\right) = s_{Y \cdot X} \sqrt{\frac{1}{m} + \frac{1}{n} + \frac{x^2}{\sum_{i=1}^{n} x_i^2}} \tag{5.37}$$

As a summary, let us illustrate the distinction between the three types of statistical intervals using the example of the typical astronaut's problem (Hahn, 1970).

> An astronaut who has been assigned a limited number of space flights is not very interested in what will happen on the average in the population of all space flights, of which his happens to be a random sample (a confidence limit), or even in what will happen in 99% of the population of such space flights (a tolerance limit). His main interest is in the worst that might happen in the one, or three, or five flights in which he will be personally involved (a prediction limit).

5.3.4 Hubaux and Vos Method

Hubaux and Vos (1970) were the first to apply the theory of statistical prediction to the problem of MDL estimation. Beginning from a calibration design in which response signals are determined for analyte-containing samples with concentrations throughout the range of L_C to L_D, they constructed a 99% prediction interval for the calibration regression line (see Figure 5.1). The prediction limit is exactly of the form given in the second equation (5.35) in the previous section on prediction intervals, and the critical level is defined as the value of the prediction limit for zero concentration (i.e., $X = 0$) which is given in the third equation (5.36) in that section (see L_C in Figure 5.1). The limit of detection is defined as the point at which we can have 99% confidence that the response signal is not L_C; therefore Hubaux and Vos suggest that it be obtained graphically by locating the abscissa corresponding to L_C on the lower prediction limit (see L_D in Figure 5.1). A somewhat more direct solution for L_D is obtained by solving a quadratic equation in X for given Y, in our case L_C. We begin by expressing $x_i = X_i - \overline{X}$ (i.e., a deviation from the average concentration); then the most computationally tractable solution is

$$x_D = \frac{\hat{x} + \left(t_{[1-\alpha, n-2]} s_{Y \cdot X}/b\right)\sqrt{[(n+1)/n](1-c^2) + \hat{x}^2/\sum x^2}}{1 - c^2}$$

where

$$c^2 = \frac{t^2 s_b^2}{b^2}$$

$$= \left(\frac{1}{\sum x^2}\right)\left(\frac{t s_{Y \cdot X}}{b}\right)^2$$

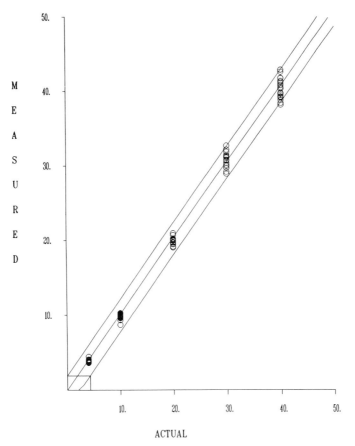

Figure 5.1 Linear calibration with constant variance: GC/MS–benzene (ppb)–prediction interval.

and

$$\hat{x} = \frac{Y_C - \overline{Y}}{b}$$

The quantity $c = ts_b/b$ is related to the significance test for b. In the present context, b will be highly significant. As such, c will be small, c^2 will become negligible, and the prediction limit becomes

$$x_D = \hat{x} + \frac{ts_{Y \cdot X}}{b} \sqrt{1 + \frac{1}{n} + \frac{\hat{x}^2}{\Sigma x^2}} \tag{5.38}$$

To transform the limit back into the original metric, simply add in the mean of the actual concentrations $\overline{X}$ to the computed limit value.

As previously mentioned, this method assumes that variability is constant throughout the range of concentrations used in the calibration design. If this assumption is violated, then a variance stabilizing transformation, such as the square root transformation, might be applied and the assumption of constant variance may be reevaluated. The choice of the square root transformation is not at all arbitrary. Since the response signals are essentially sums of ion counts, the Poisson distribution applies, and is, of course, consistent with the observation that concentration and variability are proportional (i.e., the Poisson mean and variance are identical). The square root transformation is used to bring about normality for data arising from a Poisson process (Snedecor and Cochran, 1980). Alternatively, Gibbons et al. (1991c) have shown how to compute the MDL from calibration data with nonconstant variance, as will be described in a following section.

5.3.5   Procedure Due to Clayton and Coworkers

Clayton et al. (1987) point out that the method due to Hubaux and Vos is appropriate for establishing the critical level (L_C) but not the detection limit (L_D). Their argument is that under the null hypothesis (i.e., $X = 0$) Student's t-distribution applies to the case in which σ is replaced by its sample-based estimate s and the errors of measurement are normally distributed; however, under the alternative hypothesis (i.e., $X > 0$) the appropriate distribution is the noncentral t-distribution. In light of this, they point out that the only viable method for simultaneously controlling both false positive and false negative rates at nominal levels is to derive detection limits as functions of the noncentrality parameter of the noncentral t-distribution with $n - 2$ degrees of freedom and specified values of α and β. The estimate of L_D may be found directly as

$$L_D = \frac{\phi s_{Y \cdot X}}{b} \sqrt{1 + \frac{1}{n} + \frac{\overline{X}^2}{\sum_{i=1}^{n} x_i^2}} \qquad (5.39)$$

where ϕ is the noncentrality parameter from the noncentral t-distribution with $n - 2$ degrees of freedom and specified Type I and II error rates (e.g., $\alpha = \beta = .01$). As in the case of the Hubaux–Vos method, this approach also assumes constant variance throughout the range of the calibration. As previously illustrated, this idea can also be extended to noncalibration designs.

Coleman (1993) has pointed out that this derivation ignores uncertainty in the calibration line, since only deviations from the regression line at the mean spiking concentration $\overline{X}$ are considered. In general, $\overline{X}$ is not equivalent to the concentration which corresponds to the $(1 - \alpha)100\%$ upper prediction limit for the response signal corresponding to a concentration of $X = 0$, which is the quantity considered in the derivation by Hubaux and Vos (1970). In addition, Coleman points out that as n increases, the difference between central t and noncentral t becomes negligible, rendering MDL estimates within a few percentage points of each other. These results suggest that the results of Hubaux and Vos and generalizations to be discussed may not only be computationally attractive, but also sufficiently rigorous for most applications.

5.3.6 Procedure Based on Tolerance Intervals

The previous two methods solve the problem of predicting an interval that will contain a single future measurement with specified Type I and Type II error rates. Ideally, a calibration of this type would be performed, and the corresponding MDL estimated, each time a new test sample is to be evaluated. This, however, is rarely, if ever, the case. For example, in the context of groundwater monitoring, the EPA has experimentally determined the MDLs for a variety of classes of compounds (see EPA, 1987b), and these "regulatory limits" (computed using the method of Glaser et al., 1981), are, at least in practice, used regardless of the true limit for a particular laboratory on a particular occasion. As such, the detection decision for an enormous number of test samples is being made on the basis of results obtained from a single analytical study in a single laboratory using a single analyst and a single instrument. Furthermore, the detection limits reported were computed assuming the task was to make a detection decision in a single future sample. This practice must stop and regulatory agencies must support changes so that the method by which the detection limit is computed is commensurate with the actual way in which the resulting limit is to be applied. When the number of future test samples is large, say greater than 20, and the exact number may even be unknown, the best we can do is to produce an interval that will cover a certain proportion P of the total population of measurements with a specified level of assurance $(1 - \alpha)100\%$.

As previously shown, a simultaneous tolerance interval with $P\%$ coverage and $(1 - \alpha)100\%$ confidence can be constructed for the entire calibration line using the method described by Lieberman and Miller (1963). As illustrated, a special case of computing the upper tolerance limit when $X = 0$ (i.e., the test sample does not contain the analyte in question) can provide an analogous result to the Hubaux–Vos MDL, but with the added benefit of being able to include a specified proportion P of all future measurements with confidence $(1 - \alpha)100\%$. The solution of this equation corresponds to Currie's notion of the critical level L_C. Again, solving the quadratic equation in $x = X - \bar{X}$ for $Y = L_C$ yields the detection limit

$$x_D = \frac{c + \dfrac{s_{Y \cdot X}\left(2F_{2, n-2}^{1-\alpha/2}\right)^{1/2}}{b\sqrt{\sum x^2}} \sqrt{c^2 + \dfrac{b^2 \sum x^2 - s_{Y \cdot X}^2 2F_{2, n-2}^{1-\alpha/2}}{nb^2}}}{\left(1 - \dfrac{s_{Y \cdot X}^2 2F_{2, n-2}^{1-\alpha/2}}{b^2 \sum x^2}\right)} \qquad (5.40)$$

where

$$c = \hat{x} + \frac{s_{Y \cdot X}}{b} \Phi(P) \left(\frac{n - 2}{\alpha/2 \chi^2_{n-2}} \right)^{1/2}$$

and

$$\hat{x} = \frac{Y_C - \bar{Y}}{b}$$

In order to express the detection limit in its original metric, we simply add the average absolute concentration $\bar{X}$ to the computed value of x_D.

For example, if $P = .95$ and $\alpha = .01$, we will have 99% confidence that 95% of the population of future measurements that do not contain the analyte in question will be below the MDL.

The objection raised by Clayton et al. (1987) regarding use of the central t-distribution for characterizing both false positive and false negative rates also applies to the tolerance intervals just described. As previously shown, one solution to this problem is to compute the required Type I error rate α^* necessary for the expectation interval to have coverage $1 - \beta$, for example, 99%. This solution proceeds along the lines previously described for single concentration designs. Estimates of the noncentrality parameter $\phi(\alpha^*, \beta)$ can then be substituted for $\phi(\alpha, \beta)$, and an MDL based on the tolerance interval can be obtained. In the regression framework, the degrees of freedom are now $n - 2$. Table 5.2 presents values of $\phi(\alpha^*, \beta)$ for a selection of degrees of freedom ranging from 5 to 48, and combinations of $\alpha = \beta = .05$ and .01 and coverage of 95% and 99%.

5.3.7 MDLs for Calibration Data with Nonconstant Variance

The previous detection limit calculations assume that the variance is homogeneous throughout the range of the calibration function. This assumption is rarely, if ever, justifiable (see Clayton et al., 1987; Gibbons et al., 1991). In practice, the variation in response signal is proportional to the concentration, as will be illustrated in the example in the following section. If violations of this assumption are ignored, the result is an overestimate of the variability at low levels and therefore an overestimate of the detection limit. The detection limit for nonconstant variance calibration designs can be obtained by substituting the weighted least squares (WLS) estimates b_{0w}, b_{1w}, and $(s^2_{y \cdot x})_w$ for the previously defined ordinary least squares estimates. For the simple linear model, the WLS estimators for the case in which the variance is proportional to the concentration are

$$b_{0w} = \frac{\Sigma(Y/X)\Sigma X - n\Sigma Y}{\Sigma(1/X)\Sigma X - n^2}, \tag{5.41}$$

$$b_{1w} = \frac{\Sigma(1/X)\Sigma Y - n\Sigma(Y/X)}{\Sigma(1/X)\Sigma X - n^2} \tag{5.42}$$

TABLE 5.2 Values of $\phi(\alpha^*, \beta)$ for Computing MDLs Based on Normal Tolerance Limits for Calibration Designs (df = $n - 2$)

	$\alpha = \beta = .05$ Coverage		$\alpha = \beta = .01$ Coverage	
df	95%	99%	95%	99%
5	5.512	8.292	7.254	10.984
6	5.184	7.710	6.570	9.831
7	4.950	7.304	6.102	9.051
8	4.778	7.002	5.769	8.490
9	4.644	6.770	5.515	8.062
10	4.537	6.586	5.315	7.733
11	4.449	6.436	5.156	7.468
12	4.376	6.312	5.024	7.249
13	4.314	6.206	4.913	7.072
14	4.260	6.115	4.819	6.917
15	4.214	6.038	4.737	6.786
16	4.173	5.969	4.668	6.670
17	4.136	5.909	4.605	6.570
18	4.104	5.855	4.548	6.480
19	4.074	5.805	4.501	6.400
20	4.049	5.762	4.454	6.330
21	4.024	5.723	4.412	6.264
22	4.000	5.687	4.377	6.205
23	3.980	5.654	4.341	6.153
28	3.898	5.522	4.208	5.943
33	3.837	5.427	4.112	5.797
38	3.792	5.357	4.039	5.689
48	3.725	5.256	3.963	5.532

and

$$\left(s_{y \cdot x}^2\right)_w = \frac{\Sigma(Y^2/X) - b_0 \Sigma(Y/X) - b_1 \Sigma Y}{n - 2} \qquad (5.43)$$

In those cases in which it is reasonable to assume that $b_0 = 0$, the weighted least squares estimates have the much simpler form

$$b_{1w} = \frac{\Sigma Y}{\Sigma X} = \frac{\bar{\bar{Y}}}{\bar{\bar{X}}} \qquad (5.44)$$

and

$$\left(s_{y \cdot x}^2\right)_w = \frac{\Sigma(Y^2/X) - (\Sigma Y)^2/\Sigma X}{n - 1} \qquad (5.45)$$

(see Snedecor and Cochran, 1980, page 174). In this case $(s_{y \cdot x}^2)_w$ represents the weighted mean square of the residuals from the fitted calibration line. At a given point X, on the calibration function, the variance is therefore $(s_{y \cdot x}^2)_w X$.

Substitution of the WLS estimators b_{0w}, b_{1w}, and $(s_{y \cdot x}^2)_w$ will provide the appropriate detection limit for the case in which the variance in response signal is proportional to the concentration. Results of the example, to be described in a following section, suggest that this assumption is quite reasonable.

5.3.8 Experimental Design of Detection Limit Studies

A detailed review of the principles of experimental design of method detection limit studies would easily require a chapter unto itself. These studies have been reviewed in some detail by others (see Liteanu and Rica, 1980). There are, however, several guiding concepts that are critical for producing unbiased detection limit estimates of practical relevance.

First, in analyte-present studies, the analysts must be blind to both the number of compounds in the sample as well as their spiking concentrations. To achieve this goal, the number of compounds must vary, perhaps randomly, from sample to sample. Similarly, the concentration of each constituent should also vary both within and across samples. Without ensuring that the analyst is blind to both the presence and concentration of the analyte under study, the resulting detection limit simply cannot be applied to routine practice where such uncertainty must always exist. In practice, it is often impossible to execute such studies since numerous samples would have to be prepared at widely varying concentrations. In the absence of this level of experimental control, standard calibration data, in which analysts are unaware that they are being tested, may have to suffice. The critical issue is that the analysts must *not* go back and retest samples that appear to be anomalous relative to the known spiking concentration.

Second, two or more instruments and analysts must be used, and the assignment of samples to analysts and instruments must also be random. Since in large production laboratories any one of a number of analysts and/or instruments may be called upon to analyze a test sample, this same component of variability must be included in determining the detection limit.

Third, whenever possible, the entire detection limit study should be replicated in two or more different laboratories.

Fourth, the number of samples selected should be based on statistical power criteria, such that a reasonable balance of false positive and false negative rates is achieved. For example, if we estimate σ by computing s on seven samples, our uncertainty in σ will be extremely large, and our resulting detection limit estimate L_D will also be quite large. By increasing the number of samples to, say 25, we achieve a much more reasonable estimate of σ, and the resulting L_D is greatly reduced. The cost of running a few additional samples far outweighs the drawbacks of having detection limits that are incapable of detecting anything but the largest signals.

An additional note regarding analyte-absent experiments (i.e., blank samples). Rather than running a series of blank samples at once, they should be randomly entered into the analysts' work load throughout the course of the day. Again, the purpose of this approach is to ensure that the analysts are blind to sample composition. The broader question, of course, is whether analyte-absent experiments are relevant to establishing MDLs. It can certainly be argued that the properties of the *method* can only be evaluated when the analyte is present, at least in some of the samples. The general recommendation of calibration designs over fixed concentration designs allows for a mixture of samples in which the analyte is present and absent.

There are several experimental designs that can fulfill the preceding requirements. When the number of samples is large, say $n = 30$, one possibility is to give each compound in the study a .5 probability of being in any given sample, and once selected, its concentration could also be randomly selected from a uniform distribution covering the range of 0 to $2L'_D$, where L'_D is the presumed detection limit. This design is perhaps optimal for ensuring blindness, but not necessarily for maximizing the signal-to-noise ratio which, of course, increases the amount of information that can be gained in such a study. For example, Hubaux and Vos (1970) suggest a "three values repartition," in which n_1 replicate samples with concentrations at the lowest "permissible content" (X_1) are selected, $n_2 = N - n_1 - 1$ samples at the highest "permissible content" (X_2), and a single sample at ($X_1 + X_2)/2$. In their work, they find that this design minimizes the number of required standards for a fixed level of sensitivity.

Liteanu and Rica (1980) review a wide variety of sampling designs for detection limit studies, including response surface designs, fractional factorial designs, and rotatable designs. Youden-pair-type designs are also excellent candidates for maintaining blind and unbiased detection limit studies.

Finally, in any design in which there are multiple components of variability (e.g., analyst, instrument, laboratory), appropriate steps must be taken to obtain unbiased estimates of variability. The naive estimator of σ, which is obtained by simply computing s in the usual way with all measurements, should be replaced by the unbiased variance estimate, based on the appropriate variance components model for the particular problem at hand (e.g., see Gibbons, 1987a). Methods for computing these variance components are described in a later chapter.

5.3.9   Obtaining the Calibration Data

For many constituents, for example, volatile organic priority pollutant compounds (VOCs), for which concentrations are obtained by gas chromatography/mass spectroscopy (GC/MS), estimated concentrations are obtained by applying a relative response factor (rrf) to the ratio of ion counts for the compound of interest to an internal standard. When ion count ratios of this kind are available, they are always preferable, since uncertainty in the calibration function itself can be incorporated into the MDL estimator. For analytical methods that do not calibrate this way, actual recovered concentrations can be used directly; however, the computed MDL will underestimate the true value to the extent that the computed slope of the calibration line deviates from the true value.

When using ion counts directly, a problem occurs because ion counts from different instruments or across different calibrations for the same instruments are not directly comparable. To remove this bias, the following algorithm can be used:

1. Let rrf_{ij} be the relative response factor for instrument i at concentration j.

2. Let $r\bar{r}f_i$ be the average response factor for the five concentrations for instrument i.

3. To combine data from multiple instruments and calibrations, note that the relative percentage mean deviation for instrument i at concentration j is

$$rrf'_{ij} = \left(\frac{rrf_{ij} - r\bar{r}f_i}{r\bar{r}f_i} \right) 100 \qquad (5.46)$$

which can be positive or negative.

4. To place all instruments in a common metric, compute

$$rrf^*_{ij} = \left(\frac{rrf_{ij} - r\bar{r}f_i}{r\bar{r}f_i} \right) r\bar{\bar{r}}f + r\bar{\bar{r}}f \qquad (5.47)$$

where

$$r\bar{\bar{r}}f = \sum_{i=1}^{n} \frac{r\bar{r}f_i}{n} \qquad (5.48)$$

is the grand mean of the relative response factors, and n is the number of five-point calibrations.

5. Using these transformed relative response ratios, the response factor (i.e., ion count ratio) is

$$rf^*_{ij} = rrf^*_{ij}(c_j) \qquad (5.49)$$

where c is the jth spiking concentration. These normalized response factors form the basis for all subsequent analysis.

Example 5.1

To illustrate the application of the various MDL estimators, consider the data in Figures 5.1 to 5.4, which present actual and measured concentrations of benzene in 22 five-point concentration calibrations (i.e., $n = 110$). The calibrated instrument responses are presented graphically to aid in interpretation; however, the original instrument responses (i.e., peaks are ion count ratios of analyte to internal standard) were used for computing MDLs. Figure 5.1 displays the original data with a corresponding 99% prediction interval under the assumption of constant variance as described by Hubaux and Vos (1970). Inspection of Figure 5.1 clearly reveals that variation is linearly increasing with concentration and the assumption of constant variance which underlies the Hubaux and Vos MDL estimator is untenable. Figure 5.2 displays the relation of actual and measured concentration follow-

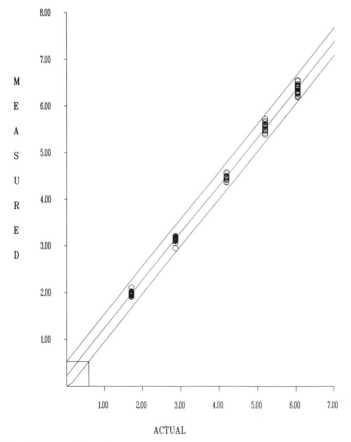

Figure 5.2 Linear calibration following square root transformation: GC/MS–benzene (ppb)–prediction interval.

ing the square root transformation suggested by Clayton et al. (1987), that is,

$$y = \sqrt{\frac{\text{peak area for compound}}{\text{peak area for internal standard}}} \tag{5.50}$$

and the concentration transformation

$$x = \sqrt{X + 0.1} - \sqrt{0.1} \tag{5.51}$$

where X is the original analytic concentration (e.g., in $\mu g/l$). The effect of this transformation is to homogenize deviations from the fitted calibration line, particularly at higher concentrations. To return to the original metric (e.g. $\mu g/l$), compute

$$X = x^2 + 0.632456x \tag{5.52}$$

Inspection of Figure 5.2 reveals that the transformation performed reasonably well for all but the lowest concentration.

Figure 5.3 displays the 99% prediction interval based on the assumption that variance is proportional to concentration (i.e., weighted least squares approach). Figure 5.3 reveals an excellent fit to the observed deviations from the regression line. Figure 5.4 displays the corresponding tolerance interval for the calibration line based on 99% confidence and 99% coverage of all future measurements. The increased width of the tolerance interval provides for the additional coverage of the distribution, whereas the prediction interval is designed to include only the next single measurement. Remarkably, the difference in width between these two intervals (i.e., Figure 5.3 versus Figure 5.4) is relatively small compared to the increased utility of the tolerance interval.

In terms of computed MDLs, estimated values for each of the four previously described methods are displayed in Table 5.3. The four methods are: (1) Hubaux and Vos (1970); (2) tolerance interval version of Hubaux and Vos (Gibbons et al., 1988a); (3) the noncentral t-method of Clayton et al. (1987); and (4) tolerance interval generalization of the method due to Clayton and coworkers (Gibbons et al., 1991c). Table 5.3 displays computed MDLs for all four methods based on the assumption of constant variance before and after variance stabilizing transformation, and based on the assumption of increasing variance for methods 3 (Clayton et al., 1987) and 4 (Gibbons et al., 1991c). Table 5.3 reveals the following. First, the assumption of constant variance leads to considerable overestimates of the MDL (i.e., estimates on the order of five times the values obtained by relaxing this assumption). Second, in all cases the variance stabilizing transformation performs well, giving estimates reasonably close to the weighted least squares values. This is a useful result, in that it permits the computational simplicity and generality of Hubaux and Vos-type calculations which only involve ordinary least squares estimation, commonly available in most computer

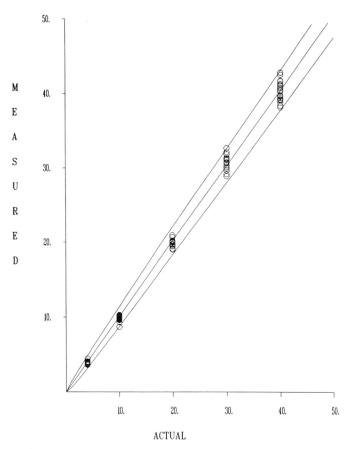

Figure 5.3 Linear calibration with increasing variance: GC/MS–benzene (ppb)–prediction interval.

packages. Third, use of the noncentral t in place of repeated application of the central t produces decreases in the MDL estimates of approximately 15% (i.e., Hubaux and Vos versus Clayton and coworkers, and comparison of the two tolerance-interval-based methods of Gibbons and coworkers), a difference which may not be trivial in practice. It should be noted, however, that this difference may be due, in part, to ignoring uncertainty in the calibration line as pointed out by Coleman (1993). Fourth, the penalty paid for the dramatically increased generalizability of the tolerance interval approach versus the prediction interval approach is quite modest. For example, MDL estimates based on the weighted least squares solution to the nonconstant variance problem were 0.75 μg/l for the next single measurement versus 0.97 μg/l for 99% of all future measurements. Both estimates provide 99% confidence of accomplishing their different objectives conditional on their assumptions' being true.

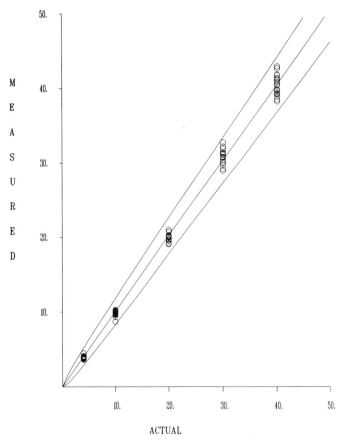

Figure 5.4 Linear calibration with increasing variance: GC/MS–benzene (ppb)–tolerance interval.

The results of this calibration study can also be used to shed light on the use of single spiking concentration designs like those required by the EPA. The relationship between concentration and variance was $s_X = .159(X)$, where X is the actual spiking concentration. As such, if we had conducted the MDL study at $X = 10$ μg/l, the EPA-estimated MDL (i.e., Glaser et al., 1981) would be

$$\text{MDL} = 3.143\sqrt{.159^2(10)} = 1.58 \ \mu\text{g/l}$$

Alternatively, had we spiked seven samples at 1 μg/l, we would have obtained an estimated MDL of

$$\text{MDL} = 3.143\sqrt{.159^2(1)} = 0.50 \ \mu\text{g/l}$$

Spiking at the new estimated MDL of 0.5 μg/l would yield a new MDL of

$$\text{MDL} = 3.143\sqrt{.159^2(0.5)} = 0.35 \ \mu\text{g/l}$$

TABLE 5.3   Comparison of the Four Types of MDL Estimators for Constant Variance before and after Transformation and Nonconstant Variance via Weighted Least Squares

Model	Constant Variance		Increasing Variance WLS
	Raw Data	Transformed	
1	4.28	0.71	—
2	5.87	1.25	—
3	3.84	0.61	.75
4	4.98	0.95	.97

Model 1—Hubaux and Vos (1970), prediction interval.
Model 2—Gibbons et al. (1988a), tolerance interval.
Model 3—Clayton et al. (1987), prediction interval.
Model 4—Gibbons et al. (1991c), tolerance interval.

and so on. These computations illustrate that if variance is proportional to concentration, any MDL can be achieved simply by lowering the spiking concentrations. As new methods are developed, analysts will naturally select lower concentrations to illustrate their sensitivity and will, of course, verify their preconception regardless of the actual detection capabilities of the new methodology. The calibration-based MDL estimators do not suffer from this critical flaw.

5.4   SUMMARY

The computation of MDLs has led to enormous controversy in the environmental literature, particularly in relation to setting drinking water and groundwater standards. In part, much of the problem is that risk-based estimates for certain carcinogenic compounds have historically been below analytical detection capabilities. These risk-based estimates are themselves statistical estimates of very questionable validity (see Chapter 14). Nevertheless, in order to detect these compounds at these levels, new analytic methods have been developed, and detection limit estimators that will yield virtually any desired MDL have been routinely used. The problem, of course, is that at these low levels, potentially hazardous constituents will be identified in trip blanks, field blanks, and wells at so-called *greenfield* sites which have had no waste disposal (see Chapter 14). At existing sites, these "false detections" will lead to costly site assessments and even corrective action for no reason whatsoever. Since variability is not homogeneous, single spiking concentration designs *cannot* be used for computing MDLs. Calibration-based MDLs represent the only viable approach. If these estimates are to be used for making detection monitoring decisions, then the effects of multiple detection decisions must also be considered as described in this chapter.

6 Practical Quantitation Limits

6.1 OVERVIEW

In the previous chapter statistical considerations regarding detection of a constituent in a new groundwater monitoring sample were discussed. The models presented were to make the binary decision of whether or not the analyte was present in the sample. Some cases may also require a quantitative estimate of the concentration of the analyte. Method detection limits provide no information regarding the quantitative value or concentration of the analyte. To have confidence in the quantitative determination, the measured concentration of the analyte must exceed a quantitation limit, often termed the *practical quantitation limit* (PQL). For example, assume that a health-based standard (e.g., maximum contaminant level, MCL) for vinyl chloride is 2 $\mu g/l$, the MDL is 1 $\mu g/l$, and the PQL is 5 $\mu g/l$. In a new sample vinyl chloride is detected and the reported concentration is 3 $\mu g/l$. Has the health-based standard been exceeded? The answer, of course, is that we have no idea whether the standard has been exceeded since we have insufficient information to accurately quantify the measurement (i.e., the measured concentration of 3 $\mu g/l$ is less than the quantitation limit of 5 $\mu g/l$). All that is known with confidence is that vinyl chloride is present in the sample. For this reason, it is often the PQL and not the MDL that is most crucial for groundwater monitoring applications.

6.2 OPERATIONAL DEFINITION

Currie (1968) defined the determination limit (L_Q) as the concentration "at which a given procedure will be sufficiently precise to yield a satisfactory quantitative estimate." This definition is similar to that used by Adams, Passmore, and Campbell (1966) who defined a "minimum working concentration" as that for which the relative standard deviation (rsd) was 10%. The determination limit has since been described by several names, most notably "practical quantitation level" (EPA, 1985), "limit of quantitation" (EPA, 1987b), and "practical quantitation limit" (EPA, 1987b). The EPA (1987b) defines the PQL as "the lowest level achievable by good laboratories within specified limits during routine laboratory operating conditions." This vague definition has been operationally defined by the EPA in three ways: first, as five or ten times the method detection limit; second, as the concentration at which 75% of the laboratories in an interlaboratory study report concentrations $\pm 20\%$ of the true value; or third, as the concentration at which 75% of the laboratories report concentrations within $\pm 40\%$ of the true value (EPA, 1987b). The first operational definition is arbitrary and depends on the validity of the corresponding method detection limit, about which serious questions were raised in the previous chapter. The second and third operational definitions are somewhat better; however, interlaboratory studies are often done at a single concentration (e.g., the MCL) in experienced government laboratories that "knew they were being tested with standard samples in distilled water without matrix interferences." The EPA (1985) points out that

> Actual day-to-day operations in a wide variety of laboratories using real samples in natural water would be expected to produce poorer results, i.e., wider performance ranges especially at the lower concentration levels.

See Koorse (1989) for an excellent review of the legal implications of these definitions. It is unclear in these definitions whether all measurements made by a single laboratory must be within $\pm 20\%$ of the true value or if this criterion can be satisfied by one or two measurements.

6.3   STATISTICAL ESTIMATE OF THE PQL

To determine the PQL of a given compound in a given laboratory using a particular methodology, begin by obtaining calibration data for a series of concentrations ranging from zero to two to five times the hypothesized PQL. These are the same types of calibration data described in the previous chapter; hence both MDLs and PQLs can be computed from the same data. Gibbons et al. (1992a) employed square root transformations originally described by Clayton et al. (1987) to homogenize variability in deviations from the fitted calibration line (see the previous chapter).

Gibbons et al. (1992a) follow Adams, Passmore, and Campbell (1966) and Currie (1968), and define the PQL operationally as the concentration at which the relative standard deviation is $\alpha\%$ (Adams, Passmore, and Campbell suggest 10%). This leads to the predicted concentration x_α^* for which

$$\hat{y} = \frac{100}{\alpha} s(\hat{y}) \qquad (6.1)$$

is the predicted response $\hat{y}$ that is $100/\alpha$ times its estimated standard deviation $s(\hat{y})$. Let y_α^* denote the value of $\hat{y}$ that satisfies (6.1). To obtain the predicted response and corresponding standard deviation, the slope of the calibration line is estimated by least squares as

$$\hat{b} = \frac{\sum_{i=1}^{n}(x_i - \bar{x})(y_i - \bar{y})}{\sum_{i=1}^{n}(x_i - \bar{x})^2} \tag{6.2}$$

Then the relationship between the predicted response $\hat{y}$ and the predicted concentration $\hat{x}$ is

$$\hat{x} = \bar{x} + \frac{(\hat{y} - \bar{y})}{b} \tag{6.3}$$

Given these two estimates, the standard deviation of $\hat{y}$ is

$$s(\hat{y}) = s_{y \cdot x}\sqrt{1 + \frac{1}{n} + \frac{(\hat{x} - \bar{x})^2}{\sum_{i=1}^{n}(x_i - \bar{x})^2}} \tag{6.4}$$

where $s_{y \cdot x}^2$ is the unbiased estimate of $\sigma_{y \cdot x}^2$, the residual variance, that is,

$$s_{y \cdot x}^2 = \frac{\sum_{i=1}^{n}(y_i - \hat{y})^2}{n - 2} \tag{6.5}$$

and

$$\hat{y}_i = \bar{y} + b(x_i - \bar{x}) \tag{6.6}$$

The PQL in the transformed metric is therefore

$$x_\alpha^* = \bar{x} + \frac{y_\alpha^* - \bar{y}}{b} \tag{6.7}$$

In order to compute the x_α^* value in (6.7), to obtain the $\alpha\%$ PQL defined by (6.1), we must know y_α^*, the predicted transformed response signal for which the relative standard deviation is $\alpha\%$. One solution to this problem is to compute $\hat{y}$ and $s(\hat{y})$ for various values of $\hat{x}$, using (6.4) to (6.6), until the ratio $\hat{y}/s(\hat{y})$ equals $100/\alpha$. To obtain a more direct solution, Gibbons et al. (1992a) solved (6.1) for y_α^* and obtained

$$y_\alpha^* = \frac{(100/\alpha)s_{y \cdot x}b\sqrt{\sum_{i=1}^{n}(x_i - \bar{x})^2\left[\bar{y}^2 + (1 + 1/n)\sum_{i=1}^{n}(x_i - \bar{x})^2 b^2 - (1 + 1/n)(100/\alpha)^2 s_{y \cdot x}^2\right]} - (100/\alpha)^2 s_{y \cdot x}^2 \bar{y}}{\sum_{i=1}^{n}(x_i - \bar{x})^2 b^2 - (100/\alpha)^2 s_{y \cdot x}^2} \tag{6.8}$$

(see the following section for the derivation of this result). Substitution of y_α^* into (6.7) yields x_α^*, the PQL in its transformed metric.

Note that $s_{y \cdot x}$ and b are estimates of the population parameters $\sigma_{y \cdot x}$ and β, based on a finite sample of size n. Repeating the study would have yielded different sample-based estimates and the corresponding PQL also would have changed. To incorporate this uncertainty into the PQL estimator, confidence limits for x_α^* must be obtained. To find the 95% confidence limits for x_α^*, we begin with the 95% confidence limits for y_α^* given x_α^*:

$$y = \bar{y} + b(x_\alpha^* - \bar{x}) \pm t_{[n-2, .05]} s_{y \cdot x} \sqrt{1 + \frac{1}{n} + \frac{(x_\alpha^* - \bar{x})^2}{\sum_{i=1}^n (x_i - \bar{x})^2}} \quad (6.9)$$

where $t_{(n-2, .05)}$ is the two-tailed 95% point of Student's t-distribution on $n - 2$ degrees of freedom. Expression (6.9) is then solved as a quadratic equation in x_α^*. The easiest solution for numerical work is obtained by expressing the two roots as

$$x = \frac{x_\alpha^* \pm (t_{[n-2, .05]} s_{y \cdot x}/b) \sqrt{[(n + 1)/n](1 - c^2) + (x_\alpha^* - \bar{x})^2/\sum_{i=1}^n (x_i - \bar{x})^2}}{1 - c^2}$$

$$(6.10)$$

where

$$c^2 = \left(\frac{t_{[n-2, .05]} s_{y \cdot x}}{b}\right)^2 \left(\frac{1}{\sum_{i=1}^n (x_i - \bar{x})^2}\right)$$

(see Snedecor and Cochran, 1980; Miller, 1966; Gibbons et al., 1992a). This result is referred to either as Fieller's (Fieller, 1940) or Paulson's (Paulson, 1942) theorem. It should be noted, however, that the quantity c is related to the test significance of b. If b is significant (which it always is in this context), c will be small and c^2 will be negligible, and the limits become

$$x_\alpha^* \pm \left(\frac{t_{[n-2, .05]} s_{y \cdot x}}{b}\right) \sqrt{1 + \frac{1}{n} + \frac{(x_\alpha^* - \bar{x})^2}{\sum_{i=1}^n (x_i - \bar{x})^2}} \quad (6.11)$$

[also see (5.38)].

In order to express the x_α^* and corresponding confidence limit in the original metric (e.g., $\mu g/l$), we compute

$$PQL = (x_\alpha^*)^2 + 0.632456(x_\alpha^*) \quad (6.12)$$

Equation (6.12) can also be used to untransform the upper and lower confidence limits in (6.11).

6.4   DERIVATION OF THE PQL

Required is

$$\frac{\hat{y}}{s(\hat{y})} = \frac{100}{\alpha}$$

where

$$\hat{y} = \bar{y} + b(\hat{x} - \bar{x})$$

and

$$\hat{x} = \bar{x} + \frac{(\hat{y} - \bar{y})}{b}$$

and

$$s(\hat{y}) = s_{y \cdot x} \sqrt{1 + \frac{1}{n} + \frac{(\hat{x} - \bar{x})^2}{\sum_{i=1}^{n}(x_i - \bar{x})^2}}$$

$$= s_{y \cdot x} \sqrt{1 + \frac{1}{n} + \frac{(\bar{x} + ((\hat{y} - \bar{y})/b) - \bar{x})^2}{\sum_{i=1}^{n}(x_i - \bar{x})^2}}$$

$$= s_{y \cdot x} \sqrt{1 + \frac{1}{n} + \frac{[(\hat{y} - \bar{y})/b]^2}{\sum_{i=1}^{n}(x_i - \bar{x})^2}}$$

Then

$$\hat{y} = \left(\frac{100}{\alpha}\right) s_{y \cdot x} \sqrt{1 + \frac{1}{n} + \frac{[(\hat{y} - \bar{y})/b]^2}{\sum_{i=1}^{n}(x_i - \bar{x})^2}}$$

To solve for $\hat{y}$, let

$$\hat{y}^2 = \left(\frac{100}{\alpha}\right)^2 s_{y \cdot x}^2 \left[1 + \frac{1}{n} + \frac{[(\hat{y} - \bar{y})/b]^2}{\sum_{i=1}^{n}(x_i - \bar{x})^2}\right]$$

and define

$$s' = \left(\frac{100}{\alpha}\right)^2 s_{y\cdot x}^2$$

$$x' = \sum_{i=1}^{n} (x_i - \bar{x})^2$$

$$n' = 1 + \frac{1}{n}$$

Then

$$x'\hat{y}^2 = s'n'x' + \frac{s'}{b^2}\left(\hat{y}^2 - 2\hat{y}\bar{y} + \bar{y}^2\right)$$

$$x'\hat{y}^2 - \frac{s'}{b^2}\hat{y}^2 + \frac{2s'}{b^2}\hat{y}\bar{y} - \frac{s'\bar{y}^2}{b^2} - s'n'x' = 0$$

$$\left(x' - \frac{s'}{b^2}\right)\hat{y}^2 + \left(\frac{2s'\bar{y}}{b^2}\right)\hat{y} - s'\left(\frac{\bar{y}^2}{b^2} + n'x'\right) = 0$$

The solution of this quadratic equation (i.e., the positive root) is therefore

$$\frac{-\dfrac{2s'\bar{y}}{b^2} + \sqrt{\left(\dfrac{2s'\bar{y}}{b^2}\right)^2 + 4s'\left(x' - \dfrac{s'}{b^2}\right)\left(\dfrac{\bar{y}^2}{b^2} + n'x'\right)}}{2\left(x' - \dfrac{s'}{b^2}\right)}$$

After a little algebra, a somewhat more computationally tractable form is

$$\hat{y} = \frac{(100/\alpha)s_{y\cdot x}b\sqrt{x'(\bar{y}^2 + n'x'b^2 - n's')} - s'\bar{y}}{x'b^2 - s'}$$

Substituting for s', x', and n' yields (6.8), which is the peak area ratio that is $100/\alpha$ times its standard deviation.

6.5   A SIMPLER ALTERNATIVE

The previous derivation is based on estimating the standard deviation of the predicted y and directly incorporates uncertainty in the calibration line. Inspection of (6.4) reveals that as n increases uncertainty in the calibration function decreases, as does $s(\hat{y})$; hence the estimated PQL will also be smaller. It can be argued that the PQL should not be smaller simply because more data have been gathered, however, since the measurement error of the process remains the same. A simple alternative is to substitute $s_{y \cdot x}$ for $s(\hat{y})$. The expectation of $s_{y \cdot x}$ does not change as more data are gathered. In this case the x_α^* in (6.7) is

$$x_\alpha^* = \bar{x} + \left(\frac{100}{\alpha} s_{y \cdot x} - \bar{y} \right) \Big/ b \tag{6.13}$$

where $s_{y \cdot x}$ is obtained in (6.5). Again, incorporation of uncertainty produced by substituting sample-based estimates for the population values can be accomplished by computing the corresponding confidence interval in (6.11).

6.6   UNCERTAINTY IN y_α^*

To this point, it has been assumed that y_α^* is a new predicted instrument response that corresponds to concentration x_α^*. This is not entirely true since y_α^* has been selected based on the ratio of $\hat{y}/s_{y \cdot x}$, that is, the ratio of a predicted value to its standard deviation. However, it is unknown what the true population standard deviation $\sigma_{y \cdot x}$ is. All that is available is a sample-based estimate $s_{y \cdot x}$ which would take on a range of possible values for a given $\sigma_{y \cdot x}$. Had the experiment been repeated many times, y_α^* would change also. For large numbers of calibration samples, this effect is negligible; however, in small samples the effect can be significant.

To explore this problem further, a confidence limit is obtained for $\sigma_{y \cdot x}$ given the estimate $s_{y \cdot x}$ obtained from a sample of size n as

$$\sqrt{\frac{(n-2) s_{y \cdot x}^2}{\chi_{(\gamma/2)}^2}} \leq \sigma_{y \cdot x} \leq \sqrt{\frac{(n-2) s_{y \cdot x}^2}{\chi_{(1-\gamma/2)}^2}} \tag{6.14}$$

where $\chi_{(1-\gamma/2)}^2$ represents the lower $100(\gamma/2)\%$ point of the chi-square distribution on $n-2$ degrees of freedom and $\chi_{\gamma/2}^2$ the corresponding upper $100(\gamma/2)\%$ point (see Snedecor and Cochran, 1980). For example, with $\gamma = .05$, there will be 95% confidence of including the true population standard deviation $\sigma_{y \cdot x}$ within the interval in (6.14). Substitution of the upper and lower confidence limits in (6.14) for $s_{y \cdot x}$ in (6.8) will yield upper and lower bounds for y_α^* and x_α^*, respectively. For convenience, let us call these lower and upper bounds $y_{\alpha 1}^*$ and $y_{\alpha 2}^*$ and $x_{\alpha 1}^*$ and $x_{\alpha 2}^*$, respectively. Substitution of $x_{\alpha 1}^*$ for x_α^* in (6.11) will yield the lower confidence for the PQL, and substitution of $x_{\alpha 2}^*$ for x_α^* in (6.11) will yield the upper confidence limit for the PQL. Alternatively, substitution of the lower and upper confidence limits for $\sigma_{y \cdot x}$ into (6.13) will yield $x_{\alpha 1}^*$ and $x_{\alpha 2}^*$ for the simpler method, and corresponding confidence limits can be obtained as before using (6.11).

6.7   EFFECT OF THE TRANSFORMATION

The previous definition of $\alpha/100$ times the standard deviation applies to the metric in which the PQL is estimated. In the present context, a square root transformation is used to stabilize variance across the calibration line while preserving linearity of the function. An rsd of 10% in the transformed response metric does not necessarily correspond to a 10% rsd in the original scale of measurement (e.g., $\mu g/l$). This is shown by simple propagation of errors as follows.

If z is the instrument response and $\sqrt{z} = y$, then by propagation of error

$$\text{var}(y) = \frac{\text{var}(z)}{4z}$$

and

$$\frac{\text{var}(y)}{y^2} = \frac{\text{var}(z)}{4z^2}$$

At the concentration where $z^* = (100/\alpha)s(z^*)$,

$$\frac{\text{var}(y_\alpha^*)}{y^{*2}} = \frac{1}{4(100/\alpha)^2}$$

therefore $y_\alpha^* = \sqrt{4(100/\alpha)^2}\,s(y_\alpha^*)$. Therefore obtaining a 10% rsd in the raw instrument response (i.e., $\alpha = 10$) would require a 5% rsd in the transformed response scale [i.e., $\sqrt{4(100/\alpha)^2} = 20$].

Example 6.1

Recall the example dataset in the previous chapter for benzene. Applying the transformations to both peak area ratios and concentrations, the summary statistics are $b = 1.02$, $s_{y \cdot x} = .11$, $\sum_{i=1}^{n}(x_i - \bar{x})^2 = 264.15$, $n = 110$, $\bar{x} = 3.98$, $\bar{y} = 4.30$, and $t_{108,.05} = 1.659$. If $\alpha\% = 10\%$ is selected, the transformed response signal that is $\alpha/100 = 20$ times its standard deviation (corresponding to a relative standard deviation of 10% in the metric of $\mu g/l$) is then computed as

$$y_\alpha^* = \frac{20(.13)1.02\sqrt{264.15[4.3^2 + (1 + 1/110)264.15(1.02^2) - (1 + 1/110)400(.13^2)]} - 400(.13^2)4.30}{264.15(1.02^2) - 400(.13^2)}$$

$$= 2.62$$

following (6.8). The PQL in the transformed metric is

$$x_\alpha^* = 3.98 + \frac{2.62 - 4.30}{1.02} = 2.33$$

following (6.7), and in the original metric,

$$PQL = 2.33^2 + 0.632456(2.33) = 6.90 \ \mu g/l$$

following (6.12). The 95% confidence interval in the transformed metric is

$$2.33 \pm \frac{1.66(.13)}{1.02} \sqrt{1 + \frac{1}{110} + \frac{(2.33 - 3.98)^2}{264.15}}$$

following (6.11), which has the roots 2.14 and 2.54, and in the original metric, 5.93 $\mu g/l$ and 8.06 $\mu g/l$ following (6.12).

In terms of the simple estimator in (6.13), for a 10% rsd in the original metric, compute

$$3.98 + \frac{20(.13) - 4.3}{1.02} = 2.03$$

or 6.80 $\mu g/l$. For a 20% rsd in the original metric, compute

$$3.98 + \frac{10(.13) - 4.3}{1.02} = 1.04$$

or 1.74 $\mu g/l$, sufficiently close to the more complete estimators of 6.90 $\mu g/l$ and 1.82 $\mu g/l$ for most practical applications.

6.8 SELECTING *n*

An important question in calibration studies is the choice of the number of spiking concentrations and the number of replicates of each. In the present context, a good choice would be a sufficient number of concentrations and replicates at each, such that our PQL estimate is primarily a function of sampling variability (e.g., between days, between analysts, between instruments, and random fluctuation of the measurement process), and not largely affected by uncertainty in the calibration function itself (i.e., *b*). Another way of saying this is that a sufficient number of measurements is desired such that the simple alternative method, which ignores uncertainty in *b*, gives a similar result to the more complete estimator, which incorporates this uncertainty. Gibbons et al. (1992a) show that with a four-point calibration, 16 replicates per concentration are required to obtain a PQL estimate relatively unaffected by uncertainty in b. In general, uncertainty in the calibration line produces a negligible increase in the PQL estimate once the sample size is approximately 50 or 60. A good minimum choice might therefore be five concentrations and twelve replicates of each. Comparison of the two PQL estimators [i.e., based on (6.12) and (6.13), respectively] will provide case-specific information on the adequacy of a particular calibration study and corresponding confidence in the use of the computationally simple alternative in (6.13).

6.9 SUMMARY

The PQL estimator described here has several advantages over existing procedures. First, it has a clear operational definition based on the percentage rsd, which has been in the literature for the last 25 years. Second, it can be determined empirically using data from a single laboratory. Third, the uncertainty in the calibration function is incorporated in both the statistical estimate of the PQL and the corresponding confidence limits. Fourth, the operational definition may be modified to reflect the degree of required, or reasonably achievable, precision, with only the most minor modifications to the estimation equation (e.g., 10% rsd versus 20% rsd). Fifth, variability throughout the working range is considered instead of simply relying on an estimate of variability obtained from a single fixed point on the calibration line. Sixth, a computationally simple alternative which will provide adequate results for calibration studies based on samples of $n = 50$ or more measurements (e.g., five concentrations and ten replicates of each) is presented.

Using this PQL estimator, it is possible to provide laboratory-specific PQLs, and in the context of groundwater detection monitoring, for example, an EPA regional administrator may now "establish facility-specific PQLs based on the technical limitations of the contracting laboratory," rather than relying on a national baseline reflecting ideal target levels that may not be achievable in routine laboratory practice.

7 Contaminant Source Analysis

7.1 OVERVIEW

To this point, focus has been on problems involving a single source, for instance, determining if a landfill has impacted groundwater quality. In some cases, however, there is concern regarding potential impact from multiple sources. As an example, consider a waste disposal facility located next to a steel mill. Wells located downgradient of the waste disposal facility are also downgradient of the steel mill; therefore changes in groundwater quality may be occurring naturally, associated with the waste disposal facility or with the steel mill (e.g., storage and disposal of slag materials). As another example, consider a Superfund investigation of a codisposal facility (i.e., a facility where both municipal and industrial waste are disposed) in which the responsible parties are the municipality that disposed of household waste and a local industry that disposed of hazardous liquids. In some cases it may be possible to determine in probability which disposal activity led to groundwater contamination. Also, consider a coastal region facility where groundwater chemistry near some downgradient wells is affected by intrusion of salt water from the ocean from a flood control channel. In this case we could assume that the difference between upgradient and downgradient groundwater quality is due to the facility when it could be due to salt water intrusion into groundwater.

In all three of these cases comparing background groundwater measurements to new monitoring measurements is complicated by multiple potential sources of contamination. If a significant difference is observed, which source is responsible?

The problem posed in this chapter is one of statistical classification. Given a p-variate vector of constituent concentrations, what is the probability it was drawn from each of Q potential sources? If constituents are properly selected to characterize the chemistry of the original sources and classification probability is larger for one of the Q sources than the others, it may be reasonable to assign the impact to that source. In the extreme case, if there is no overlap between constituents in the Q sources, merely the presence or absence of those compounds in downgradient groundwater samples may be sufficient to identify the source. Conversely, when the same constituents are found at the same concentrations in all sources, classification of new ground-

132

water samples may not be possible since the original sources are not differentiable. In practice, there will likely be overlap between sources in the presence or absence and concentration of relevant constituents; however, patterns of concentrations across multiple constituents (i.e., covariation) may provide reliable information to differentiate contaminant sources and produce probability estimates for new groundwater samples. In the following sections statistical models for classification are discussed and relevant examples are provided.

7.2   STATISTICAL CLASSIFICATION PROBLEMS

To begin, statistical classification problems can be classified into two general types. First, there are problems in which multivariate measurements are available (e.g., concentration measurements for a suite of constituents), but there is no a priori grouping of the measurements. Not only are statistical models such as cluster analysis, factor analysis (Harman, 1970), and finite mixture distributions (Everitt and Hand, 1981) used to derive classification weights, but they also create an empirical grouping of the measurements. These models are not the focus of this chapter, since, in general, sample origin is known. For example, in the Superfund illustration, separate leachate samples from identifiable municipal and industrial waste cells may be available. Alternatively, in the example of a coastal landfill, we may have leachate samples for the facility as well as surface water samples from the bay or flood control channel. The chemical measurements from these samples define the patterns for each group (i.e., source). Statistical methods such as discriminant function analysis (Hand, 1981) or logistic regression analysis are useful for analysis of classification problems in which group membership is known in advance. In this chapter, focus is on discriminant function analysis since the chemical measurements used to differentiate the sources are continuous.

7.2.1   Classical Discriminant Function Analysis

Fisher (1936) developed the classical discriminant function analysis model for use of multiple measurements in taxonomic problems. An early application of this model was classification of archaeological skull specimens as English or Eskimo on the basis of a series of morphological measurements made on each skull. An example of a more recent application is classification of crops from high-altitude photographs to estimate total acreage. In the context of groundwater monitoring problems, classification of new downgradient samples into one of several possible sources (e.g., surface water, leachate, background, off-site industrial activity) is desired.

To develop the theory, consider the following notation:

$\mathbf{x}$ = the p-dimensional observation vector
$\mathbf{S}$ = the pooled covariance matrix
j = a subscript to distinguish groups
n_j = the number of observations in group j
m_j = the p-dimensional mean vector
$\mathbf{S}_j$ = the covariance matrix in group j
$|\mathbf{S}_j|$ = the determinant of $\mathbf{S}_j$
q_j = the prior probability of membership in group j
$P(j|\mathbf{x})$ = the posterior probability of observation $\mathbf{x}$ in group j
$f_j(\mathbf{x})$ = the group-specific density estimate of $\mathbf{x}$ from group j
$f(\mathbf{x})$ = the unconditional density at $\mathbf{x}$ $[\Sigma_j q_j f_j(\mathbf{x})]$

Using Bayes theorem, assume q_j are known and $f_j(\mathbf{x})$ can be estimated. Then the probability that the new measurement vector (i.e., constituent concentrations for a new groundwater sample) was drawn from source j is

$$P(j|\mathbf{x}) = \frac{q_j f_j(\mathbf{x})}{f(\mathbf{x})} \tag{7.1}$$

The idea is to partition the p-dimensional space into regions R_j, which contain all vectors $\mathbf{x}$, such that $P(j|\mathbf{x})$ is largest among all groups. A new measurement vector $\mathbf{x}$ (i.e., groundwater sample) is classified in group j if it lies in region R_j.

Fisher (1936) proceeded under the assumptions that the variables are distributed multivariate normally (MVN) within each group and that prior probabilities q_j are known. Given these assumptions, the objective is to place each observation in the group that has the smallest generalized distance

$$d_j^2(\mathbf{x}) = (\mathbf{x} - \mathbf{m}_j)'\mathbf{V}_j^{-1}(\mathbf{x} - \mathbf{m}_j) \tag{7.2}$$

where $\mathbf{V}_j = \mathbf{S}_j$ (i.e., each group has a unique covariance matrix describing association among the p variables, in this case chemical constituents), or $\mathbf{V}_j = \mathbf{S}$ which reflects those cases in which covariation among variables is sufficiently homogeneous across groups that a pooled estimate of $\mathbf{V}$ can be used. The group-specific density estimate is then

$$f_j(\mathbf{x}) = (2\pi)^{-p/2}|\mathbf{V}_j|^{-1/2}\exp\left(-.5d_j^2\mathbf{x}\right) \tag{7.3}$$

and from Bayes theorem the classification probability of the vector observation $\mathbf{x}$ into group j is

$$P(j|\mathbf{x}) = \frac{q_j f_j(\mathbf{x})}{\Sigma_i q_i f_i(\mathbf{x})} \tag{7.4}$$

In practice, multivariate measurement vectors will not fit any group perfectly. To determine the amount of this deviation, the generalized squared distance from $\mathbf{x}$ to group j can be computed as

$$D^2(\mathbf{x}) = d_j^2(\mathbf{x}) + g_1(j) + g_2(j) \tag{7.5}$$

where $g_1(j) = \log_e|\mathbf{S}_j|$ for the case of group-specific covariance estimates, otherwise $g_1(j) = 0$ and $g_2(j) = -2\log_e(q_j)$ if prior probabilities for group membership differ, otherwise $g_2(j) = 0$. In terms of D, the posterior probability of $\mathbf{x}$ belonging in group j is

$$P(j|\mathbf{x}) = \frac{\exp(-.5D^2(\mathbf{x}))}{\Sigma_i \exp(-.5D^2(\mathbf{x}))} \tag{7.6}$$

which can be used to classify $\mathbf{x}$ in group i if setting $j = i$ maximizes $P(j|\mathbf{x})$ or minimizes $D_j^2(\mathbf{x})$.

7.2.2 Parameter Estimation

Fisher (1936) proposed a method for estimating coefficients of the discriminant function model that is now termed Fisher's criterion. The method is best illustrated using two groups, but is generalizable to more than two groups. The idea is to identify the linear surface that best discriminates two groups, equivalent to finding the surface normal vector that best separates the two groups.

Find the direction $\mathbf{v}$ such th $(\mathbf{v}'\bar{\mathbf{x}}_1 - \mathbf{v}'\bar{\mathbf{x}}_2)$ is maximized relative to the standard deviation $\sqrt{\mathbf{v}'\mathbf{S}\mathbf{v}}$, where $\bar{\mathbf{x}}_j$ is the mean vector for group j and $\mathbf{S}$ is the assumed common covariance matrix. Choose $\mathbf{v}$ to maximize

$$C = \frac{\text{distance between sample means}}{\text{standard deviation within samples}} = \frac{\mathbf{v}'\bar{\mathbf{x}}_1 - \mathbf{v}'\bar{\mathbf{x}}_2}{\sqrt{\mathbf{v}'\mathbf{S}\mathbf{v}}} \tag{7.7}$$

Differentiating with respect to $\mathbf{v}$ and equating to $\mathbf{0}$ yields

$$\bar{\mathbf{x}}_1 - \bar{\mathbf{x}}_2 = \frac{\mathbf{v}'(\bar{\mathbf{x}}_1 - \bar{\mathbf{x}}_2)\mathbf{S}\mathbf{v}}{\sqrt{\mathbf{v}'\mathbf{S}\mathbf{v}}} \tag{7.8}$$

Since we only need the direction of $\mathbf{v}$, multiplication by a scalar makes no difference. In particular,

$$\frac{\mathbf{v}'(\bar{\mathbf{x}}_1 - \bar{\mathbf{x}}_2)}{\sqrt{\mathbf{v}'\mathbf{S}\mathbf{v}}} \tag{7.9}$$

is a scalar; therefore

$$\mathbf{v} = \mathbf{S}^{-1}(\bar{\mathbf{x}}_1 - \bar{\mathbf{x}}_2) \tag{7.10}$$

Hand (1981) describes a variety of other approaches to parameter estimation for discriminant function models.

7.3   NONPARAMETRIC METHODS

In some cases it may be unreasonable to assume multivariate normality for the underlying measurements, but more often it is the presence of nondetects or occasional elevated values which make this assumption unreasonable. Fortunately, deviations from multivariate normality have been shown to have little effect on classification accuracy and associated probability estimates (Hand, 1981). Even analysis of binary data often yields reasonable results despite the obvious departure from multivariate normality. Nonparametric alternatives to the classical discriminant function model are available. The two most common approaches are k-nearest-neighbor methods and kernel methods (i.e., nonparametric density estimation). Nonparametric approaches are based on computing squared distance between any two observation vectors $\mathbf{x}$ and $\mathbf{y}$ in group j as

$$d_j^2(\mathbf{x}, \mathbf{y}) = (\mathbf{x} - \mathbf{y})'\mathbf{V}_j^{-1}(\mathbf{x} - \mathbf{y}) \tag{7.11}$$

Classification of $\mathbf{x}$ is based on estimated group-specific densities from which $P(j|\mathbf{x})$ is computed. Specific approaches are described in the following sections.

7.3.1   Kernel Methods

To compute a discriminant function analysis using a kernel method, begin by defining the following quantities. Let r be a fixed radius, K_j be a specified kernel, and $\mathbf{z}$ be a p-dimensional vector. The volume of a unit sphere bounded by $\mathbf{z}'\mathbf{z} = 1$ is

$$v_0 = \frac{\pi^{p/2}}{\Gamma(p/2 + 1)} \tag{7.12}$$

For group j, the volume of a p-dimensional ellipsoid bounded by $\{z | z' V_j^{-1} z = r\}$ is

$$v_r(j) = r^p |V_j|^{1/2} v_0 \tag{7.13}$$

A variety of group-specific densities can be used. For a uniform kernel

$$K_j(z) = \frac{1}{v_r(j)} \tag{7.14}$$

if $z' V_j^{-1} z \le r^2$, otherwise $K_j(z) = 0$. For a normal kernel

$$K_j(z) = \frac{1}{c_0(j)} \exp\left(- \frac{.5 z' V_j^{-1} z}{r^2}\right) \tag{7.15}$$

where $c_0(j) = (2\pi)^{p/2} r^p |V_j|^{1/2}$. For an Epanechnikov kernel

$$K_j(z) = c_1(j)\left(1 - \frac{z' V_j^{-1} z}{r^2}\right) \tag{7.16}$$

if $z' V_j^{-1} z \le r^2$, otherwise $K_j(z) = 0$, where $c_1(j) = (1 + p/2)/v_r(j)$. For a biweight kernel

$$K_j(z) = c_2(j)\left(1 - \frac{z' V_j^{-1} z}{r^2}\right)^2 \tag{7.17}$$

if $z' V_j^{-1} z \le r^2$, otherwise $K_j(z) = 0$, where $c_2(j) = (1 + p/4) c_1(j)$. The group density at x is

$$f_j(x) = \frac{1}{n_j} \sum_y K_j(x - y) \tag{7.18}$$

where the summation is over all observations in group j. The posterior probability of membership in group j is

$$P(j|x) = \frac{q_j f_j(x)}{f(x)} \tag{7.19}$$

and if $f(x) = 0$ the observation cannot be classified. Note that selection of r is tricky. One approach is to vary r and plot group densities for fixed kernel shape. Another approach is to assume multivariate normality and minimize mean integrated square error of the estimated density. Then

$$r = \left(\frac{A(K_j)}{n_j} \right)^{1/(p+4)}$$

(7.20)

where for a uniform kernel

$$A(K_j) = \frac{2^{p+1}(p+2)\Gamma(p/2)}{p}$$

(7.21)

or for a normal kernel

$$A(K_j) = \frac{4}{2p+1}$$

(7.22)

or for an Epanechnikov kernel

$$A(K_j) = \frac{2^{p+2}p^2(p+2)(p+4)\Gamma(p/2)}{2p+1}$$

(7.23)

7.3.2   The *k*-Nearest-Neighbor Method

To compute a nonparametric discriminant function analysis based on the *k*-nearest-neighbor method, begin by saving *k* smallest distances from **x**. Let k_j represent the number of distances associated with group *j*. The estimated group *j* density at **x** is then

$$f_j(\mathbf{x}) = \frac{k_j}{n_j v_j(\mathbf{x})}$$

(7.24)

where $v_j(\mathbf{x})$ is the volume of the ellipsoid bounded by $\{\mathbf{z} | \mathbf{z}' \mathbf{V}_j^{-1} \mathbf{z}\}$. When $k = 1$, **x** is classified in the group with the **y** point that yields the smallest $d_j^2(\mathbf{x}, \mathbf{y})$.

Example 7.1

One question underlying development of regulations for municipal solid waste facilities is whether or not municipal solid waste leachate (MSWLF) is differentiable from hazardous waste leachate in terms of types, frequencies, and concentrations of hazardous constituents. It has been suggested that leachate from sites with histories of accepting hazardous waste is virtually the same as leachate from sites that accepted nonhazardous solid waste (Brown and Donnelly, 1988). For example, the preamble of the new Subtitle D

TABLE 7.3   Standardized Discriminant Function Coefficients Leachate Comparison Study (16 Priority Pollutant List Volatile Organics)

Constituent	Function 1	Function 2
Ethylbenzene	−0.85373	1.33903
1,2-Dichloroethane	0.20917	−0.31996
Toluene	−0.27676	−1.01725
Chlorobenzene	−0.43472	−1.12535
Tetrachloroethene	−0.30367	1.34204
trans-1,2-Dichloroethene	−1.12865	−0.61367
Chloroform	0.59268	0.04392
Benzene	0.77089	−0.40560
1,1,1-Trichloroethane	0.33018	−0.39083
Vinyl chloride	0.95020	0.09952
Methylene chloride	0.31670	−0.02633
1,1-Dichloroethane	−0.51178	1.62448
1,1-Dichloroethene	0.53929	−0.03210
1,1,2-Trichloroethane	0.04071	1.14297
Trichloroethene	1.52181	−1.53949
1,1,2,2-Trichloroethane	−1.23001	0.17053

From Gibbons et al. (1992b).

hazardous waste disposal facility that took waste prior to 1987, what is the average concentration? Table 7.2 reveals that, when detected, the average concentration of VOCs was one to three orders of magnitude higher in old hazardous waste disposal facilities relative to the old MSWLFs. In general, average concentrations for codisposal facility leachate fell between hazardous and MSWLF levels.

To determine if 16 VOCs detected in at least one type of facility could be used to accurately classify facility type, Gibbons and coworkers applied linear discriminant function analysis on natural-log-transformed data (i.e., assuming the log transforms of the original measurements were multivariate normal). The results (see Tables 7.3 and 7.4) and reanalysis of these data using alternate nonparametric methods are reported.

TABLE 7.4   Standardized Discriminant Function Group Means Leachate Comparison Study (16 Priority Pollutant List Volatile Organics)

Facility Type	Function 1	Function 2
Municipal	−1.27622	−0.60351
Codisposal	−0.93824	0.64565
Hazardous	1.86047	−0.06002

From Gibbons et al. (1992b).

regulation (40 CFR Parts 257 and 258) of the EPA states that,

> Technical data gathered by the Agency and available in the docket to this rulemaking do not reveal significant differences in the number of toxic constituents and their concentrations in the leachates of the two categories of facilities.

Gibbons et al. (1992b) have constructed a large national database consisting of volatile organic priority pollutant compounds from 1490 leachate samples obtained from 283 sample points in 93 landfill waste cells from 48 separate sites with MSWLF, codisposal, or hazardous waste site histories. Gibbons and coworkers further classified hazardous and municipal waste landfills as new or old on the basis of whether or not they had accepted waste prior to new government regulations banning disposal of certain substances. They demonstrate that MSWLFs and hazardous waste landfills are *easily* differentiable based on detection frequency and concentration, with codisposal facilities generally falling in the middle. This analysis is used to illustrate the methods described in this chapter.

To begin, landfill operation and sample point information was listed. Based on site histories, leachate sample points were categorized as codisposal, hazardous, or MSWLF. Waste cells that could not be accurately categorized or for which insufficient documentation was available were not used in the analysis.

In terms of constituents, Gibbons and coworkers began by examining leachate data for 56 Appendix IX volatile organic compounds (VOCs), because the majority of these leachate samples had been analyzed for volatile organic compounds and because of the greater environmental mobility of these compounds relative to the other classes of compounds (i.e., semivolatiles, pesticides, PCBs, and metals). In addition, these compounds do not occur naturally; therefore natural variation over geographical regions was not confounded with facility type. Gibbons and coworkers observed that the majority of detections were on the sublist of 29 Priority Pollutant List VOCs. Detection frequencies for each facility type are listed in Table 7.1.

Inspection of Table 7.1 reveals that numerous constituents are only or predominantly detected in hazardous waste landfill leachate. Perhaps the best illustration is 1,2-DCA, detected almost half of the time (48%) in hazardous waste landfill leachate samples, but in only 3% of codisposal samples, 4% of old MSWLF samples, and 6% of new MSWLF samples. In contrast, chloroform was detected in 39% of old hazardous waste samples, but in only 2% of the new hazardous waste samples, with low detection frequencies for codisposal and MSWLF samples. From these data Gibbons and coworkers concluded that hazardous waste facilities *do* differ from MSWLFs in the number of toxic constituents found, and that for some, but not all, of these compounds hazardous waste treatment standards imple-

TABLE 7.1 Detection Frequencies by Site Type Leachate Comparison Study (Priority Pollutant List Volatile Organics)

| | Codisposal | | Hazardous | | | | MSWLF | | | |
| | | | Old | | New | | Old | | New | |
Constituent	$P(D)$	N	$P(D)$	N	$P(D)$	N	$P(D)$	N	$P(D)$	N
1,1,1-Trichloroethane	.02	195	.35	849	.04	94	.02	261	.23	78
1,1,2,2-Tetrachloroethane	.00	191	.10	847	.00	94	.00	261	.00	78
1,1,2-Trichloroethane	.01	191	.09	847	.00	94	.00	261	.01	78
1,1-Dichloroethane	.27	195	.52	847	.15	94	.17	261	.41	78
1,1-Dichloroethene	.02	195	.28	848	.09	94	.02	261	.01	78
1,2-Dichloroethane	.03	195	.48	848	.49	94	.04	261	.06	78
1,2-Dichloropropane	.03	189	.01	845	.00	94	.05	260	.00	76
Acrolein	.00	191	.00	821	.02	94	.00	247	.00	76
Acrylonitrile	.00	191	.00	823	.00	94	.00	247	.00	76
Benzene	.30	195	.31	847	.72	94	.49	261	.13	80
Bromodichloromethane	.00	189	.00	845	.00	94	.00	261	.00	76
Bromoform	.00	191	.00	845	.00	94	.00	261	.00	78
Bromomethane	.00	191	.00	842	.00	94	.00	261	.00	76
Carbon tetrachloride	.00	191	.02	847	.00	94	.00	261	.03	76
Chlorobenzene	.21	191	.20	847	.34	94	.35	263	.04	77
Chloroethane	.05	191	.08	843	.00	94	.08	261	.08	76
Chloroform	.01	191	.39	849	.02	94	.03	261	.04	76
Chloromethane	.01	191	.00	845	.00	94	.02	261	.00	76
cis-1,3-Dichloropropene	.00	191	.00	845	.00	94	.00	261	.00	77
Dibromochloromethane	.00	191	.00	845	.00	94	.00	261	.00	78
Dichlorodifluoromethane	.00	186	.06	834	.09	94	.02	247	.09	76
Ethylbenzene	.67	198	.39	840	.38	94	.57	267	.39	79
Methylene chloride	.58	195	.86	828	.83	94	.59	261	.78	78
Tetrachloroethene	.05	193	.38	849	.34	94	.05	260	.04	80
Toluene	.72	194	.75	824	.77	94	.69	261	.89	80
trans-1,2-Dichloroethene	.30	195	.34	847	.26	94	.33	257	.32	78
Trichloroethene	.11	195	.65	849	.30	94	.13	261	.22	79
Trichlorofluoromethane	.01	193	.02	851	.02	94	.01	247	.09	76
Vinyl chloride	.17	195	.40	846	.51	94	.10	261	.15	78

New Hazardous = accepted waste after 1986; New MSWLF = accepted waste after 1984; $P(D)$ = proportion detected; N = total number of measurements.

From Gibbons et al. (1992b).

mented since 1987 have decreased the frequency of detection in new hazardous waste disposal sites.

Table 7.2 presents the arithmetic mean concentration for those compounds that were detected in 10% or more of the samples for each facility type. Average concentrations were computed from detected concentrations only. These data address the question: When a constituent is detected in a

TABLE 7.2 Average Concentration in mg/l by Site Type Leachate Comparison Study (Priority Pollutant List Volatile Organics)

| | | Hazardous | | MSWLF | |
Constituent	Codisposal	Old	New	Old	New
Compounds Rarely Detected Regardless of Facility Type					
1,2-Dichloropropane					
Acrolein					
Acrylonitrile					
Bromodichloromethane					
Bromoform					
Bromomethane					
Carbon tetrachloride					
Chloroethane					
Chloromethane					
cis-1,3-Dichloropropene					
Dibromochloromethane					
Dichlorodifluoromethane					
Trichlorofluoromethane					
Compounds Only Detected in Hazardous Waste Facilities*					
1,1,2,2-Tetrachloroethane		30.7			
1,1,2-Trichloroethane		39.6			
1,1-Dichloroethene		20.6	0.140		
1,2-Dichloroethane		200.2	4.25		
Chloroform		97.5			
Tetrachloroethene		67.4	0.112		
Compounds Primarily Detected in Hazardous Waste Facilities					
1,1,1-Trichloroethane		200.6			0.1?
1,1-Dichloroethane	0.739	33.2	0.132	0.400	0.1?
Trichloroethene	0.837	95.5	0.038	0.051	0.0?
Vinyl chloride	0.841	8.96	40.1	0.107	0.0?
Compounds Frequently Found in all Facility Types					
Benzene	2.31	4.03	0.529	0.065	0.0?
Chlorobenzene	130.5	9.11	0.156	0.736	
Ethylbenzene	1.86	152.7	0.185	0.198	0.0?
Methylene chloride	2.38	385.4	1.97	0.898	1.3?
Toluene	3.97	79.7	1.67	0.583	0.4?
trans-1,2-Dichloroethene	1.35	1.67	0.081	0.492	0.1?

New Hazardous = accepted waste after 1986; New MSWLF = accepted waste after 1984; Me? values reported for detection frequencies $\geq$ 10%.

*$P(D) > 6\%$.

From Gibbons et al. (1992b).

Classification results for the original analysis are as follows:

Actual Facility Type	Number of Cases	Predicted Facility Type		
		Sanitary—Old	Codisposal	Hazardous—Old
Sanitary—Old	245	74%	24%	2%
Codisposal	184	30%	60%	10%
Hazardous—Old	747	8%	9%	83%
Sanitary—New	76	45%	47%	8%
Hazardous—New	47	26%	4%	70%

Inspection of the classification table reveals that only 2% (i.e., 6 out of 245) of old MSWLF leachate measurements were classified as old hazardous waste leachate. If the EPA's statement was correct, 33% should have been found, consistent with chance if leachate from old MSWLFs and old hazardous waste landfills were the same. Conversely, only 8% of old hazardous waste landfill samples were classified as old MSWLF. These findings reveal that using 16 VOCs alone, we can accurately determine whether the facility accepted MSWLF or hazardous waste. In terms of misclassification, both rates were low; however, it was more probable to misclassify a hazardous sample point as MSWLF than the reverse.

In codisposal samples, 30% were misclassified as MSWLF, 60% were correctly classified, and 10% were misclassified as hazardous. This result supports previous findings based on average concentrations, that codisposal leachate is more similar to leachate from MSWLFs than from hazardous landfills.

In newer facilities, 70% of new hazardous waste landfill leachate samples were classified as old hazardous waste leachate, 4% as codisposal, and 26% as old MSWLF. This increase from 8% misclassification of old hazardous leachate as old MSWLF leachate to 26% of new hazardous leachate misclassified as old MSWLF leachate is a testimony to the effectiveness of new treatment standards since 1987. Nevertheless, leachate from new hazardous waste landfills still looks a lot more like leachate from old hazardous facilities than leachate from old MSWLFs. Conversely, leachate from new MSWLFs does not look like leachate from old hazardous facilities (8% misclassified), but does appear to be split equally between old MSWLFs and codisposal in terms of classification rates. Given the small sample sizes for these two "new" groups, these findings should be interpreted at an "exploratory" level.

The first function, which discriminates between hazardous and the combination of MSWLF and codisposal leachate, is characterized by the relative ratio of TCE, PCE, and vinyl chloride to trans-1,2-DCE and ethylbenzene (see Table 7.3). The larger this ratio, the greater is the probability that the sample point is hazardous (see Table 7.4). The second function, which differentiates MSWLF from codisposal leachate, is characterized by the relative ratio of ethylbenzene and toluene to chlorobenzene, benzene, and methylene chloride. The larger this ratio, the greater is the probability that the sample point is codisposal. The directions of these differences are determined by the locations of the function means (see Table 7.4).

A valid criticism of this analysis is that comparisons are made at the level of the individual sample and not at the sample point, waste cell, or site. To begin, it should be noted that hazardous, codisposal, and MSWLF waste cells may be at a single facility; therefore analysis at the level of the site can be misleading and inappropriate. For a given waste cell at a given site, there are some cases in which there is more than one sample point; however, these waste cells are often heterogeneous in composition, and these sample points have been selectively installed to incorporate this heterogeneity. For a given sample point, repeat samples are often taken to incorporate the effects of an ever-changing waste stream over time. The "multilevel" nature of this problem calls into question use of traditional statistical models. For example, it can be argued that for hazardous waste disposal facilities, sample points with the greatest number of constituents are sampled most often. Therefore, elevated detection frequencies become an artifact of repeated sampling, and the classification function is biased in this direction.

To shed some light on this question, data were aggregated to the level of the sample point. Average concentrations were computed for each sample point (i.e., averaging repeated measurements for each sample point) assuming that nondetected constituents were present at one-half of the respective detection limits, and the classification functions were recomputed and accuracy reestablished.

In terms of classification accuracy, the aggregated data yielded improved results as follows:

Actual Facility Type	Number of Cases	Predicted Facility Type		
		Sanitary—Old	Codisposal	Hazardous—Old
Sanitary—Old	42	86%	14%	0%
Codisposal	44	32%	66%	2%
Hazardous—Old	51	10%	10%	80%
Sanitary—New	13	46%	54%	0%
Hazardous—New	4	50%	0%	50%

Inspection of these classification results reveals that when data are analyzed at the level of the sample point, *none* of the MSWLF sample points are classified as originating from old hazardous waste landfills (in contrast to 2% for old MSWLF and 8% for new MSWLF from the previous analysis). In addition, the percentage of instances in which a codisposal sample was classified as hazardous also fell from 10% to 2% when data were aggregated to the sample point level.

It is of interest to determine the extent to which classification results depend on model assumptions. For example, relaxing the assumption of homogeneous covariance matrices across the three facility types yielded

Actual Facility Type	Number of Cases	Predicted Facility Type		
		Sanitary—Old	Codisposal	Hazardous—Old
Sanitary	42	45%	43%	12%
Codisposal	44	25%	59%	16%
Hazardous	51	2%	6%	92%

which appears to be better at differentiating hazardous waste from codisposal and MSW, but poorer at differentiating MSW and codisposal. In terms of nonparametric approaches, one nearest neighbor yielded

Actual Facility Type	Number of Cases	Predicted Facility Type		
		Sanitary—Old	Codisposal	Hazardous—Old
Sanitary	42	53%	45%	2%
Codisposal	44	32%	64%	4%
Hazardous	51	4%	8%	88%

and two nearest neighbors yielded

Actual Facility Type	Number of Cases	Predicted Facility Type		
		Sanitary—Old	Codisposal	Hazardous—Old
Sanitary	42	76%	24%	0%
Codisposal	44	64%	36%	0%
Hazardous	51	10%	12%	78%

The nearest-neighbor methods yield results similar to the normal linear discriminant function method for discrimination between MSW and hazardous waste landfill leachate, but both appear to be less able to differentiate codisposal from MSW landfill leachate. The normal kernel estimator yielded the following results:

Actual Facility Type	Number of Cases	Predicted Facility Type		
		Sanitary—Old	Codisposal	Hazardous—Old
Sanitary	42	64%	36%	0%
Codisposal	44	20%	80%	0%
Hazardous	51	8%	21%	71%

which are also quite similar to the classical discriminant function model, although more hazardous facilities are classified as codisposal but no codisposal facilities are classified as hazardous. Neither the uniform kernel nor the Epanechnikov kernel was able to classify all of the cases, although for those cases that were classifiable using these estimators, none of the MSW or codisposal samples was classified as hazardous, and none of the hazardous samples was classified as MSW.

The results reveal that leachate of MSWLFs and hazardous waste disposal facilities are not the same and, in fact, can be classified with accuracy based on chemical composition alone. Furthermore, codisposal facilities can be distinguished from both purely hazardous waste landfills and purely MSWLFs with accuracy, although they appear to be more similar to MSWLFs than hazardous waste landfills. The best overall classification accuracy for this problem was achieved using the traditional normal linear discriminant function model.

Example 7.2

As a second example, consider data from a facility that accepted two distinct waste streams, one termed the "Tire Pile" where millions of tires and industrial liquid waste were disposed of, and a second termed "South Central" where a mixture of industrial liquid waste and municipal solid waste were disposed of. Collection wells were drilled directly into waste pits in each area. In addition, background (i.e., upgradient detection monitoring wells) water quality samples were available. Based solely on the geochemistry, the questions are: (1) Are the two waste streams differentiable from background? (2) Are the two waste streams differentiable from each other?

To examine these questions, the six primary VOCs detected at the facility, 1,1,1-trichloroethane (1,1,1-TCA), 1,1-dichloroethane (1,1-DCA), tetra-chloroethene (PCE), trichloroethene (TCE), benzene, and toluene, were used in constructing parametric and nonparametric discriminant functions. Inspection of the distribution of these data revealed lognormal distributions for the six constituents. Observed source geometric means are displayed in Table 7.5. Inspection of Table 7.5 reveals that (1) background water quality has concentrations at the detection limit; (2) Tire Pile area samples are characterized by high levels of 1,1,1-TCA, 1,1-DCA, PCE, and TCE; and (3) waste pits in the South Central portion of the landfill are high in benzene and toluene relative to the other four constituents. Table 7.6 displays the standardized discriminant function coefficients which support previous observations for geometric mean values.

TABLE 7.5   Observed Source Geometric Means in μg/l

Source	111-TCA	11-DCA	PCE	TCE	Benzene	Toluene
Background	4	5	4	2	4	6
Tire Pile	1998	602	299	148	30	37
South Central	20	22	12	5	74	1212

TABLE 7.6   Standardized Discriminant Function Coefficients Natural-Log-Transformed Data

Parameter	Function 1	Function 2
111-TCA	0.59621	−0.09486
11-DCA	0.26869	0.00006
PCE	0.42360	−0.77628
TCE	−0.02044	0.65911
Benzene	0.03821	0.18924
Toluene	−0.37423	1.07026

The first function, which contrasts Tire Pile areas from background and South Central areas (see the function means in Table 7.7), is based on the ratio of 1,1,1-TCA, 1,1-DCA, and PCE to toluene. The second function, which contrasts South Central waste pits from background and Tire Pile areas, is dominated by the ratio of toluene to PCE which is large in the South Central area but small in the background and Tire Pile areas.

Classification accuracy for several methods is displayed in Table 7.8. Table 7.8 reveals that parametric and nonparametric discriminant function models performed comparably in terms of differentiating among the three sources based on the six constituents. Of the two parametric models, the quadratic model (i.e., unequal variances and covariances among the six constituents across the three sources) performed better in contrasting the two waste pit areas but misclassified four background samples as Tire Pile. No other methods misclassified background groundwater samples as Tire Pile or South Central waste pit samples. Of the nonparametric methods, the first- and second-nearest-neighbor methods and the normal kernel estimator yielded comparable results, slightly better than the normal linear model in differentiating waste pit area samples from background samples. The uniform kernel and Epanechnikov kernel estimators were not able to classify six Tire Pile and ten South Central waste pit samples. For those cases that were classifiable using these estimators, classification was nearly perfect.

To better understand the properties of these estimators and their corresponding assumptions, analysis was focused on 1,1,1-TCA and toluene, the two best discriminators of the three groups. Using only two variables, estimated source densities and classification surfaces can be depicted graphi-

TABLE 7.7   Discriminant Function Source Means

Source	Function 1	Function 2
Background	−1.44578	−0.82968
Tire Pile	2.62824	−0.04821
South Central	−1.28060	2.37803

TABLE 7.8   Classification Accuracy for Parametric and Nonparametric Models

	Number of	Predicted Source		
Actual Source	Cases	Background	Tire Pile	South Central
Background				
Normal linear	55	55	0	0
Normal quadratic	55	51	4	0
NN $k = 1$	55	55	0	0
NN $k = 2$	55	55	0	0
Normal kernel	55	55	0	0
Tire Pile				
Normal linear	41	3	36	2
Normal quadratic	41	0	38	3
NN $k = 1$	41	1	38	2
NN $k = 2$	41	1	37	3
Normal kernel	41	1	38	2
South Central				
Normal linear	20	4	3	13
Normal quadratic	20	1	2	17
NN $k = 1$	20	1	3	16
NN $k = 2$	20	1	2	17
Normal kernel	20	3	3	14

cally as contours or three-dimensional surfaces. Figure 7.1 displays a plot of natural-log-transformed 1,1,1-TCA (LNTCA) versus natural-log-transformed toluene (LNTOL) using different symbols for background groundwater samples, Tire Pile waste pits, and South Central waste pits.

Figure 7.1 reveals that the three sources are well differentiated by these two constituents, with only a few anomalous values. Figures 7.2a to c display estimated densities for background, Tire Pile, and South Central areas based on the assumption of bivariate normality with common variances and covariance. These densities are bell shaped and there is little overlap in probability. The classification surface is linear (see Figure 7.2d) as assumed by the model, with distinct classification regions.

In contrast, Figures 7.3a to d display the same results for the quadratic model, which assumes bivariate normality with different variances and covariance for each source. Figure 7.3a reveals that there is small variability for the background area, more for the Tire Pile area (see Figure 7.3b), and even more for the South Central area (see Figure 7.3c). Overlap, though minimal, nevertheless accounts for misclassification of four background samples as Tire Pile. The classification surface is now curvilinear (i.e., quadratic) as assumed by the model (see Figure 7.3d).

For completeness, Figures 7.4a to d display the same results for the normal kernel estimator, which makes no assumption regarding joint distri-

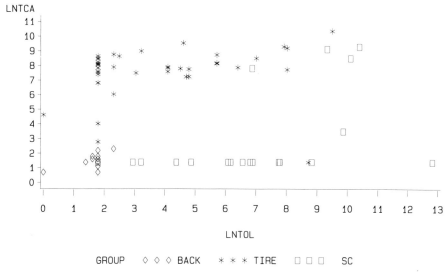

Figure 7.1   Plot of LNTCA versus LNTOL by group.

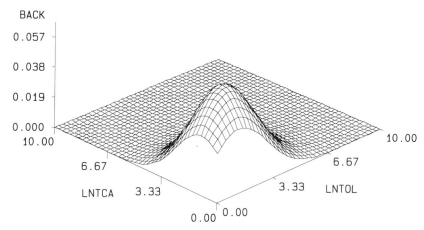

Figure 7.2*a*   Plot of estimated density for background (normal linear discriminant function).

bution of 1,1,1-TCA and toluene other than a continuous bivariate function. Figures 7.4*b* and *c* do a particularly nice job of illustrating how the nonparametric estimator remains true to the observed form of bivariate distributions (see Figure 7.1). Figure 7.4*c* shows how the nonparametric kernel estimator incorporates information from the two misclassified results by creating a density with two discrete regions. Figure 7.4*d* displays a nonlinear classification surface.

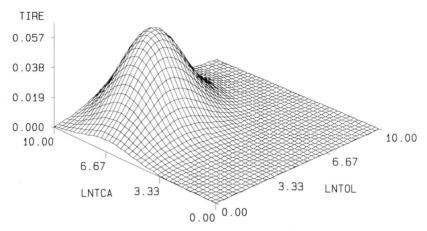

Figure 7.2b Plot of estimated density for Tire Pile (normal linear discriminant function).

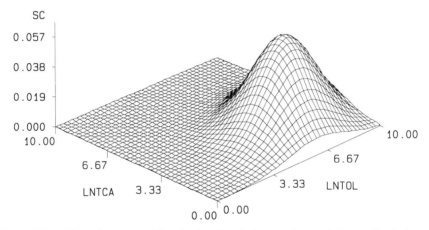

Figure 7.2c Plot of estimated density for South Central (normal linear discriminant function).

Similar results are seen for the one-, two-, and three-nearest-neighbor estimators (see Figures 7.5a to c). With one nearest neighbor, both regions of misclassification (i.e., the Tire Pile and South Central waste pits) are shown as distinct classification subregions. With two nearest neighbors, classification boundaries are nonlinear and quite irregular, but there are no distinct subregions. Finally, using three nearest neighbors appears to provide very little additional information.

These graphical depictions of estimated source densities and classification regions help to illustrate important differences between alternative paramet-

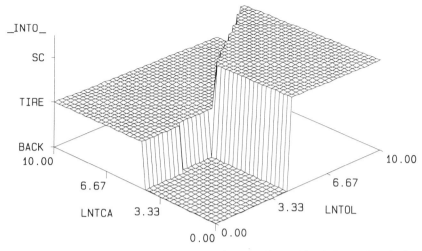

Figure 7.2d   Plot of classification results (normal linear discriminant function).

ric and nonparametric approaches to discriminant function analysis. The most restrictive normal linear model appears to do quite well in this problem, in part due to the other methods' being more sensitive to incorporating information from misclassified or outlying values (i.e., measurements that look more like a different source than the one they were sampled from). Of the nonparametric approaches, the two-nearest-neighbor estimator appears to give the most reasonable classification surface in that it accommodates nonlinearity without creating "islands" of classification. These results may be specific to this example and consideration of alternatives is encouraged.

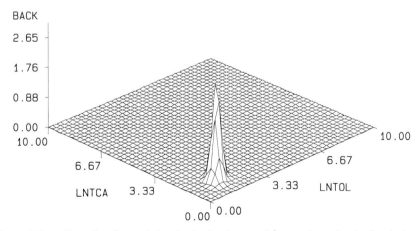

Figure 7.3a   Plot of estimated density for background (normal quadratic discriminant function).

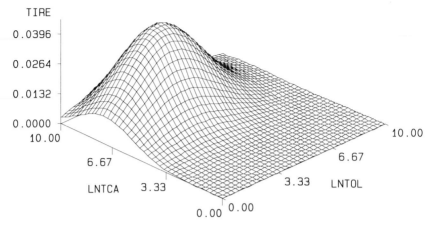

Figure 7.3*b* Plot of estimated density for background (normal quadratic discriminant function).

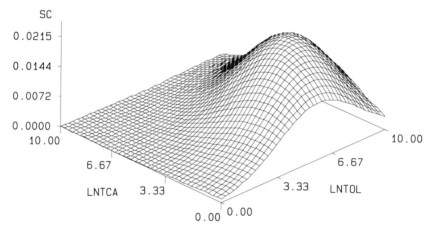

Figure 7.3*c* Plot of estimated density for background (normal quadratic discriminant function).

Example 7.3

As a final example, a multiple-source detection monitoring problem is examined. In this example groundwater quality was questionable for a site on an ocean bay surrounded by enormous salt ponds harvested for human consumption. In the middle of the facility is a flood control channel filled with varying levels of salt water from the bay. The waste disposal facility is monitored by a series of shallow and deep wells located downgradient of the landfill. Four potential groundwater impacts are (1) leachate from the waste

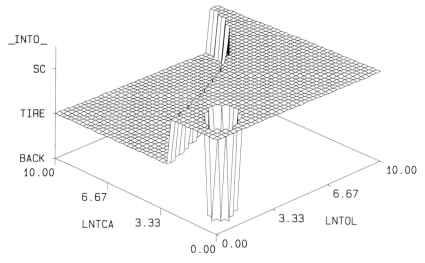

Figure 7.3d   Plot of classification results (normal quadratic discriminant function).

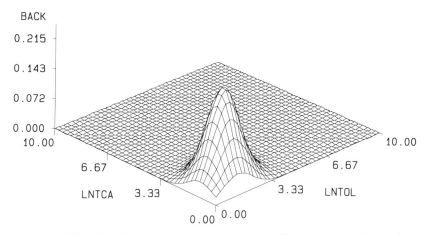

Figure 7.4a   Plot of estimated density for background [normal kernel ($r = .5$) nonparametric discriminant function].

disposal facility, (2) surface water from the bay, (3) liquid from the salt ponds, and (4) background groundwater obtained from upgradient wells. The primary detection monitoring constituents were alkalinity, total organic carbon (TOC), and total dissolved solids (TDS). Observed source means and shallow and deep downgradient well means are presented in Table 7.9.

Leachate from the facility is high in alkalinity and TOC, whereas liquid from the salt ponds is high in TDS. Surface water from the bay is similar to

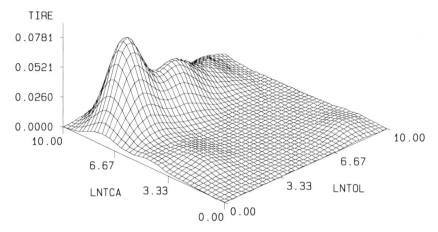

Figure 7.4b   Plot of estimated density for Tire Pile [normal kernel (r = .5) nonparametric discriminant function].

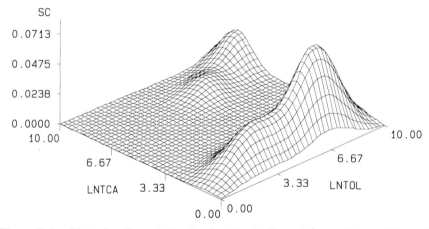

Figure 7.4c   Plot of estimated density for South Central [normal kernel (r = .5) nonparametric discriminant function].

salt pond liquid but has much lower TDS and TOC levels. Background groundwater quality is low in all three constituents (i.e., alkalinity, TOC, and TDS), whereas shallow and deep downgradient wells have levels more similar to surface water for all three constituents. Standardized discriminant function coefficients are presented in Table 7.10, and their corresponding source means are found in Table 7.11. The first function is based on high levels of TOC and TDS, and differentiates leachate and salt ponds from background and surface water measurements. The second function is based on the ratio of alkalinity to TDS, and differentiates leachate (and to a lesser extent background) from salt pond and surface water (i.e., a function sensitive to

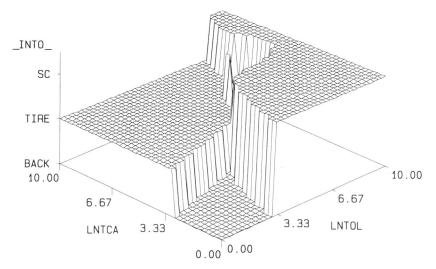

Figure 7.4d   Plot of classification results [normal kernel ($r = .5$) nonparametric discriminant function].

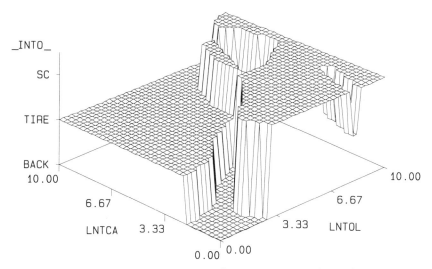

Figure 7.5a   Plot of classification results [nearest-neighbor ($K = 1$) nonparametric discriminant function].

salt water). The third function is based on the ratio of alkalinity and TDS to TOC, and differentiates surface water and leachate from background and salt ponds. Each discriminant function represents an independent contribution to the classification function. Note that the sign of the TDS coefficient is always the same as the sign of the salt pond source mean, and the same relationship between TOC and leachate is observed.

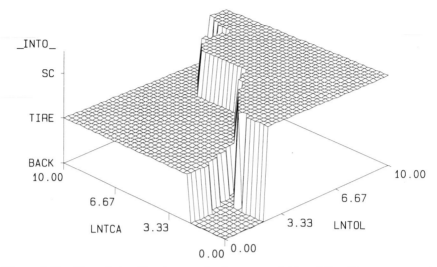

Figure 7.5b Plot of classification results [nearest-neighbor ($K = 2$) nonparametric discriminant function].

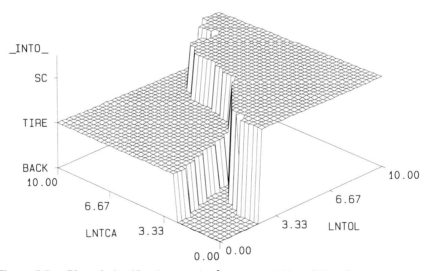

Figure 7.5c Plot of classification results [nearest-neighbor ($K = 3$) nonparametric discriminant function].

TABLE 7.9   Observed Source Means

Source	Alkalinity	TOC	TDS
Reference	584	3.27	1,502
Leachate	2,462	312.06	21,721
Salt pond	517	81.06	260,125
Surface	218	8.10	16,362
DG Deep	429	2.44	41,589
DG Shallow	672	13.63	54,252

TABLE 7.10   Standardized Discriminant Function Coefficients

Parameter	Function 1	Function 2	Function 3
Alkalinity	0.22104	1.17944	0.37193
TOC	1.06229	0.04695	−0.41237
TDS	0.72343	−1.02691	0.45264

TABLE 7.11   Discriminant Function Source Means

Source	Function 1	Function 2	Function 3
Background	−7.15827	3.30829	0.40684
Leachate	7.15461	6.58559	−0.38676
Salt pond	5.12936	−3.72771	0.35699
Surface water	−4.45031	−3.55290	−0.62666

Classification accuracy is displayed in Table 7.12. Note that 100% of samples used in constructing the classification function were correctly classified. In new downgradient monitoring results, all but six shallow well measurements were classified as surface water. These two wells (each consisting of three measurements) were classified as being impacted by leachate from the facility. In deeper downgradient monitoring wells, one-third were classified as background and two-thirds as surface water. These results indicate that in all but two monitoring wells, differences in upgradient versus downgradient groundwater quality were attributable to impact from surface water and not impact from the waste disposal facility. The facility may be responsible for impact to the other two wells, but this conclusion would require a more detailed assessment using constituents that are characteristics of the leachate (e.g., VOCs).

To obtain a feel for the magnitude of the differences between sources and the strength of the classification of groundwater samples, the classification function is displayed graphically in Figure 7.6. Figure 7.6 presents the source means as large transparent spheres with diameters of four standard deviation

TABLE 7.12   Classification Accuracy Normal Linear Model

Actual Source	N	Predicted Source			
		Reference	Leachate	Salt Pond	Surface Water
Background	13	13	0	0	0
		100.0%	0.0%	0.0%	0.0%
Leachate	9	0	9	0	0
		0.0%	100.0%	0.0%	0.0%
Salt pond	16	0	0	16	0
		0.0%	0.0%	100.0%	0.0%
Surface	12	0	0	0	12
		0.0%	0.0%	0.0%	100.0%
DG shallow	31	0	6	0	25
		0.0%	19.4%	0.0%	80.6%
DG deep	30	10	0	0	20
		33.3%	0.0%	0.0%	66.7%

Percentage of "grouped" cases correctly classified: 100.00%

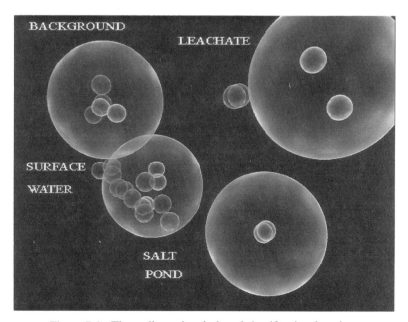

Figure 7.6   Three-dimensional plot of classification function.

units. The x axis refers to the first function, the y axis the second function, and the z axis (i.e., into the picture) the third function. Downgradient monitoring wells are displayed as small transparent spheres. The small highlighted transparent spheres are the well means that gave rise to the group centroids (i.e., the large transparent spheres).

Figure 7.6 reveals that (a) sources are well differentiated with background and leachate means 14 standard deviation units apart, (b) the majority of monitoring wells are consistent with either background groundwater or have been impacted by surface water perhaps from the flood control channel, and (c) two of the downgradient wells appear to be similar to leachate from the facility in terms of these three constituents. Conventional statistical evaluation of the site would have concluded that most downgradient wells were significantly different from background and thus potentially impacted by the facility. Application of discriminant function analysis, however, has revealed that if a site impact has occurred, it is restricted to only two of the monitoring wells.

7.4 SUMMARY

Discriminant function analysis is one of several multivariate statistical methods of use in groundwater monitoring applications. Other pattern recognition procedures such as cluster analysis, factor analysis, multivariate mixture models, and canonical correlation may also prove useful. Multivariate prediction limits (e.g., see Guttman, 1970; Bock, 1975) may be useful in simultaneously comparing constituent profiles or patterns between background and point-of-compliance monitoring wells. Discriminant function analysis is used since contaminant sources are known in advance and each sample obtained from each source is routinely analyzed for the same set of constituents. Both parametric and nonparametric discriminant function analysis models are well suited to taxonomic problems of this kind. The three examples illustrate the flexibility of the approach but do not exhaust potential groundwater monitoring applications.

8 Intrawell Comparisons

8.1 OVERVIEW

There are two general approaches to groundwater detection monitoring at waste disposal facilities. In the first, new downgradient monitoring measurements are compared to a series of n water quality measurements obtained from well(s) that are located hydraulically upgradient of the facility. In the second, new downgradient measurements are compared to their own history, hence the name *intrawell* comparisons. There are advantages and disadvantages to both approaches.

Upgradient versus downgradient comparisons assume that the only difference between upgradient and downgradient water quality is the site, which sits in between. This is often not the case. Large spatial fluctuations in water quality across the facility and surrounding area often exist. Considerable differences in water quality can be seen simply by drilling two holes in the ground, regardless of their positions relative to the site (i.e., upgradient or downgradient). This is further complicated by the fact that there are almost always far more downgradient wells then upgradient wells. Just by chance alone, spatial variation will be greater downgradient than upgradient simply due to the imbalance in the number of wells. Also, in some areas, groundwater moves so slowly that upgradient water may not pass below downgradient wells for years. This also confounds the integrity of this approach to testing groundwater quality. Finally, previous industrial activity (e.g., disposal of slag) may also contribute to widespread spatial variability in groundwater quality, often invalidating upgradient versus downgradient comparisons.

In contrast, intrawell comparisons completely remove the spatial component of variation from the comparison (i.e., each well is compared to its own history). The problem here, however, is that if previous contamination exists, the method will not detect it unless it significantly increases. For this reason, intrawell comparisons are useful for new sites or for those sites in which it can be documented that previous contamination does not exist.

An intermediate solution can be obtained through the use of control charts. Perhaps the most useful are combined Shewart–CUSUM control charts that detect releases both in terms of their absolute magnitude and cumulative increases (i.e., trends). Although background data are collected for each well and summary statistics computed, the cumulative sum includes these data; hence even gradual trends in groundwater quality are detected.

Further confidence can be placed in these methods by removing outliers and existing trends from the background database for each well prior to computing the historical mean and variance from which the limits are derived. The actual data, however, are then compared to these limits, so that if trends are present even in background data they will be detected. In some cases more powerful tests can be obtained by pooling background variance estimates from several wells (i.e., an unbiased estimate of the temporal variance component obtained from several monitoring wells) which can be used in constructing prediction limits for the next future measurement in each of k monitoring wells.

In the following sections several approaches to performing intrawell comparisons are described.

8.2 SHEWART CONTROL CHARTS

The Shewart control chart (Shewart, 1931) is perhaps the oldest and easiest quality control procedure available. The chart plots time (x-axis) versus concentration (y-axis). A horizontal line is drawn to intersect the vertical axis at the point $\mu + Z\sigma$, where Z is an upper percentage point of the normal distribution and μ and σ are the population values of the mean and standard deviation, typically established from years of historical observation. When such long-run data are not available, the sample-based estimates $\bar{x}$ and s are used, and a larger value of Z is selected. In the present context, the line defined by $\bar{x} + Zs$ is called the control limit, and new measurements that exceed the control limit are declared out of control. Often, $Z = 3$ which corresponds to a confidence level of $1 - \alpha = .9987$ for a single new comparison where μ and σ are known. In contrast, however, when only the sample-based estimates $\bar{x}$ and s are available from $n = 8$ historical measurements (e.g., two years of quarterly sampling), the overall confidence level is only 95% for five new comparisons and decreases as the number of future comparisons increases (see Table 1.2). For this reason, Lucas (1982) and the EPA (1989) have suggested a control limit of $\bar{x} + 4.5s$ for routine groundwater monitoring applications. Overall confidence levels for this control limit are 95% with $n = 8$ and 35 future comparisons; however, verification resampling further reduces false positive rates to acceptable levels for most monitoring programs.

8.3 CUMULATIVE SUMMATION (CUSUM) CONTROL CHARTS

The CUSUM control chart was introduced by Page (1954) and is more complicated to compute and implement than the Shewart control chart (see Lucas, 1985; Starks, 1988). Unlike the Shewart control chart that focuses solely on the current monitoring value, the CUSUM control chart incorporates information from previous measurements. The CUSUM scheme involves computation of the cumulative sum S which for the ith sample is given by

$$S_i = \max[0, z_i - k + S_{i-1}] \qquad (8.1)$$

where $z_i = (x_i - \mu)/\sigma$ and k is a parameter selected to be approximately one-half the size of a difference worth detecting. The EPA (1989) has suggested a value of $k = 1$, so that a cumulative increase of two standard deviation units per sampling event would be detected. The EPA (1989) suggests a control limit of $h = 5$ for the values of S_i. The advantage of the CUSUM control chart over the Shewart control chart is sensitivity to small, gradual changes. The advantage of the Shewart control chart over the CUSUM control chart is immediate sensitivity to large releases.

As an illustration, Starks (1988) presents simulated data drawn from a normal distribution with mean $\mu = 10$ and $\sigma = 2$. To illustrate an out-of-control situation, the value 2 was added to each random number drawn after the fourth sampling event (i.e., $i = 4$). The in-control situation has data distributed $N(10, 4)$, so that

$$z_i = \frac{x_i - \mu}{\sigma} = \frac{x_i - 10}{2} \qquad (8.2)$$

Table 8.1 presents results of the computation. Note that if the Shewart control limit is set at $\mu + 3\sigma$, then a value of $z_i = (x_i - \mu)/\sigma > 3$ will exceed the control limit (i.e., the new value is three standard deviation units above the background mean). Table 8.1 reveals that at $i = 10$, the CUSUM

TABLE 8.1   CUSM Quality Control Scheme ($k = .5$, $h = 5$)

	In Control			Out of Control at $i = 5$		
i	x_i	z_i	s_i	x_i	z_i	s_i
0			0			0
1	14.504	2.252	1.752	14.504	2.252	1.752
2	11.108	0.554	1.806	11.108	0.554	1.806
3	7.594	−1.203	0.103	7.594	−1.203	0.103
4	7.580	−1.210	0.000	7.580	−1.210	0.000
5	11.588	0.794	0.294	13.588	1.794	1.294
6	12.002	1.001	0.795	14.002	2.001	2.595
7	10.434	0.217	0.512	12.434	1.217	3.312
8	9.378	−0.311	0.000	11.378	0.689	3.501
9	10.708	0.354	0.000	12.708	1.354	4.355
10	11.278	0.639	0.139	13.278	1.639	5.494*

*Exceeds control limit.
From Starks (1988).

limit is exceeded, but the Shewart limit is never even close to being exceeded, even after the simulated release.

For this reason, the most effective approach to control charting in groundwater monitoring applications is to combine the two approaches as described in the following section.

8.4   COMBINED SHEWART–CUSUM CONTROL CHARTS

The combined Shewart–CUSUM control chart (Lucas, 1982) combines two traditional approaches to quality control: the Shewart control chart (Shewart, 1931) and the cumulative sum control chart (Page, 1954). In this way both immediate and gradual releases are rapidly detected.

8.4.1   Assumptions

The combined Shewart–CUSUM control chart procedure assumes that the data are *independent* and *normally* distributed with a *fixed* mean μ and constant variance σ^2. The most important assumption is independence, and as a result wells should be sampled *no more* frequently than quarterly. The assumption of normality is somewhat less of a concern, and, if problematic, natural log or square root transformation of the observed data should be adequate for most practical applications. For this method nondetects can be replaced by the method detection limit without serious consequence. This procedure should *only* be applied to those constituents that are detected at least in 25% of all samples, otherwise, σ^2 is not adequately defined.

8.4.2   Procedure

In general, at least eight historical independent samples must be available to provide reliable estimates of the mean μ and standard deviation σ of the constituent's concentration in each well.

1. Select the three Shewart–CUSUM parameters h (the value against which the cumulative sum will be compared), k (a parameter related to the displacement that should be quickly detected), and SCL (the upper Shewart limit which is the number of standard deviation units for an immediate release). Lucas (1982) and Starks (1988) suggest that $k = 1$, $h = 5$, and $SCL = 4.5$ are most appropriate for groundwater monitoring applications. This sentiment is echoed by the EPA (1989) in its interim final guidance document. For ease of application, however, $h = SCL = 4.5$ has been selected, which is slightly more conservative than the value of $h = 5$ suggested by the EPA.

2. Denote the new measurement at time point t_i as x_i.

3. Compute the standardized value z_i:

$$z_i = \frac{x_i - \bar{x}}{s} \tag{8.3}$$

where $\bar{x}$ and s are the mean and standard deviation of at least eight historical measurements for that well and constituent (collected in a period of no less than one year).

4. At each time period t_i, compute the cumulative sum S_i as

$$S_i = \max\left[0, (z_i - k) + S_{i-1}\right] \tag{8.4}$$

where $\max[A, B]$ is the maximum of A and B, starting with $S_0 = 0$.

5. Plot the values of S_i (y axis) versus t_i (x axis) on a time chart. Declare an "out-of-control" situation on sampling period t_i if, for the first time, $S_i \geq h$ or $z_i \geq SCL$. Any such designation, however, must be verified on the next round of sampling before further investigation is indicated.

The reader should note that, unlike prediction limits which provide a fixed confidence level (e.g., 95%) for a given number of future comparisons, control charts do not provide explicit confidence levels, and they do not adjust for the number of future comparisons. The selection of $h = SCL = 4.5$ and $k = 1$ is based on the EPA's own review of the literature and simulations (see Lucas, 1982; Starks, 1988; EPA, 1989). The EPA indicates that these values "allow a displacement of two standard deviations to be detected quickly." Since 1.96 standard deviation units correspond to 95% confidence on a normal distribution, we can have approximately 95% confidence for this method as well.

In terms of plotting the results, it is more intuitive to plot values in their original metric (e.g., $\mu g/l$) rather than in standard deviation units. In this case $h = SCL = \bar{x} + 4.5s$ and the S_i are converted to the concentration metric by the transformation $S_i(s) + \bar{x}$, noting that when normalized (i.e., in standard deviation units) $\bar{x} = 0$ and $s = 1$, so that $h = SCL = 4.5$ and $S_i(1) + 0 = S_i$.

8.4.3   Detection of Outliers

From time to time, inconsistently large or small values (outliers) can be observed due to sampling, laboratory, transportation, transcription errors, or even by chance alone. The verification resampling procedure that we have proposed will tremendously reduce the probability of concluding that an impact has occurred if such an anomalous value is obtained for any of these reasons. However, nothing has eliminated the chance that such errors might be included in the historical measurements for a particular well and constituent. If such erroneous values (either too high or too low) are included in the historical database, the result would be an artificial increase in the magnitude of the control limit and a corresponding increase in the false negative rate of the statistical test (i.e., concluding that there is no site impact when, in fact, there is). To remove the possibility of this type of error, the historical data should be screened for each well and constituent for the existence of outliers using the well-known method described by Dixon (1953) (see Chapter 13). These outlying data points should be indicated on the control charts (using a different symbol), but should be excluded from the measurements that are used to compute the background mean and standard deviation. In the future new measurements that turn out to be outliers, in that they exceed the control limit, should be dealt with by verification resampling in downgradient wells only.

8.4.4   Existing Trends

If contamination is preexisting, trends will often be observed in the background database from which the mean and variance are computed. This will lead to upward biased estimates and grossly inflated control limits. To remove this possibility, we first remove outliers and then obtain an estimate of trend using one of the methods described in Chapter 9. In the presence of significant trend, we can remove its effect by the transformation

$$x_i^* = \alpha + (x_i - \alpha + \beta t) \tag{8.5}$$

where α and β are the intercept and slope of the trend line and t represents the measurement occasion recorded as $1, \ldots, n$. In words, this transformation sets the new values equal to the estimated initial level measurement plus the deviation of the actual measurement from the trend line. In this way, unbiased estimates of the background mean, variance, and control limits are obtained even in the presence of a trend potentially caused by a historical release. Of course, the actual measured values (x_i) are then compared to the control limits obtained from the de-trended data (x_i^*). In this way, even preexisting trends in the background dataset will be detected.

Example 8.1

To illustrate the combined Shewart–CUSUM control chart procedure, consider the following hypothetical example.

The example in Table 8.2 illustrates a case in which previous historical measurements were found to be distributed with a mean of 50 μg/l and a standard deviation of 10 μg/l. The data listed in Table 8.2 represent the new levels obtained from eight monitoring events following the establishment of background. Columns 5 and 7 of Table 8.2 reveal that the process is out of control both in terms of trend and absolute value on the third quarter

TABLE 8.2   Example Dataset Following Collection of Eight Background Sample Constituents in $\mu g/l$ (Background Levels $\bar{x} = 50$ and $s = 10$)

Quarter	Year	Period t_i	Concentration x_i	Standardized z_i	$z_i - k$	CUSUM S_i
1	90	1	50	0	−1	0
2	90	2	40	−1	−2	0
3	90	3	60	1	0	0
4	90	4	50	0	−1	0
1	91	5	70	2	1	1
2	91	6	80	3	2	3
3	91	7	100	5^a	4	7^b
4	91	8	120	7^a	6	13^b

[a]Shewart "out-of-control" limit exceeded ($z_i > 4.5$).
[b]CUSUM "out-of-control" limit exceeded ($S_i > 4.5$).

of 1991. The result is confirmed on the fourth quarter of 1991, and further investigation would be indicated. These same data and procedure are illustrated graphically in Figure 8.1.

Note that in Figure 8.1 the CUSUM and absolute measurements are reported in the original metric of $\mu g/l$. To do this, we transform the S_i in Table 8.2 by $S_i(s) + \bar{x}$. Using the example in Table 8.2 with $\bar{x} = 50$ and $s = 10$, the final value of $S_i = 13$ in standard deviation units relates to a concentration of $13(10) + 50 = 180$ $\mu g/l$.

8.4.5   A Note on Verification Sampling

It should be noted that when a new monitoring value is an outlier, perhaps due to a transcription error, sampling error, or analytical error, the Shewart and CUSUM portions of the control chart compares each individual new measurement to the control limit; therefore the next monitoring event measurement constitutes an independent verification of the original result. In contrast, however, the CUSUM procedure incorporates *all* historical values in the computation; therefore the effect of the outlier will be present for both the initial and verification sample, and the statistical test will be invalid.

For example, assume $\bar{x} = 50$ and $s = 10$. On quarter 1, the new monitoring value is 50, so $z = (50 - 50)/10 = 0$ and $S_i = \max[0, (z - 1) + 0] = 0$. On quarter 2, a sampling error occurs and the reported value is 200, yielding $z = (200 - 50)/10 = 15$ and $S_i = \max[0, (15 - 1) + 0] = 14$, which is considerably larger than 4.5; hence an initial exceedance is recorded. On the next round of sampling, the previous result is not confirmed, because the result is back to 50. Inspection of the CUSUM, however, yields $z = (50 - 50)/10 = 0$ and $S_i = \max[(0, (0 - 1) + 14] = 13$, which would be taken as a confirmation of the exceedance, when, in fact, no such confirmation was

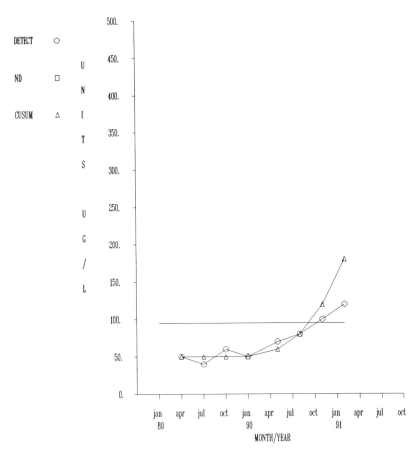

Figure 8.1 Combined Shewart–CUSUM control chart. Example in Table 8.2

observed. For this reason, the verification must *replace* the suspected result in order to have an unbiased confirmation.

8.4.6 Updating the Control Chart

As monitoring continues and the process is shown to be in control, the background mean and variance should be updated periodically to incorporate these new data. Every two years, all new data that are *in control* should be pooled with the initial samples and $\bar{x}$ and s recomputed. These new values of $\bar{x}$ and s are then used in constructing future control charts.

8.4.7 Statistical Power

To better understand the operating characteristics of this procedure, two simulation studies were conducted. In the first study, a release was introduced on month 1, at a rate of 0.1 to 5 standard deviation (sd) units per month. For example, following eight quarters of background monitoring at a particular well for a hypothetical constituent with a mean of 100 $\mu g/l$ and an sd of 10 $\mu g/l$, a release of 1 sd unit per month is introduced. On month 1, the mean of the sampled distribution is 110 $\mu g/l$, on month 2, the mean is 120 $\mu g/l$, and so on. The question of interest is: How long does it take to detect this release using this statistical method? The answer is obtained by simulating this process 1000 times and computing the median time to detection in months. As the rate of the release increases (e.g., 1 sd unit per month versus 0.1 sd units per month), the time it takes to detect the release decreases. The simulation used one verification resample, a background sample of eight measurements, $k = 1$, and $h = SCL = 4.5$. As a comparison, these same simulated data were evaluated using a 99% prediction limit and corresponding median detection times recorded.

In the second study, all conditions were identical, except that the release was constant (i.e., not increasing over time) in the amount of 0.1 to 5 sd units. For those simulations that did not result in a verified exceedance in 5 years, the value of 5 years was recorded (i.e., 5 years or more).

Figure 8.2 displays the results of the first simulation study. Inspection of Figure 8.2 reveals that for a tiny release of only 0.1 sd units per month (i.e., for our previous example, a difference of 100 $\mu g/l$ versus 101 $\mu g/l$ in month 1 following the release), it would take 30 months (over 3 years) to detect it using a 99% prediction limit and 27 months using the combined Shewart–CUSUM control chart procedure. In contrast, a release of 1 sd unit per month would be detected on the second quarterly sampling event, and a release of 2 sd units per month on the very next quarterly sampling event. Results for the two methods were identical, with the exception of a 1.5-sd-unit release which was detected in one quarter by the prediction limit and two quarters by the control chart.

Results of the second simulation study (i.e., constant release) are displayed in Figure 8.3. For a constant release, the control chart outperforms the prediction limit for small releases (i.e., 0.1 to 2 sd units), whereas the prediction limit appears to be slightly more sensitive to larger releases (i.e., releases in the range of 3 to 4 sd units are detected about one quarter faster by the prediction limit). Overall, constant releases in the amount of 2.5 sd units are detected within a year and releases of approximately 3 sd units within 6 months.

8.5 PREDICTION LIMITS

An alternative approach to the intrawell comparison problem involves the use of prediction limits as described in Chapter 1. Here, the prediction limit is computed separately in each monitoring well, using the first available n measurements as background. Since different prediction limits are used for

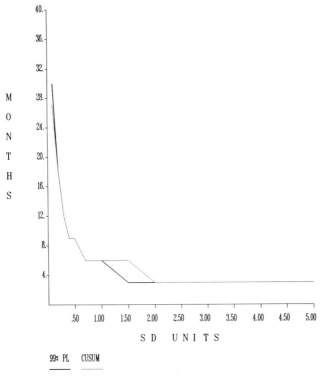

Figure 8.2 Power of combined Shewart–CUSUM control chart under gradual release.

each monitoring well, comparisons across wells are independent and the multiple-comparison problem is greatly simplified. Here, the only complicated statistical adjustment involves the repeated comparison of the resamples to the same prediction limit for that well which is required when the initial sample exceeds the prediction limit. The reader should note that this is simply a repeated application of the method described by Davis and McNichols (1987) for the case of including the next r of m samples in a single future well. Since the wells are independent, to adjust for multiple wells, we simply set

$$\alpha^* = \frac{\alpha}{k}$$

where k is the number of future comparisons (i.e., monitoring wells and constituents) and α is the desired sitewide false positive rate (e.g., $\alpha = .05$). Table 8.3 presents factors for $\alpha^* = .001, .005,$ and $.0001,$ for one of two, one of three, and first or both of two resampling strategies. Using Table 8.3,

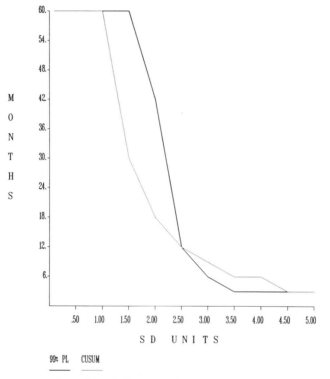

M
O
N
T
H
S

S D   U N I T S

99% PL   CUSUM

Figure 8.3 Power of combined Shewart–CUSUM control chart under constant release.

factors can identified for numbers of future comparisons ranging from 10 (i.e., .05/10 = .005) to 500 (i.e., .05/500 = .0001). Interpolation is best performed on a logarithmic scale.

When tabled values are not available, an approximate solution can be obtained by assuming all measurements are independent and adjusting the individual test false positive rate via the Bonferroni inequality. For the case in which all resamples must pass the limit to fail (i.e., $r = m - 1$), the adjusted individual test false positive rate is approximately

$$\alpha = \sqrt{1 - [.95]^{1/k}} \sqrt{\frac{1}{m - 1}} \qquad (8.6)$$

where $m - 1$ is the number of resamples and k is the total number of comparisons (i.e., both wells and constituents). This result was first suggested by Neil Willits of the University of California at Davis and is part of the California Subtitle 15, Chapter 5 rule for groundwater monitoring. For

TABLE 8.3   Simultaneous Normal Prediction Limit Factors for Intrawell Comparisons

Plan Alpha N	1 of 2			1 of 3			First or next 2		
	.0050	.0010	.0001	.0050	.0010	.0001	.0050	.0010	.0001
4	3.60	6.38	13.95	2.40	4.34	9.55	4.33	7.61	16.59
8	2.14	3.02	4.56	1.44	2.13	3.29	2.45	3.40	5.05
12	1.87	2.52	3.51	1.25	1.77	2.55	2.12	2.79	3.81
16	1.75	2.32	3.13	1.17	1.63	2.28	1.98	2.55	3.37
20	1.69	2.21	2.93	1.12	1.55	2.14	1.90	2.42	3.14
24	1.65	2.15	2.81	1.09	1.50	2.05	1.86	2.34	3.01
28	1.63	2.10	2.74	1.07	1.47	1.99	1.82	2.29	2.91
32	1.61	2.07	2.68	1.06	1.44	1.95	1.80	2.25	2.85
36	1.59	2.04	2.64	1.04	1.43	1.92	1.78	2.22	2.80
40	1.58	2.03	2.60	1.04	1.41	1.89	1.77	2.20	2.76
44	1.57	2.01	2.58	1.03	1.40	1.87	1.75	2.18	2.73
48	1.56	2.00	2.55	1.02	1.39	1.85	1.74	2.17	2.71
52	1.55	1.99	2.54	1.02	1.38	1.84	1.74	2.15	2.69
56	1.55	1.98	2.52	1.01	1.37	1.83	1.73	2.14	2.67
60	1.54	1.97	2.51	1.01	1.37	1.82	1.72	2.13	2.66
64	1.54	1.96	2.50	1.00	1.36	1.81	1.72	2.13	2.64
68	1.53	1.96	2.49	1.00	1.36	1.80	1.71	2.12	2.63
72	1.53	1.95	2.48	1.00	1.35	1.79	1.71	2.11	2.62
76	1.53	1.95	2.47	1.00	1.35	1.79	1.71	2.11	2.61
80	1.53	1.94	2.46	0.99	1.35	1.78	1.70	2.10	2.60
84	1.52	1.94	2.46	0.99	1.34	1.78	1.70	2.10	2.60
88	1.52	1.93	2.45	0.99	1.34	1.77	1.70	2.09	2.59
92	1.52	1.93	2.44	0.99	1.34	1.77	1.70	2.09	2.58
96	1.52	1.93	2.44	0.99	1.34	1.77	1.69	2.09	2.58
100	1.52	1.92	2.43	0.99	1.33	1.76	1.69	2.08	2.57

Prepared by Charles Davis.

example, with $k = 50$ comparisons and two verification resamples,

$$\alpha = \sqrt{1 - [.95]^{1/50}} \sqrt{\frac{1}{2}} = .02264 \qquad (8.7)$$

We can verify this result by computing the sitewide false positive rate using (1.11) which yields

$$\alpha^* = 1 - \left[1 - .02264 + .02264(1 - .02264)^2\right]^{50} = .0494$$

which is almost identical to the intended 5% sitewide false positive rate.

For example, with $n = 8$ background measurements, 10 constituents, and 5 monitoring wells (i.e., $k = 50$ future comparisons), the prediction limit for each well and constituent would be

$$\bar{x} + 2.432s\sqrt{1 + \frac{1}{8}} = \bar{x} + 2.580s$$

Had we instead used Table 8.3, we would have obtained a factor of 3.40. Note that the large difference in these two factors is due to the correlation introduced by comparing the initial sample and all of the verification resamples to the same historical background for that well and constituent. This correlation is inversely proportional to sample size, that is, it is large for small samples (e.g., $n = 8$); therefore the approximation should only be used for background samples of 20 or more which are often unavailable for intrawell comparisons.

8.6 POOLING VARIANCE ESTIMATES

In the previous discussion, prediction limits and control charts have been computed separately for each well and constituent using only those data available in each well. In general, the historical record for each well will be limited, and uncertainty in the sampling variability σ^2 as estimated by s^2 will be large. One way to increase the power of the intrawell comparison is to obtain an unbiased estimator of the intrawell variance pooling information from all wells (both upgradient and downgradient) adjusted for spatial differences. The details of such a procedure have been described by Davis (1994) and can result in substantial increases in statistical power.

To begin, let the measurements x_{ij} follow a mixed-effects linear model

$$x_{ij} = \mu + w_i + t_j + e_{ij} \tag{8.8}$$

where μ is the overall mean, w_i is a fixed effect due to well location, t_j is the random effect due to the common time variability at all monitoring locations, and e_{ij} is the residual variability due to the sampling, transportation, measurement, an so on. The t_j are assumed to be independent and normally distributed $N(0, \sigma_t^2)$ and independent of the residuals e_{ij} which are assumed independent and normally distributed $N(0, \sigma_e^2)$.

Davis (1994) illustrates the procedure using fluoride measurements made during quarterly measurements from 16 wells at a hazardous waste disposal facility with 8 historical measurements. The variance components analysis of the background data yielded an analysis of variance (ANOVA) table (Table 8.4).

Specific details on how to compute a mixed-effects ANOVA are provided in numerous sources (e.g., Winer, 1971; Snedecor and Cochran, 1980) and are available in most comprehensive statistical computer programs (e.g., SPSS, SAS, BMDP, and SYSTAT). In addition, computation of variance

TABLE 8.4 Analysis of Variance Table Variance Component Analysis

Source	DF	SS	MS	F	p <	EMS	Variance
Wells	15	4.967	.331	24.4	.0001		
Up–down	1	0.732	.732				
Other	14	4.235	.302				
Time	7	2.426	.347	25.6	.0001	$\sigma_e^2 + 16\sigma_t^2$	.021
Residual	105	1.424	.014			σ_e	.014
Total	127	8.817					.035

DF = degrees of freedom; SS = sum of squared deviations; MS = mean squared deviations; F = F-statistic; p = probability value associated with F-statistic; EMS = expected value of mean squares; Variance = variance component estimate.

components models is described in Chapter 12. Results of this analysis reveal that both temporal and spatial variability are quite pronounced. The pooled intrawell variance is given by

$$\hat{\sigma}_{\text{pooled}}^2 = \hat{\sigma}_t^2 + \hat{\sigma}_e^2 = .021 + .014 = .035 \qquad (8.9)$$

which yields a standard deviation of $\hat{\sigma}_{\text{pooled}} = \sqrt{.035} = 0.187$ mg/l.

Note that this pooled estimate of the standard deviation is used in computing prediction limits for all downgradient wells, that is,

$$\bar{x}_i + t_{D,\alpha}\hat{\sigma}_{\text{pooled}}\sqrt{1 + \frac{1}{n}} \qquad (8.10)$$

In fact, the usual naive estimator of the sample variance is the correct unbiased estimator, but the degrees of freedom D are now given by

$$D = \frac{(Q-1)(F+W-1)^2}{F^2+W-1} \qquad (8.11)$$

where Q represents the number of quarters, W the number of wells, and F the F-ratio for the test of the main effect of time. Note that when $F = 1$ (i.e., clearly no significant time effect), the degrees of freedom are equal to those of the naive estimator [i.e., $(Q-1)W$]. As F increases, the amount of independent information decreases and in the extreme of $F = \infty$, the degrees of freedom are simply $Q - 1$ which is equivalent to treating each well individually. In the fluoride example, with $F = 25.5$, $Q = 8$, and $W = 16$, the approximate degrees of freedom are $D = 17.24$ which is a modest increase over the 7 degrees of freedom available for each well but considerably short of the naive 112 pooled degrees of freedom. The user should note that the same homogeneity of variance assumption that applies to the ANOVA also applies to this variance estimator and its associated degrees of freedom. Also note that in the presence of censored data, a value must be imputed for each missing data point. Several approaches to this problem are described in Chapter 10. As a general rule, however, only datasets with low levels of censoring (e.g., 20% or less) should be considered for this type of pooling procedure.

8.7 SUMMARY

In many ways intrawell comparisons are the method of choice for environmental monitoring applications. Intrawell comparisons completely remove the potentially confounding effect of spatial variability which is the primary threat to the validity of traditional upgradient versus downgradient comparisons. Even if the spatial variability upgradient is representative of spatial variability downgradient, the very presence of spatial variability compromises the independence assumption inherent in the prediction limit estimators described in Chapters 1 through 3. In some cases this problem can be eliminated by computing the variance estimate based on the correct variance components model (see Chapter 12); however, this may not be feasible unless there are several upgradient wells and the data are not censored. Often these requirements are not met in routine practice. In contrast, the validity of intrawell comparisons is threatened by the possibility of previous site impact. In general, it should be possible to empirically test for previous site impact in two ways. First, the historical data for each well should not exhibit significant trend (see Chapter 9). Second, the wells should not contain verified detections of nonnaturally occurring constituents (e.g., VOCs) that are present in leachate. The two conditions also eliminate the possibility of a long-term steady-state release that might not show evidence of significant trend in the available historical database. In this case we would expect to see evidence of leachate impact in the well through the presence of anthropogenic substances (e.g., VOCs) that are characteristic of facility leachate.

9 Trend Analysis

9.1 OVERVIEW

Detecting trend in environmental data is a broad area which could easily encompass a separate volume. Issues of seasonality, autocorrelation, and corrections for flow substantially complicate trend detection application. Good sources of information include Gilbert (1987), McCleary and Hay (1980), Chatfield (1984), and Box and Jenkins (1976). Environmental applications can be found in McMichael and Hunter (1972), Fuller and Tsokos (1971), Carlson MacCormick, and Watts (1970), Hsu and Hunter (1976), and McCollister and Wilson (1975). However, these papers generally focus on groundwater monitoring trend estimates in stream, river, and air monitoring applications. Typically, they require more historical measurements than are available in groundwater monitoring wells. For example, time-series methods such as the Box–Jenkins method (Montgomery and Johnson, 1976) are often used, which usually require 50 to 100 measurements at equally spaced intervals to properly evaluate the autocorrelation function. Additionally, trace level or nondetected measurements must be treated as missing data, complicating application of traditional estimation methods and often invalidating parametric methods. For example, if a series of measurements is reported at the detection limit, deviations from the trend line will not be normally distributed and the standard error of the usual least squares trend estimator will no longer apply, thus invalidating the test of the null hypothesis that the trend is zero. In many cases outliers in the data will produce biased estimates of the least squares estimated slope itself. As such, the most reasonable trend estimator for groundwater monitoring data is nonparametric. Fortunately, there are several estimators available which are described in the following sections.

9.2 SEN'S TEST

Sen (1968) developed a simple nonparametric estimator of trend which is particularly useful for groundwater monitoring applications. Gilbert (1987) points out that the method is an extension of an earlier work by Theil (1950). The method is robust to outliers, missing data, and nondetects.

175

To compute Sen's trend estimator, begin by obtaining the N' slope estimates Q for each well as

$$Q = \frac{x'_i - x_i}{i' - i} \tag{9.1}$$

where x'_i and x_i are the measured concentrations on monitoring events i' and i where $i' > i$ and N' is the number of data pairs for which $i' > i$. The median value of the N' values of Q is Sen's estimator of trend (i.e., slope of the time by concentration regression line). With a single measurement per monitoring event (which is always recommended),

$$N' = \frac{n(n - 1)}{2} \tag{9.2}$$

where n is the number of monitoring events. For nondetects, the quantitation limit may be used for x_i since it represents the lowest quantifiable concentration. To obtain the median value of Q, denoted as S, the N' values of Q are ranked from smallest to largest (i.e., $Q_1 \le Q_2 \cdots Q_{n-1} \le Q_n$) and the median slope is computed as

$$S = Q_{[(N'+1)/2]} \quad \text{if } N' \text{ is odd} \tag{9.3}$$

or

$$S = \frac{Q_{[N'/2]} + Q_{[(N'+2)/2]}}{2} \quad \text{if } N' \text{ is even} \tag{9.4}$$

To test the null hypothesis of zero slope (i.e., no trend), the lower $(1 - \alpha)$ 100% confidence limit for the true slope is computed. To compute the confidence limit, an estimate of the variance of S is required. Gilbert (1987) suggests employing the estimator due to Kendall (1975) which can be used with small samples (i.e., $n \le 10$) and in the presence of ties (i.e., measurements with the same value or nondetects). The variance estimate is given by

$$\text{var}(S) = \frac{1}{18}\left[n(n - 1)(2n + 5) - \sum_{p=1}^{q} t_p(t_p - 1)(2t_p + 5) \right] \tag{9.5}$$

where q is the number of values for which there are ties and t_p is the number of tied measurements for a particular value. The approximate normal theory lower confidence limit is the M_1th largest value of Q where

$$M_1 = \frac{N' - Z_{1-\alpha}[\text{var}(S)]^{1/2}}{2} \tag{9.6}$$

TABLE 9.1   Intrawell Monitoring Data for TOC in mg / l

Year	Quarter	Concentration
1991	1	1
	2	5
	3	4
	4	8
1992	1	8
	2	12
	3	16
	4	30

and $Z_{1-\alpha}$ is the $(1 - \alpha)100\%$ point of the normal distribution (e.g., $Z_{.95} = 1.645$). If $Q_{M_1} > 0$ then the null hypothesis that the trend is zero can be rejected.

The upper confidence limit for S is given by the $(M_2 + 1)$th largest value of Q where

$$M_2 = \frac{N' + Z_{1-\alpha}[\text{var}(S)]^{1/2}}{2} \tag{9.7}$$

Note that in order to compute a 95% two-sided confidence interval for S, $Z_{1-\alpha/2} = 1.965$ is required.

Example 9.1

Consider the data in Table 9.1 for TOC in a single monitoring well, sampled quarterly for the past 2 years. There are $N' = 28$ pairs for which $i' > i$. Individual slope estimates for these pairs are obtained by dividing the differences by $i' - i$ as in Table 9.2.

TABLE 9.2   Individual Slope Estimates for Example 9.1

Time Period	1	2	3	4	5	6	7	8
Concentration	1	5	4	8	8	12	16	30
		4.00	1.33	2.33	1.75	2.20	2.50	4.14
			-1.00	1.50	1.00	1.75	2.20	4.17
				4.00	2.00	2.67	3.00	5.20
					0.00	2.00	2.67	5.50
						4.00	4.00	7.33
							4.00	9.00
								14.00

Ranking these values from smallest to largest gives

$-1, 0, 1, 1.33, 1.5, 1.75, 1.75, 2.00, 2.00, 2.20, 2.20, 2.33, 2.50, 2.67, 2.67, 3,$
$4, 4, 4, 4, 4, 4.14, 4.17, 5.20, 5.50, 7.33, 9.14$

Since $N' = 28$ is even, the median is the average of the 14th and 15th largest values (2.67) which is the Sen slope estimate. Since there is one pair of tied values, the variance estimate is

$$\text{var}(S) = \tfrac{1}{18}[8(7)(21) - 2(1)(9)] = 64.33$$

The 95% confidence interval is therefore given by the values corresponding to the order statistics

$$M_1 = \frac{28 - 1.96\sqrt{64.33}}{2} = 6.14$$

and

$$M_2 = \frac{28 + 1.96\sqrt{644.33}}{2} = 21.86$$

which correspond to values of 1.75 and 4.12, obtained by interpolating between the 6th and 7th and 21st and 22nd largest values. This interval does not contain the value zero; therefore the null hypothesis of no trend (i.e., slope equals zero) can be rejected. Note that if the concern had been with increasing trends, a one-sided interval or limit should have been used. In this case the one-sided lower confidence limit is computed as

$$M_1 = \frac{28 - 1.65\sqrt{64.33}}{2} = 7.38$$

which has a value of 1.85 mg/l which is also greater than zero.

9.3   MANN–KENDALL TEST

Another trend estimator well suited to environmental data is the Mann–Kendall test (or Kendall's test) for trend (Mann, 1945; Theil, 1950; Kendall, 1975). As in Sen's test, there are no distributional assumptions and missing data (i.e., nondetects) or irregularly spaced measurement periods are permitted. Nondetects are assigned a value smaller than the smallest measured value, typically the PQL, since no quantitative value can or should be reported below the PQL.

The version of the Mann–Kendall test presented here is recommended for 40 or fewer measurements (Gilbert, 1987), adequate for virtually all ground-

TABLE 9.3 Mann–Kendall Test

Measurements Ordered by Time							
x_1	x_2	x_3	$\cdots$	x_{n-1}	x_n	No. +	No. −
	$x_2 - x_1$	$x_3 - x_1$		$x_{n-1} - x_1$	$x_n - x_1$		
		$x_3 - x_2$		$x_{n-1} - x_2$	$x_n - x_2$		
				$x_{n-1} - x_3$	$x_n - x_3$		
				$\vdots$	$\vdots$		
				$x_{n-1} - x_{n-2}$	$x_n - x_{n-2}$		
					$x_n - x_{n-1}$		
						Sum of +	Sum of −

water monitoring applications. First, order the data by sampling date $x_1, x_2, \ldots, x_n$ where x_i is the measured value on occasion i. Second, record the signs of each of the N' possible differences $x_{i'} - x_i$, where $i' > i$. For example, let

$$\operatorname{sgn}(x_{i'} - x_i) = \begin{cases} 1 & \text{if } x_{i'} - x_i > 0 \\ 0 & \text{if } x_{i'} - x_i = 0 \\ -1 & \text{if } x_{i'} - x_i < 0 \end{cases} \tag{9.8}$$

The Mann–Kendall statistic is then computed as

$$S = \sum_{i=1}^{n-1} \sum_{i'=k+1} \operatorname{sgn}(x_{i'} - x_i) \tag{9.9}$$

which is the number of positive differences minus the number of negative differences, easily obtained from the last two columns of Table 9.3. The values of S, n, and the associated probability for the test of $S = 0$ (i.e., no increasing trend) are given in Table 9.4 (see Kendall, 1975). Negative trends can also be tested (e.g., to determine if the effects of remediation are significant) by reversing the sign of S (i.e., for decreasing trends $S < 0$). A two-sided test (i.e., either increasing or decreasing trend) can also be obtained by doubling the probability values listed in Table 9.4. For $n > 10$, Gilbert (1987) suggests that the normal approximation

$$Z = \begin{cases} \dfrac{S - 1}{[\operatorname{var}(S)]^{1/2}} & \text{if } S > 0 \\ 0 & \text{if } S > 0 \\ \dfrac{S + 1}{[\operatorname{var}(S)]^{1/2}} & \text{if } S < 0 \end{cases} \tag{9.10}$$

works adequately for most purposes.

TABLE 9.4 Values of S, n and Associated Probability for Mann–Kendall Test

	Values of n					Values of n		
S	4	5	8	9	S	6	7	10
0	0.625	0.592	0.548	0.540	1	0.500	0.500	0.500
2	0.375	0.408	0.452	0.460	3	0.360	0.386	0.431
4	0.167	0.242	0.360	0.381	5	0.235	0.281	0.364
6	0.042	0.117	0.274	0.306	7	0.136	0.191	0.300
8	—	0.042	0.199	0.238	9	0.068	0.119	0.242
10	—	$0.0^2 83$	0.138	0.179	11	0.028	0.068	0.190
12	—		0.089	0.130	13	$0.0^2 83$	0.035	0.146
14	—		0.054	0.090	15	$0.0^2 14$	0.015	0.108
16	—		0.031	0.060	17		$0.0^2 54$	0.078
18	—		0.016	0.038	19		$0.0^2 14$	0.054
20	—		$0.0^2 71$	0.022	21		$0.0^3 20$	0.036
22	—		$0.0^2 28$	0.012	23			0.023
24	—		$0.0^3 87$	$0.0^2 63$	25			0.014
26	—		$0.0^3 19$	$0.0^2 29$	27			$0.0^2 83$
28	—		$0.0^4 25$	$0.0^2 12$	29			$0.0^2 46$
30	—			$0.0^3 43$	31			$0.0^2 23$
32	—			$0.0^3 12$	33			$0.0^2 11$
34	—			$0.0^4 25$	35			$0.0^3 47$
36	—			$0.0^5 28$	37			$0.0^3 18$
					39			$0.0^4 58$
					41			$0.0^4 15$
					43			$0.0^5 28$
					45			$0.0^6 28$

Repeated zeros are indicated by powers; for example, $0.0^3 47$ stands for 0.00047.
Reprinted with permission from Kendall (1975), *Rank Correlation Methods*, Charles Griffin, London.

Example 9.2

Using the example data in Table 9.1, the Mann–Kendall test statistic is computed as shown in Table 9.5. The value of S is therefore $26 - 1 = 25$. From Table 9.4, entering at $n = 8$ and $S = 25$ yields a probability of .00053, indicating a significant positive trend. Using the normal approximation,

$$Z = \frac{25 - 1}{\sqrt{64.33}} = 2.99$$

is obtained, which has an associated probability of .0014, slightly overestimating the true value.

TABLE 9.5   Computation of Mann–Kendall Test

Time Period	1	2	3	4	5	6	7	8	No. + Signs	No. − Signs
Concentration	1	5	4	8	8	12	16	30		
		4	3	7	7	11	15	29	7	0
			−1	3	3	7	11	25	5	1
				4	4	8	12	26	5	0
					0	4	8	22	3	0
						4	8	22	3	0
							4	18	2	0
								14	1	0
									26	1

9.4   SEASONAL KENDALL TEST

When data are influenced by seasonal effects, the previously described estimators will yield biased results, requiring a trend estimator adjustable for seasonal variation. The seasonal Kendall test was developed by Smith, Hirsch, and Slack (1982) and is discussed further by Gilbert (1987). As in the previous two tests, the seasonal Kendall test is free of distributional assumptions and will work with a mixture of quantifiable and nonquantifiable measurements.

Let x_{il} be the measurement for the ith season in the lth year, K the number of seasons, and L the number of years as in Table 9.6. In this illustration $K = 4$ seasons, but, there is no reason the test could not consider monthly effects (i.e., $K = 12$) as long as independent measurements are possible.

For each season, data are accumulated over the L years to compute the Mann–Kendall test statistic S. For season i, compute

$$S_i = \sum_{k=1}^{n_i-1} \sum_{l=k+1}^{n_i} \text{sgn}(x_{il} - x_{ik}) \qquad (9.11)$$

TABLE 9.6   Dataset for Computing Seasonal Kendall Test

	Season			
Year	Spring	Summer	Fall	Winter
1	x_{11}	x_{21}	x_{31}	x_{41}
2	x_{12}	x_{22}	x_{32}	x_{42}
$\vdots$				
L	x_{1L}	x_{2L}	x_{3L}	x_{4L}

where $l > k$, n_i is the number of measurements available for the season i (i.e., over the L years) and

$$\text{sgn}(x_{il} - x_{ik}) = \begin{cases} 1 & \text{if } x_{il} - x_{ik} > 0 \\ 0 & \text{if } x_{il} - x_{ik} = 0 \\ -1 & \text{if } x_{il} - x_{ik} < 0 \end{cases} \tag{9.12}$$

The variance of S_i is computed as

$$\begin{aligned} \text{var}(S_i) = \frac{1}{18} & \left[n_i(n_i - 1)(2n_i + 5) \right. \\ & \left. - \sum_{p=1}^{g_i} t_{ip}(t_{ip} - 1)(2t_{ip} + 5) - \sum_{q=1}^{h_i} \mu_{iq}(\mu_{iq} - 1)(2\mu_{iq} + 5) \right] \\ & + \frac{\sum_{p=1}^{g_i} t_{ip}(t_{ip} - 1)(t_{ip} - 2)\sum_{q=1}^{h_i} \mu_{iq}(\mu_{iq} - 1)(\mu_{iq} - 2)}{9n_i(n_i - 1)(n_i - 2)} \\ & + \frac{\sum_{p=1}^{g_i} t_{ip}(t_{ip} - 1)\sum_{q=1}^{h_i} \mu_{iq}(\mu_{iq} - 1)}{2n_i(n_i - 1)} \end{aligned} \tag{9.13}$$

where g_i is the number of groups of values tied in season i, t_{ip} is the number of ties in the pth group for season i, h_i is the number of sampling times in season i that contains more than one measurement, and μ_{iq} is the number of measurements in the qth time period in season i. Note that in practice there is generally a single measurement per sampling event (e.g., quarterly measurements); therefore $h_i = 0$ for $i = 1, \ldots, 4$ and $\mu_{iq} = 0$ for all i and q. In this case the variance estimator is simply

$$\text{var}(S_i) = \frac{1}{18} \left[n_i(n_{i-1})(2n_i + 5) - \sum_{p=1}^{g_i} t_{ip}(t_{ip} - 1)(2t_{ip} + 5) \right] \tag{9.14}$$

which is the usual estimator of variance of the Mann–Kendall test computed separately for the data in season i. Pooling across the K seasons,

$$S' = \sum_{i=1}^{K} S_i \tag{9.15}$$

and

$$\text{var}(S') = \sum_{i=1}^{K} \text{var}(S_i) \tag{9.16}$$

The quantity

$$
Z = \begin{cases}
\dfrac{S' - 1}{[\mathrm{var}(S')]^{1/2}} & \text{if } S' > 0 \\[2ex]
0 & \text{if } S' = 0 \\[2ex]
\dfrac{S' + 1}{[\mathrm{var}(S')]^{1/2}} & \text{if } S' < 0
\end{cases}
\tag{9.17}
$$

can be compared to standard normal cumulative distribution probabilities to test the null hypotheses of no trend. For 95% confidence, the critical one-tailed value is 1.65 (i.e., test of an upward trend) or, for a two-tailed test, the value is 1.96 (i.e., test of either an upward or downward trend). The corresponding 99% confidence values are 2.33 and 2.57, respectively.

While the previous method provides a test of the null hypothesis of no trend, it does not provide an estimate of the slope of the trend line or corresponding interval estimates. The seasonal Kendall slope estimator is obtained by computing the individual N_i' slope estimates for season i as

$$
Q_i = \frac{x_{il} - x_{ik}}{l - k}
\tag{9.18}
$$

where x_{il} is the concentration for the ith season of the lth year and x_{ik} is the concentration for the ith season of the kth year, where $l > k$. This computation is repeated for each season. The median of the $N' = N_1' + N_2' + \cdots + N_K'$ individual slope estimates is then the Kendall slope estimator. Confidence limits are obtained in the same way as for Sen's estimator, substituting S' for S.

Example 9.3

To illustrate the seasonal Kendall test, consider the data in Table 9.7, covering three years and four seasons. In this example,

$$
n_1 = n_2 = n_3 = n_4
$$

$$
g_1 = g_2 = g_3 = g_4
$$

$$
t_{ip} = 0 \quad \text{for all } i \text{ and } p
$$

$$
N_1' = N_2' = N_3' = N_4' = 3
$$

$$
N' = 12
$$

TABLE 9.7   Seasonal Intrawell Monitoring Data for TOC in mg/l

Year	Season	Concentration
1990	Spring	3
	Summer	4
	Fall	5
	Winter	6
1991	Spring	4
	Summer	5
	Fall	6
	Winter	7
1992	Spring	5
	Summer	6
	Fall	7
	Winter	8

Individual slope estimates are computed as

	Spring			Summer			Fall			Winter		
Year	1	2	3	1	2	3	1	2	3	1	2	3
TOC	3	4	5	4	5	6	5	6	7	6	7	8
		+1	+1		+1	+1		+1	+1		+1	+1
			+1			+1			+1			+1
$S_i =$		3			3			3			3	

The seasonal Kendall test statistic is therefore

$$S' = S_1 + S_2 + S_3 + S_4 = 3 + 3 + 3 + 3 = 12$$

with seasonal variances

$$\text{var}(S_1) = \text{var}(S_2) = \text{var}(S_3) = \text{var}(S_4) = \tfrac{1}{18}[3(2)(11) - 0] = 3.67$$

and overall variance

$$\text{var}(S') = 3.67 + 3.67 + 3.67 + 3.67 = 14.68$$

The test statistic is therefore

$$Z = \frac{12 - 1}{\sqrt{14.68}} = 2.87$$

which has probability .002, indicating rejection of the null hypothesis of no trend.

The seasonal Kendall slope estimator is the median of the 12 individual slope estimates which is 1.0 mg/l, with a 95% lower confidence limit

$$M_1 = \frac{12 - 1.65\sqrt{14.68}}{2} = 2.84$$

which corresponds to a concentration of 1.0 mg/l since in this fabricated example all slope estimates were equal to 1.0 mg/l.

9.5 SOME STATISTICAL PROPERTIES

Recently, El-Shaarawi and Niculescu (1992) have considered the statistical properties of the Mann–Kendall test of trend in environmental time-series data. They showed that (1) the statistic is asymptotically normal, thus justifying use of the approximate normal theory confidence limits derived in this chapter, and (2) the statistic is strongly affected by statistical dependence among the reported measurements. They derive the appropriate variance estimators for moving average correlated time-series data for both seasonal and nonseasonal models. The interested reader is referred to their paper, particularly equations (15) and (40). In their illustration, using monthly chloride data in the South Saskatchewan River, they show that the amount of dependence is sufficient to generate incorrect conclusions using the original test (which assumes independence). This result adds support to the statement that groundwater should not be sampled more than quarterly to minimize dependence and to avoid use of more complicated estimators.

9.6 SUMMARY

The detection of trend in environmental data is a field unto itself, and this chapter has presented a small selection of possible approaches that are potentially useful in this area. In general, trend analysis is useful as a global screening tool and in justifying use of intrawell comparisons. In the presence of significant trend, intrawell comparisons can be adjusted to eliminate historical trend thereby providing unbiased comparisons. In general, however, the presence of significant trend is an indication of a problem that should be investigated fully prior to establishing a groundwater detection monitoring program.

10 Censored Data

One of the most difficult problems in the analysis of groundwater monitoring data involves the incorporation of "nondetects" into estimates of summary statistics (e.g., mean and standard deviation) and their corresponding tests of hypotheses and interval estimates. More often than not, environmental monitoring data consist of a mixture of results that can and cannot be accurately quantified. In practice, the "censoring" mechanism is the detection limit, values below which are reported as "ND" or " < MDL" to signify that they were not found in the sample. All other values are reported as a concentration. Based on previous chapters, one should immediately note that this is the wrong procedure. The PQL and not the MDL should be the censoring mechanism since values above the MDL and below the PQL are detected but not quantifiable. Using the MDL as the censoring point produces data with widely varying levels of uncertainty, violating the assumption of homoscedasticity (i.e., constant measurement variation) which is assumed by all of the previous statistical theory and methods. Even with an agreed-upon censoring point, there is considerable controversy regarding the appropriate method or methods for incorporating the censored data in computing summary statistics, testing hypotheses, and computing interval estimates. This is not at all surprising since the correct choice of method depends on both the degree of censoring (e.g., 20% versus 80% nondetects) and the type of application (e.g., computing the mean versus computing a prediction limit from data that are a mixture of quantifiable and nonquantifiable measurements).

10.1 CONCEPTUAL FOUNDATION

Assume that there is a population of true concentrations from which we have drawn a sample of size n. For convenience, also assume that variation in the sampled population can be represented by a continuous probability distribution, for which a fraction of the true concentration is essentially zero. This partial loss of information occurs because of censoring imposed by limits of detection and/or quantification.

For example, Davis (1994) points out that we may assume an underlying distribution as in Figure 10.1, but what we observe is the distribution in Figure 10.2, where the vertical line represents a point mass at MDL/2

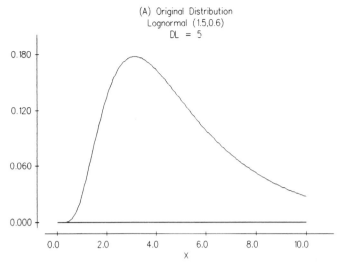

Figure 10.1   Assumed underlying distribution, before censoring (Davis, 1994).

containing the probability content of the region $< MDL$. In practice, the measurements are often coarsely rounded, so that the observed frequency distribution looks like Figure 10.3. Davis (1994) points out that in real-world application the true underlying model in Figure 10.3 is unknown; therefore different approaches will yield widely varying results, depending on the degree to which they rely on the assumed distribution. This is even more

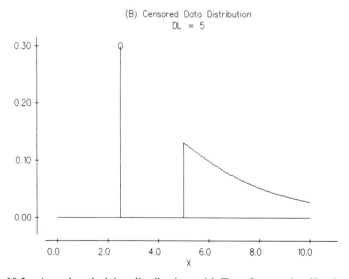

Figure 10.2   Actual underlying distribution, with Type I censoring (Davis, 1994).

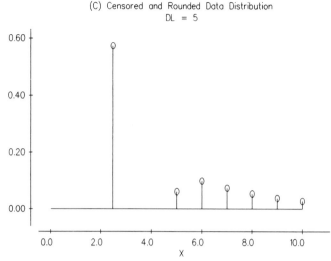

Figure 10.3   The effects of rounding, with censoring and coarse rounding (Davis, 1994).

critical in groundwater monitoring in which repeated application of tail probabilities is used to control the overall sitewide false positive rate (i.e., prediction limits). How well a censored data estimator works in the center of the distribution (e.g., to estimate mean concentration) is often a poor index of how well that method will work in the tails of the distribution (e.g., to estimate a 99% confidence prediction limit for a new single measurement). In the following sections several methods are described, and some general recommendations are provided.

10.2   SIMPLE SUBSTITUTION METHODS

Historically, the EPA has advocated that nondetected measurements be replaced by one-half the detection limit (i.e., MDL/2) and data be analyzed as if all measurements were observable with equal precision. Other substitution values for nondetects required or advocated by the EPA include zero (for drinking water contaminants in public water supplies) and one-half the Superfund quantitation limit (SQL) (for contaminants being evaluated in Superfund risk assessments). Of course, this assumption is demonstrably false, but the method is computationally simple and is often adequate for most practical purposes if the detection frequency is 80% or more. For example, Haas and Scheff (1990) found that the simple substitution method was negatively biased (i.e., too low) for low degrees of censoring but positively biased thereafter (i.e., too high). Bias became large for samples of size

10 when detection frequencies fell below 80%. Kushner (1976) found similar results for bias in estimated mean concentrations when 20% or more of the samples were nondetects. Several other studies have also shown that simple substitution methods perform poorly in recovering true population statistics as compared to other statistical procedures (Gilliom and Helsel, 1986, Gleit, 1985; Helsel and Cohn, 1988, Helsel and Gilliom, 1986; Helsel, 1990).

Surprisingly, however, McNichols and Davis (1988) and Davis (1994) suggest that the simple substitution technique (e.g., MDL/2) maintains reasonable control over sitewide false positive rates evaluated through repeated application of prediction limits, for nondetect proportions up to 80%. They report that adding verification resamples not only aids in controlling the sitewide false positive rate, but also in increasing the robustness of the resulting tests. As they point out, this result is vastly different from the conclusions of others and more recent guidelines, which have focused on estimation rather than hypothesis testing.

The simple substitution method may be enhanced in a couple of ways. Gleit (1985) suggests substituting nondetects with expected order statistics from a fitted distribution. This approach involves initially fitting the distribution to the quantifiable measurements substituting MDL/2 for the nondetects. Gleit suggests iteratively refitting and resubstituting until the parameter estimates converge. An alternative approach suggested by Gilliom and Helsel (1986) involves substitution of random variates drawn from a right-truncated normal distribution with working parameters. Both of these estimates should lead to improved results over the simple substitution method to the extent that the assumed distribution is correct.

10.3 MAXIMUM LIKELIHOOD ESTIMATORS

For a left singly censored normal distribution, Cohen (1959, 1961) derived the maximum likelihood estimator (MLE) for the mean and variance of the overall distribution as

$$\hat{\mu} = \bar{x}' - (\bar{x}' - x_0)\lambda(g, h) \tag{10.1}$$

$$\sigma^2 = s^{2\prime} + (\bar{x}' - x_0)^2 \lambda(g, h) \tag{10.2}$$

where

$\bar{x}'$ = the mean of the n_1 measured values

$$\bar{x}' = \frac{1}{n_1} \sum_{i=1}^{n_1} x_i$$

$s^{2\prime}$ = the variance of the n_1 measured values

$$s^{2\prime} = \frac{1}{n_1} \sum_{i=1}^{n_1} (x_i - \bar{x}')^2$$

h = the proportion of nondetects
x_0 = the censoring point (e.g., MDL/2)

and

$$g = \frac{s^{2\prime}}{(\bar{x}' - x_0)^2}$$

Cohen (1961) provided values of $\lambda(g, h)$ for values of g up to 1, Schneider (1986) up to $g = 1.48$, and Schmee et al. (1985) up to $g = 10$, which are reproduced in Table 10.1.

TABLE 10.1   Values of $\lambda(g, h)$ for Values of g up to 10

				h (Proportion Censored)					
g	0.1	0.2	0.3	0.4	0.5	0.6	0.7	0.8	0.9
0.0	0.1102	0.2426	0.4021	0.5961	0.8368	1.145	1.561	2.176	3.283
0.1	0.1180	0.2574	0.4233	0.6234	0.8703	1.185	1.608	2.229	3.345
0.2	0.1247	0.2703	0.4422	0.6483	0.9012	1.222	1.651	2.280	3.405
0.3	0.1306	0.2819	0.4595	0.6713	0.9300	1.257	1.693	2.329	3.464
0.4	0.1360	0.2926	0.4755	0.6927	0.9570	1.290	1.732	2.376	3.520
0.5	0.1409	0.3025	0.4904	0.7129	0.9826	1.321	1.770	2.421	3.575
0.6	0.1455	0.3118	0.5045	0.7320	1.0070	1.351	1.806	2.465	3.628
0.7	0.1499	0.3207	0.5180	0.7502	1.0300	1.380	1.841	2.507	3.679
0.8	0.1540	0.3290	0.5308	0.7676	1.0530	1.408	1.875	2.548	3.730
0.9	0.1579	0.3370	0.5430	0.7844	1.0740	1.435	1.908	2.588	3.779
1.0	0.1617	0.3447	0.5548	0.8005	1.0950	1.461	1.940	2.626	3.827
2.0	0.1932	0.4093	0.6547	0.9382	1.2740	1.686	2.217	2.968	4.258
3.0	0.2182	0.4609	0.7349	1.0490	1.4200	1.870	2.447	3.255	4.625
4.0	0.2395	0.5052	0.8038	1.1460	1.5460	2.031	2.649	3.508	4.952
5.0	0.2585	0.5450	0.8653	1.2310	1.6590	2.175	2.829	3.736	5.249
6.0	0.2757	0.5803	0.8912	1.3090	1.7620	2.307	2.995	3.945	5.522
7.0	0.2916	0.6134	0.9729	1.3820	1.8570	2.428	3.149	4.140	5.778
8.0	0.3065	0.6442	1.0210	1.4490	1.9460	2.543	3.293	4.322	6.018
9.0	0.3205	0.6733	1.0670	1.5130	2.0310	2.650	3.430	4.495	6.245
10.0	0.3337	0.7009	1.1100	1.5730	2.1100	2.753	3.559	4.660	6.462

Reprinted with permission from J. Schmee, D. Gladstein, and W. Nelson, *Technometrics*, **27** 2(1985):119–128. ©1985 by the American Statistical Association and the American Society for Quality Control.

Haas and Scheff (1990) have developed the following power series expansion that fits these tabled values to within 6% relative error:

$$\log_e \lambda(g, y) = 0.182344 - \frac{0.3756}{g + 1}$$

$$+ 0.10017g + 0.78079y - 0.00581g^2$$

$$- 0.06642y^2 - 0.0234gy + 0.000174g^3$$

$$+ 0.001663g^2y - 0.00086gy^2 - 0.00653y^3 \quad (10.3)$$

where

$$y = \log_e \left(\frac{h}{1 - h} \right)$$

This approximation is most useful for routine computer applications of this method.

Saw (1961) noted that the maximum likelihood estimators proposed by Cohen (1959) were biased, and he derived the first-order bias correction terms, for the case in which a constant proportion of observations is censored rather than all observations below the censoring point (e.g., MDL or PQL). Schneider (1986) provided simple computational formulas

$$B_\mu = -\exp\left[2.692 - \frac{5.439(n - n_0)}{n + 1}\right] \quad (10.4)$$

$$B_\sigma = -\left[0.312 + \frac{0.859(n - n_0)}{n + 1}\right]^2 \quad (10.5)$$

where n is the total number of measurements, n_0 is the number of censored observations, and the unbiased mean and standard deviation are given by

$$\hat{\mu}_u = \mu - \frac{\hat{\sigma}B_\mu}{n + 1} \quad (10.6)$$

$$\hat{\sigma}_u = \sigma - \frac{\hat{\sigma}B_\sigma}{n + 1} \quad (10.7)$$

Haas and Scheff (1990) suggest that this correction applies to the fixed censoring point case as well.

Example 10.1

As an example, consider the data given in Table 10.2 for TOC from a single background monitoring well over 10 quarterly monitoring events. Inspection

TABLE 10.2 Historical TOC Example

Year	Quarter	TOC in mg/l
1990	1	5
	2	7
	3	< 1
	4	3
1991	1	< 1
	2	4
	3	6
	4	5
1992	1	< 1
	2	6

of Table 10.2 reveals that 30% of the data are censored at the detection limit of 1 mg/l. Simple substitution of MDL/2 for the three censored values yields $\bar{x} = 3.75$ and $s = 2.50$. To compute the maximum likelihood estimates, note that $h = .3$ (i.e., the proportion censored) and the mean and standard deviation of the seven uncensored values are $\bar{x}' = 5.14$ and $s' = 1.35$, respectively. Therefore

$$g = \frac{1.35^2}{(5.14 - 1)^2} = .1063$$

Inspection of Table 10.1 reveals that $\lambda(g, h) = .425$ (i.e., interpolating between $g = .1$ and $g = .2$). Using the approximation suggested by Haas and Scheff (1990) yields

$$y = \log_e\left(\frac{.3}{1 - .3}\right) = -.8473$$

$$
\begin{aligned}
\log_2 \lambda(g, y) = {} & 0.182344 - \frac{.3756}{.1063 + 1} \\
& + .10017(.1063) + .78079(-.8473) - .00581(.1063)^2 \\
& - .06642(-.8473)^2 - .0234(.1063)(-.8473) \\
& + .000174(.1063)^3 + .001663(.1063)^2(-.8473) \\
& - .00086(.1063)(-.8473)^2 - .00653(-.8473)^3 \\
= {} & .427
\end{aligned}
$$

which is in close agreement with the tabled valued. The maximum likelihood estimators are

$$\hat{\mu} = 5.14 - (5.14 - 1.0).427 = 3.37$$
$$\hat{\sigma} = 1.35^2 + (5.14 - 1.0)^2.427 = 3.02$$

Adjusting for bias in the estimators yields

$$B_\mu = -\exp\left[2.692 - \frac{5.439(10 - 3)}{10 + 1}\right] = -.463$$

$$B_\sigma = -\left[.312 + \frac{.859(10 - 3)}{10 + 1}\right]^2 = -.737$$

and

$$\hat{\mu}_\mu = 3.37 - \frac{3.02(-.463)}{10 + 1} = 3.50$$

$$\hat{\sigma}_\mu = 3.02 - \frac{3.02(-.737)}{10 + 1} = 3.22$$

Interestingly, the simple substitution method yielded an overestimate of the mean (i.e., 3.75 versus 3.50) and an underestimate of the standard deviation (i.e., 2.50 versus 3.22) relative to the bias-adjusted maximum likelihood estimates. The biased estimates were approximately 5% lower than the bias-adjusted estimates.

10.4   RESTRICTED MAXIMUM LIKELIHOOD ESTIMATORS

To produce a computationally simple estimator for singly censored samples, Persson and Rootzen (1977) combined the method of maximum likelihood and the method of moments. Their estimator is quite close to the maximum likelihood estimator of Cohen (1959) but somewhat simpler to compute. Use of this method in analysis of environmental data was first considered by Haas and Scheff (1990). The restricted maximum likelihood estimators (RMLEs) are given by

$$\lambda = \Phi^{-1}\frac{n_0}{n}$$

$$k = n - n_0$$

$$a = \sum_{i=n_0+1}^{n}(x_i - x_0)$$

$$b = \sum_{i=n_0+1}^{n}(x_i - x_0)^2$$

$$\hat{\sigma} = \frac{1}{2}\left(\frac{\lambda a}{k} + \left[\left(\frac{\lambda a}{k}\right)^2 + 4\frac{b}{k}\right]^{1/2}\right)$$

$$\hat{\mu} = x_0 - \lambda\hat{\sigma}$$

where $\Phi^{-1}(n_0/n)$ is the inverse normal (i.e., the value of the standardized normal deviate for cumulative probability n_0/n). These estimators are biased at low levels of censoring; therefore the previously described correction (Saw, 1961) can also be applied to the restricted maximum likelihood estimators.

All that is required to compute these estimators is a method of evaluating the inverse normal cumulative density function Φ^{-1} which can be approximated as

$$\Phi^{-1}(p) \sim 1.238T(1 + .0262T) \qquad \text{if } p \geq .5$$
$$\sim -1.238T(1 + .0262T) \quad \text{if } p < .5$$

where

$$T = (-\log_e[4p(1 - p)])^{1/2}$$

(see Maindonald, 1984). For example, with $p = .95$, the approximation yields $\Phi^{-1}(.95) = 1.648$ where the correct value is 1.645. In general, two decimal places of accuracy are guaranteed which should be sufficient for this purpose.

Example 10.2

Using the TOC data from Table 10.2, we obtain

$$\sum_{i=n_0+1}^{n} x_i = 36$$

$$\sum_{i=n_0+1}^{n} x_i^2 = 196$$

$$\sum_{i=n_0+1}^{n} x_i - x_0 = 29$$

$$\sum_{i=n_0+1}^{n} (x_i - x_0)^2 = 131$$

$$k = 10 - 3 = 7$$

and

$$\lambda = -.523$$

leading to the estimates

$$\hat{\sigma} = \frac{1}{2}\left(\frac{-.523(29)}{7} + \left[\left(\frac{-.523(29)}{7} \right)^2 + 4\left(\frac{131}{7} \right) \right]^{1/2} \right) = 3.38$$

and

$$\hat{\mu} = 1 - (-.523)(3.38) = 2.77$$

The nearly unbiased estimators are

$$\hat{\mu}_\mu = 2.91$$

and

$$\hat{\sigma}_\mu = 3.60$$

which are reasonably close to the maximum likelihood estimates of $\hat{\mu}_\mu = 3.50$ and $\hat{\sigma}_\mu = 3.22$.

10.5   LINEAR ESTIMATORS

Although maximum likelihood estimators are optimal with respect to estimating variances, they are biased, particularly in small samples. To eliminate bias in small samples, the weighted least squares method is used to develop the best linear unbiased estimator (BLUE) of the mean and variance of a censored normal distribution (see Cohen, 1991, Chapter 4, for a review). The problem has been considered in detail by Gupta (1952) and Sarhan and Greenberg (1962). The estimates are calculated as sums of products of the uncensored observations and the appropriate normal order statistics as

$$\hat{\mu} = \sum_{i=n_0+1}^{n} a_{1i} x_{i|n} \tag{10.8}$$

$$\hat{\sigma} = \sum_{i=n_0+1}^{n} a_{2i} x_{i|n} \tag{10.9}$$

where the x_i, $i = n_0 + 1, \ldots, n$, represent the $k = n - n_0$ uncensored observations. Gupta (1952) has tabulated the coefficients (a_{1i} and a_{2i}) for sample sizes up to $n = 10$ (reproduced here in Tables 10.3 and 10.4), and Sarhan and Greenberg (1962) have tabulated the coefficients for sample sizes up to $n = 20$. For sample sizes greater than 20, maximum likelihood estimators are preferred. The tables are designed principally for right-censored distributions. Gupta (1952) notes that for left-censored distributions the observations should be ranked from largest to smallest and the coefficients for the standard deviation should be of reversed sign.

TABLE 10.3   Coefficients for BLUE of the Mean

n	n_1	n_r	$x(1)$	$x(2)$	$x(3)$	$x(4)$	$x(5)$	$x(6)$	$x(7)$
3	0	1	0.0000	1.0000	—	—	—	—	—
4	0	1	0.1161	0.2408	0.6431	—	—	—	—
	0	2	−0.4056	1.4056	—	—	—	—	—
	1	1	—	−1.6834	1.6834	—	—	—	—
5	0	1	0.1252	0.1830	0.2147	0.4771	—	—	—
	0	2	−0.0638	0.1498	0.9139	—	—	—	—
	0	3	−0.7411	1.7411	—	—	—	—	—
	1	1	—	−1.0101	0.0000	1.0101	—	—	—
	1	2	—	−2.0201	2.0201	—	—	—	—
6	0	1	0.1183	0.1510	0.1680	0.1828	0.3799	—	—
	0	2	0.0185	0.1226	0.1761	0.6828	—	—	—
	0	3	−0.2159	0.0649	1.1511	—	—	—	—
	0	4	−1.0261	2.0261	—	—	—	—	—
	1	1	—	0.3198	0.1802	0.1802	0.3198	—	—
	1	2	—	0.1539	0.1781	0.6680	—	—	—
	1	3	—	−0.4578	1.4578	—	—	—	—
	2	2	—	—	0.5000	0.5000	—	—	—
7	0	1	0.1088	0.1295	0.1400	0.1487	0.1571	0.3159	—
	0	2	0.0465	0.1072	0.1375	0.1626	0.5462	—	—
	0	3	−0.0738	0.0677	0.1375	0.8686	—	—	—
	0	4	−0.3474	−0.0135	1.3609	—	—	—	—
	0	5	−1.2733	2.2733	—	—	—	—	—
	1	1	—	0.2718	0.1520	0.1524	0.1520	0.2718	—
	1	2	—	0.1748	0.1432	0.1634	0.5186	—	—
	1	3	—	−0.0592	0.1270	0.9321	—	—	—
	1	4	—	−0.8716	1.8716	—	—	—	—
	2	2	—	—	0.4157	0.1686	0.4157	—	—
	2	3	—	—	0.0000	1.0000	—	—	—
8	0	1	0.0997	0.1139	0.1208	0.1265	0.1318	0.1370	0.2704
	0	2	0.0569	0.0962	0.1153	0.1309	0.1451	0.4555	—
	0	3	−0.0167	0.0677	0.1084	0.1413	0.6993	—	—
	0	4	−0.1549	0.0176	0.1001	1.0372	—	—	—
	0	5	−0.4632	−0.0855	1.5487	—	—	—	—
	0	6	−1.4915	2.4915	—	—	—	—	—
	1	1	—	0.2367	0.1315	0.1319	0.1319	0.1315	0.2367
	1	2	—	0.1716	0.1222	0.1338	0.1442	0.4282	—
	1	3	—	0.0431	0.1061	0.1406	0.7102	—	—
	1	4	—	−0.2519	0.0741	1.1778	—	—	—
	1	5	—	−1.2462	2.2462	—	—	—	—
	2	2	—	—	0.3569	0.1431	0.1431	0.3569	—
	2	3	—	—	0.1742	0.1429	0.6829	—	—
	2	4	—	—	−0.4761	1.4761	—	—	—
	3	3	—	—	—	0.5000	0.5000	—	—

TABLE 10.3 (*Continued*)

n	n_1	n_r	x(1)	x(2)	x(3)	x(4)	x(5)	x(6)	x(7)	x(8)	x(9)
9	0	1	0.0915	0.1018	0.1067	0.1106	0.1142	0.1177	0.1212	0.2365	—
	0	2	0.0602	0.0876	0.1006	0.1110	0.1204	0.1294	0.3909	—	—
	0	3	0.0104	0.0660	0.0923	0.1133	0.1320	0.5860	—	—	—
	0	4	−0.0731	0.0316	0.0809	0.1199	0.8408	—	—	—	—
	0	5	−0.2272	−0.0284	0.0644	1.1912	—	—	—	—	—
	0	6	−0.5664	−0.1521	1.7185	—	—	—	—	—	—
	0	7	−1.6868	2.6868	—	—	—	—	—	—	—
	1	1	—	0.2097	0.1159	0.1162	0.1163	0.1162	0.1159	0.2097	—
	1	2	—	0.1626	0.1074	0.1148	0.1214	0.1275	0.3663	—	—
	1	3	—	0.0799	0.0936	0.1140	0.1321	0.5804	—	—	—
	1	4	—	−0.0768	0.0699	0.1153	0.8916	—	—	—	—
	1	5	—	−0.4272	0.0218	1.4054	—	—	—	—	—
	1	6	—	−1.5874	2.5874	—	—	—	—	—	—
	2	2	—	—	0.3134	0.1243	0.1246	0.1243	0.3134	—	—
	2	3	—	—	0.2040	0.1191	0.1330	0.5440	—	—	—
	2	4	—	—	−0.0527	0.1098	0.9429	—	—	—	—
	2	5	—	—	−0.9229	1.9229	—	—	—	—	—
	3	3	—	—	—	0.4315	0.1370	0.4315	—	—	—
	3	4	—	—	—	0.0000	1.0000	—	—	—	—
10	0	1	0.0843	0.0921	0.0957	0.0986	0.1011	0.1036	0.1060	0.1085	0.2101
	0	2	0.0605	0.0804	0.0898	0.0972	0.1037	0.1099	0.1161	0.3424	—
	0	3	0.0244	0.0636	0.0818	0.0962	0.1089	0.1207	0.5045	—	—
	0	4	−0.0316	0.0383	0.0707	0.0962	0.1185	0.7078	—	—	—
	0	5	−0.1240	−0.0016	0.0549	0.0990	0.9718	—	—	—	—
	0	6	−0.2923	−0.0709	0.0305	1.3327	—	—	—	—	—
	0	7	−0.6596	−0.2138	1.8734	—	—	—	—	—	—
	0	8	−1.8634	2.8634	—	—	—	—	—	—	—
	1	1	—	0.1884	0.1036	0.1040	0.1041	0.1041	0.1040	0.1036	0.1884
	1	2	—	0.1525	0.0961	0.1013	0.1057	0.1098	0.1138	0.3209	—
	1	3	—	0.0942	0.0846	0.0979	0.1095	0.1204	0.4933	—	—
	1	4	—	−0.0043	0.0665	0.0938	0.1179	0.7261	—	—	—
	1	5	—	−0.1866	0.0351	0.0892	1.0623	—	—	—	—
	1	6	—	−0.5877	−0.0289	1.6166	—	—	—	—	—
	1	7	—	−1.9000	2.9000	—	—	—	—	—	—
	2	2	—	—	0.2798	0.1099	0.1103	0.1103	0.1099	0.2798	—
	2	3	—	—	0.2050	0.1038	0.1122	0.1198	0.4592	—	—
	2	4	—	—	0.0606	0.0935	0.1178	0.7281	—	—	—
	2	5	—	—	−0.2648	0.0735	1.1914	—	—	—	—
	2	6	—	—	−1.3406	2.3406	—	—	—	—	—
	3	3	—	—	—	0.3807	0.1193	0.1193	0.3807	—	—
	3	4	—	—	—	0.1871	0.1198	0.6930	—	—	—
	3	5	—	—	—	−0.4847	1.4847	—	—	—	—
	4	4	—	—	—	—	0.5000	0.5000	—	—	—

Reproduced with permission from Sarhan and Greenberg, *Annals of Mathematical Statistics*, **27**(1956):427–457.

TABLE 10.4   Coefficients for BLUE of the Variance

n	n_1	n_r	$x(1)$	$x(2)$	$x(3)$	$x(4)$	$x(5)$	$x(6)$	$x(7)$
3	0	1	-1.1816	1.1816	—	—	—	—	—
4	0	1	-0.6971	-0.1268	0.8239	—	—	—	—
	0	2	-1.3654	1.3654	—	—	—	—	—
	1	1	—	-1.6834	1.6834	—	—	—	—
5	0	1	-0.5117	-0.1668	0.0274	0.6511	—	—	—
	0	2	-0.7696	-0.2121	0.9817	—	—	—	—
	0	3	-1.4971	1.4971	—	—	—	—	—
	1	1	—	-1.0101	0.0000	1.0101	—	—	—
	1	2	—	-2.0201	2.0201	—	—	—	—
6	0	1	-0.4097	-0.1685	-0.0406	0.0740	0.5448	—	—
	0	2	-0.5528	-0.2091	-0.0290	0.7909	—	—	—
	0	3	-0.8244	-0.2760	1.1004	—	—	—	—
	0	4	-1.5988	1.5988	—	—	—	—	—
	1	1	—	-0.7531	-0.0829	0.0829	0.7531	—	—
	1	2	—	-1.1438	-0.0878	1.2317	—	—	—
	1	3	—	-2.2717	2.2717	—	—	—	—
7	0	1	-0.3440	-0.1610	-0.0681	0.0114	0.0901	0.4716	—
	0	2	-0.4370	-0.1943	-0.0718	0.0321	0.6709	—	—
	0	3	-0.5848	-0.2428	-0.0717	0.8994	—	—	—
	0	4	-0.8682	-0.3269	1.1951	—	—	—	—
	0	5	-1.6812	1.6812	—	—	—	—	—
	1	1	—	-0.6108	-0.1061	0.0000	0.1061	0.6108	—
	1	2	—	-0.8288	-0.1258	0.0248	0.9298	—	—
	1	3	—	-1.2483	-0.1548	1.4030	—	—	—
	1	4	—	-2.4712	2.4712	—	—	—	—
	2	2	—	—	-1.4176	0.0000	1.4176	—	—
	2	3	—	—	-2.8352	2.8352	—	—	—
8	0	1	-0.2978	-0.1515	-0.0796	-0.0200	0.0364	0.0951	0.4175
	0	2	-0.3638	-0.1788	-0.0881	-0.0132	0.0570	0.5868	—
	0	3	-0.4586	-0.2156	-0.0970	0.0002	0.7709	—	—
	0	4	-0.6110	-0.2707	-0.1061	0.9878	—	—	—
	0	5	-0.9045	-0.3690	1.2735	—	—	—	—
	0	6	-1.7502	1.7502	—	—	—	—	—
	1	1	—	-0.5184	-0.1115	-0.0361	0.0361	0.1115	0.5184
	1	2	—	-0.6608	-0.1319	-0.0318	0.0630	0.7615	—
	1	3	—	-0.8894	-0.1605	-0.0197	1.0696	—	—
	1	4	—	-1.3337	-0.2086	1.5423	—	—	—
	1	5	—	-2.6357	2.6357	—	—	—	—
	2	2	—	—	-1.0357	-0.0674	0.0674	1.0357	—
	2	3	—	—	-1.5661	-0.0678	1.6338	—	—
	2	4	—	—	-3.1220	3.1220	—	—	—
	3	3	—	—	—	-3.2784	3.2784	—	—

TABLE 10.4   (*Continued*)

n	n_1	n_r	x(1)	x(2)	x(3)	x(4)	x(5)	x(6)	x(7)	x(8)	x(9)
9	0	1	−0.2633	−0.1421	−0.0841	−0.0370	0.0062	0.0492	0.0954	0.3757	—
	0	2	−0.3129	−0.1647	−0.0938	−0.0364	0.0160	0.0678	0.5239	—	—
	0	3	−0.3797	−0.1936	−0.1048	−0.0333	0.0317	0.6797	—	—	—
	0	4	−0.4766	−0.2335	−0.1181	−0.0256	0.8537	—	—	—	—
	0	5	−0.6330	−0.2944	−0.1348	1.0622	—	—	—	—	—
	0	6	−0.9355	−0.4047	1.3402	—	—	—	—	—	—
	0	7	−1.8092	1.8092	—	—	—	—	—	—	—
	1	1	—	−0.4527	−0.1107	−0.0532	0.0000	0.0532	0.1107	0.4527	—
	1	2	—	−0.5544	−0.1291	−0.0563	0.0109	0.0775	0.6514	—	—
	1	3	—	−0.7015	−0.1535	−0.0578	0.0299	0.8828	—	—	—
	1	4	—	−0.9399	−0.1896	−0.0558	1.1852	—	—	—	—
	1	5	—	−1.4057	−0.2534	1.6591	—	—	—	—	—
	1	6	—	−2.7753	2.7753	—	—	—	—	—	—
	2	2	—	—	−0.8817	−0.0885	0.0000	0.0885	0.8317	—	—
	2	3	—	—	−1.1222	−0.1023	0.0223	1.2022	—	—	—
	2	4	—	—	−1.6894	−0.1227	1.8122	—	—	—	—
	2	5	—	—	−3.3620	3.3620	—	—	—	—	—
	3	3	—	—	—	−1.8213	0.0000	1.8213	—	—	—
	3	4	—	—	—	−3.6426	3.6426	—	—	—	—
10	0	1	−0.2364	−0.1334	−0.0851	−0.0465	−0.0119	0.0215	0.0559	0.0937	0.3423
	0	2	−0.2753	−0.1523	−0.0947	−0.0488	−0.0077	0.0319	0.0722	0.4746	—
	0	3	−0.3252	−0.1758	−0.1058	−0.0502	−0.0006	0.0469	0.6107	—	—
	0	4	−0.3930	−0.2063	−0.1192	−0.0501	0.0111	0.7576	—	—	—
	0	5	−0.4919	−0.2491	−0.1362	−0.0472	0.9243	—	—	—	—
	0	6	−0.6520	−0.3150	−0.1593	1.1263	—	—	—	—	—
	0	7	−0.9625	−0.4357	1.3981	—	—	—	—	—	—
	0	8	−1.8608	1.8608	—	—	—	—	—	—	—
	1	1	—	−0.4034	−0.1074	−0.0616	−0.0201	0.0201	0.0616	0.1074	0.4034
	1	2	—	−0.4803	−0.1235	−0.0674	−0.0166	0.0325	0.0827	0.5726	—
	1	3	—	−0.5842	−0.1440	−0.0734	−0.0097	0.0514	0.7599	—	—
	1	4	—	−0.7359	−0.1719	−0.0797	0.0031	0.9844	—	—	—
	1	5	—	−0.9831	−0.2145	−0.0859	1.2835	—	—	—	—
	1	6	—	−1.4678	−0.2918	1.7595	—	—	—	—	—
	1	7	—	−2.8960	2.8960	—	—	—	—	—	—
	2	2	—	—	−0.7021	−0.0947	−0.0310	0.0310	0.0947	0.7021	—
	2	3	—	—	−0.8898	−0.1101	−0.0262	0.0549	0.9711	—	—
	2	4	—	—	−1.1952	−0.1318	−0.0144	1.3415	—	—	—
	2	5	—	—	−1.7947	−0.1688	1.9635	—	—	—	—
	2	6	—	—	−3.5677	3.5677	—	—	—	—	—
	3	3	—	—	—	−1.2832	−0.0559	0.0559	1.2832	—	—
	3	4	—	—	—	−1.9791	−0.0553	2.0344	—	—	—
	3	5	—	—	—	−3.9511	3.9511	—	—	—	—
	4	4	—	—	—	—	−4.0761	4.0761	—	—	—

Reproduced with permission from Sarhan and Greenberg, *Annals of Mathematical Statistics*, **27**(1956):427–457.

Example 10.3

Again, using the TOC data in Table 10.2 and the coefficients in Tables 10.3 and 10.4, we begin by arranging the data from largest to smallest (i.e., $x_{1|n} > x_{2|n} > \cdots > x_{k|n}$, where $k = n - n_0$) as

$$7, 6, 6, 5, 5, 4, 3$$

$$\hat{\mu} = .0244(7) + .0636(6) + .0818(6) + .0962(5)$$

$$+ .1089(5) + .1207(4) + .5045(3) = 4.07$$

and

$$\hat{\sigma} = .3252(7) + .1758(6) + .1058(6) + .0502(5)$$

$$+ .0006(5) - .0469(4) - .6107(3) = 2.20$$

which are, in fact, more similar to the simple substitution estimates than the ML or RML estimates, at least for this example. Note that for this estimator, there is no information regarding the value of the censoring point as in the ML or RML estimators, only knowledge that $n - k$ observations have been censored. This distribution has sometimes been termed Type II censoring to distinguish it from those cases in which there is a known censoring point (i.e., Type I censoring).

10.6 ALTERNATIVE LINEAR ESTIMATORS

Gupta (1952) suggested an alternative linear estimator for the mean and standard deviation of the censored normal distribution which is only slightly less efficient than the best linear estimators. The estimators are somewhat easier to compute since they only require the expected values of the order statistics from a standard normal distribution (see Table 10.5). The estimators are

$$\hat{\mu} = \sum_{i=n_0+1}^{n} b_i x_{(i)} \tag{10.10}$$

and

$$\hat{\sigma} = \sum_{i=n_0+1}^{n} c_i x_{(i)} \tag{10.11}$$

TABLE 10.5   Expected Values of Normal Order Statistics

Rank	2	3	4	5	6	7	8
1	0.5642	0.8463	1.0294	1.1630	1.2672	1.3522	1.4236
2	− 0.5642	0.0000	0.2970	0.4950	0.6418	0.7574	0.8522
3	—	− 0.8463	− 0.2970	0.0000	0.2015	0.3527	0.4728
4	—	—	− 1.0294	− 0.4950	− 0.2015	0.0000	0.1525

Rank	9	10	11	12	13	14	15
1	1.4850	1.5388	1.5864	1.6292	1.6680	1.7034	1.7359
2	0.9323	1.0014	1.0619	1.1157	1.1641	1.2079	1.2479
3	0.5720	0.6561	0.7288	0.7928	0.8498	0.9011	0.9477
4	0.2745	0.3758	0.4620	0.5368	0.6028	0.6618	0.7149
5	0.0000	0.1227	0.2249	0.3122	0.3883	0.4556	0.5157
6	− 0.2745	− 0.1227	0.0000	0.1026	0.1905	0.2673	0.3353
7	− 0.5720	− 0.3758	− 0.2249	− 0.1026	0.0000	0.0882	0.1653

Rank	16	17	18	19	20	21	22
1	1.7660	1.7939	1.8200	1.8445	1.8675	1.8892	1.9097
2	1.2847	1.3188	1.3504	1.3799	1.4076	1.4336	1.4582
3	0.9903	1.0295	1.0657	1.0995	1.1309	1.1605	1.1882
4	0.7632	0.8074	0.8481	0.8859	0.9210	0.9538	0.9846
5	0.5700	0.6195	0.6648	0.7066	0.7454	0.7815	0.8153
6	0.3962	0.4513	0.5016	0.5477	0.5903	0.6298	0.6667
7	0.2338	0.2952	0.3508	0.4016	0.4483	0.4915	0.5316
8	0.0773	0.1460	0.2077	0.2637	0.3149	0.3620	0.4056
9	− 0.0773	0.0000	0.0688	0.1307	0.1870	0.2384	0.2858
10	− 0.2338	− 0.1460	− 0.0688	0.0000	0.0620	0.1184	0.1700
11	− 0.3962	− 0.2952	− 0.2077	− 0.1307	− 0.0620	0.0000	0.0564

Rank	23	24	25	26	27	28	29
1	1.9292	1.9477	1.9653	1.9822	1.9983	2.0137	2.0285
2	1.4814	1.5034	1.5243	1.5442	1.5633	1.5815	1.5989
3	1.2144	1.2392	1.2628	1.2851	1.3064	1.3267	1.3462
4	1.1036	1.0409	1.0668	1.0914	1.1147	1.1370	1.1582
5	0.8470	0.8768	0.9050	0.9317	0.9570	0.9812	1.0041
6	0.7012	0.7335	0.7641	0.7929	0.8202	0.8461	0.8708
7	0.5690	0.6040	0.6369	0.6679	0.6973	0.7251	0.7515
8	0.4461	0.4839	0.5193	0.5527	0.5841	0.6138	0.6420
9	0.3297	0.3705	0.4086	0.4444	0.4780	0.5098	0.5398
10	0.2175	0.2616	0.3027	0.3410	0.3771	0.4110	0.4430
11	0.1081	0.1558	0.2001	0.2413	0.2798	0.3160	0.3501
12	0.0000	0.0518	0.0995	0.1439	0.1852	0.2239	0.2602
13	− 0.1081	− 0.0518	0.0000	0.0478	0.0922	0.1336	0.1724
14	− 0.2175	− 0.1558	− 0.0995	− 0.0478	0.0000	0.0444	0.0859

TABLE 10.5   (*Continued*)

Rank	30	31	32	33	34	35	36
1	2.0428	2.0565	2.0697	2.0824	2.0947	2.1066	2.1181
2	1.6156	1.6317	1.6471	1.6620	1.6764	1.6902	1.7036
3	1.3648	1.3827	1.3999	1.4164	1.4323	1.4476	1.4624
4	1.1786	1.1980	1.2167	1.2347	1.2520	1.2686	1.2847
5	1.0261	1.0471	1.0672	1.0865	1.1051	1.1229	1.1402
6	0.8944	0.9169	0.9384	0.9590	0.9789	0.9979	1.0162
7	0.7767	0.8007	0.8236	0.8455	0.8666	0.8868	0.9062
8	0.6688	0.6944	0.7187	0.7420	0.7643	0.7857	0.8063
9	0.5683	0.5955	0.6213	0.6460	0.6695	0.6921	0.7138
10	0.4733	0.5021	0.5294	0.5555	0.5804	0.6043	0.6271
11	0.3824	0.4129	0.4418	0.4694	0.4957	0.5208	0.5449
12	0.2945	0.3269	0.3575	0.3867	0.4144	0.4409	0.4662
13	0.2088	0.2432	0.2757	0.3065	0.3358	0.3637	0.3903
14	0.1247	0.1613	0.1957	0.2283	0.2592	0.2886	0.3166
15	0.0415	0.0804	0.1169	0.1515	0.1842	0.2152	0.2446
16	-0.0415	0.0000	0.0389	0.0755	0.1101	0.1428	0.1739
17	-0.1247	-0.0804	-0.0389	0.0000	0.0366	0.0712	0.1040
18	-0.2088	-0.1613	-0.1169	-0.0755	-0.0366	0.0000	0.0346

Rank	37	38	39	40	41	42	43
1	2.1293	2.1401	2.1506	2.1608	2.1707	2.1803	2.1897
2	1.7166	1.7291	1.7413	1.7351	1.7646	1.7757	1.7865
3	1.4768	1.4906	1.5040	1.5170	1.5296	1.5419	1.5538
4	1.3002	1.3151	1.3296	1.3437	1.3573	1.3705	1.3833
5	1.1568	1.1728	1.1883	1.2033	1.2178	1.2319	1.2456
6	1.0339	1.0509	1.0674	1.0833	1.0987	1.1136	1.1281
7	0.9250	0.9430	0.9604	0.9772	0.9935	1.0092	1.0245
8	0.8260	0.8451	0.8634	0.8811	0.8983	0.9148	0.9308
9	0.7346	0.8547	0.7740	0.7926	0.8106	0.8279	0.8447
10	0.6490	0.6701	0.6904	0.7099	0.7287	0.7469	0.7645
11	0.5679	0.5900	0.6113	0.6318	0.6515	0.6705	0.6889
12	0.4904	0.5136	0.5359	0.5574	0.5780	0.5979	0.6171
13	0.4158	0.4401	0.4635	0.4859	0.5075	0.5283	0.5483
14	0.3434	0.3689	0.3934	0.4169	0.4394	0.4611	0.4820
15	0.2727	0.2995	0.3252	0.3498	0.3734	0.3960	0.4178
16	0.2034	0.2316	0.2585	0.2842	0.3089	0.3326	0.3553
17	0.1351	0.1647	0.1929	0.2199	0.2457	0.2704	0.2942
18	0.0674	0.0985	0.1282	0.1564	0.1835	0.2093	0.2341
19	0.0000	0.0328	0.0640	0.0936	0.1219	0.1490	0.1749
20	-0.0674	-0.0328	0.0000	0.0312	0.0608	0.0892	0.1163
21	-1.1351	-0.0985	-0.0640	-0.0312	0.0000	0.0297	0.0580

TABLE 10.5 (*Continued*)

Rank	44	45	46	47	48	49	50
1	2.1988	2.2077	2.2164	2.2249	2.2331	2.2412	2.2491
2	1.7971	1.8073	1.8173	1.8271	1.8366	1.8458	1.8549
3	1.5653	1.5766	1.5875	1.5982	1.6086	1.6187	1.6286
4	1.3957	1.4078	1.4196	1.4311	1.4422	1.4531	1.4637
5	1.2588	1.2717	1.2842	1.2964	1.3083	1.3198	1.3311
6	1.1421	1.1558	1.1690	1.1819	1.1944	1.2066	1.2185
7	1.0392	1.0536	1.0675	1.0810	1.0942	1.1070	1.1195
8	0.9463	0.9614	0.9760	0.9902	1.0040	1.0174	1.0304
9	0.8610	0.8767	0.8920	0.9068	0.9212	0.9353	0.9489
10	0.7815	0.7979	0.8139	0.8294	0.8444	0.8590	0.8732
11	0.7067	0.7238	0.7405	0.7566	0.7723	0.7875	0.8022
12	0.6356	0.6535	0.6709	0.6877	0.7040	0.7198	0.7351
13	0.5676	0.5863	0.6044	0.6219	0.6388	0.6552	0.6712
14	0.5022	0.5217	0.5405	0.5586	0.5763	0.5933	0.6099
15	0.4389	0.4591	0.4787	0.4976	0.5159	0.5336	0.5508
16	0.3772	0.3983	0.4187	0.4383	0.4573	0.4757	0.4935
17	0.3170	0.3390	0.3602	0.3806	0.4003	0.4194	0.4379
18	0.2579	0.2808	0.3029	0.3241	0.3446	0.3644	0.3836
19	0.1997	0.2236	0.2465	0.2686	0.2899	0.3105	0.3304
20	0.1422	0.1671	0.1910	0.2140	0.2361	0.2575	0.2781
21	0.0851	0.1111	0.1360	0.1599	0.1830	0.2051	0.2265
22	0.0283	0.0555	0.0814	0.1064	0.1303	0.1534	0.1756
23	−0.0283	0.0000	0.0271	0.0531	0.0781	0.1020	0.1251
24	−0.0851	−0.0555	−0.0271	0.0000	0.0260	0.0509	0.0749
25	−0.1422	−0.1111	−0.0814	−0.0531	−0.0260	0.0000	0.0250

where

$$b_i = \frac{1}{n - n_0} - \frac{\bar{u}_k(u_i - \bar{u}_k)}{\sum_{j=n_0+1}^{n}(u_j - \bar{u}_k)^2} \tag{10.12}$$

$$c_i = \frac{u_i - \bar{u}_k}{\sum_{j=n_0+1}^{n}(u_j - \bar{u}_k)^2} \tag{10.13}$$

and

$$\bar{u}_k = \frac{1}{n - n_0} \sum_{j=n_0+1}^{n} u_j \tag{10.14}$$

or the arithmetic mean of the expected values of the uncensored sample elements. The values u_i are the expected values of the order statistics from a standard normal distribution [i.e., $N(0, 1)$]. In Table 10.5 the expected values of the normal order statistics (u_j) are tabulated for sample sizes ranging from $n = 2$ to $n = 50$, so that this alternative linear estimator can be computed for most practical groundwater monitoring problems.

Example 10.4

Returning to the TOC example and ranking from smallest to largest

$$3, 4, 5, 5, 6, 6, 7$$

we obtain the equation for the mean as

$$\hat{\mu} = .2838(3) + .2413(4) + .2001(5) + .1576(5)$$
$$+ .1105(6) + .0525(6) - .0457(7) = 4.26$$

and the standard deviation as

$$\hat{\sigma} = -.3041(3) - .2124(4) - .1235(5) - .0318(5)$$
$$+ .0698(6) + .1950(6) + .4070(7) = 1.90$$

which are similar to the BLU estimates of 4.07 and 2.20, respectively. Again, application of these Type II censoring estimators to a Type I censoring problem yields different results than the ML or RML estimators, because the linear estimator does not incorporate knowledge of the censoring point. As the distance between the censoring point and the smallest measured value increases, as in this example, the discrepancy between Type I and Type II estimators will increase. In general, the linear estimators will have higher mean and lower standard deviation values than the ML or RML estimators.

10.7   DELTA DISTRIBUTIONS

An alternative approach to the censored data problem involves the so-called delta distribution (Aitchison, 1955) in which the parameters of a continuous probability distribution with some probability mass at zero are estimated. Owen and DeRouen (1980) have shown that the lognormal delta distribution is optimal for measuring exposure to air contaminants. The distribution is well suited to the current problem because it accommodates both the problem of nondetects as well as the lognormality of the detected constituent concentrations. Usually, the concentration of contaminants in environmental media is lognormally distributed (Ott, 1990). If detection and/or quantitation limits are close to zero, there is typically little loss of information in assuming that the censored portion of the distribution is at zero.

Statistically, the delta distribution is a two-parameter lognormal distribution in which some proportion of the probability mass is located at zero. Computationally, the mean and variance of the delta distribution are obtained as follows.

Denoting the number of observations that are not detected as n_0, the number of detected measurements as n_1, and the total number of measurements as n, the mean and variance of the lognormal delta distribution are given by

$$\hat{\mu} = \frac{n_1}{n} \exp(\bar{y}) \Delta_{n_1}\!\left(\frac{s_y^2}{2}\right) \tag{10.15}$$

$$\hat{\sigma}^2 = \frac{n_1}{n} \exp(2\bar{y})\left[\Delta_{n_1}\!\left(2s_y^2\right) - \frac{n_1 - 1}{n - 1}\Delta_{n_1}\!\left(\frac{n_1 - 2}{n_1 - 1}s_y^2\right)\right] \tag{10.16}$$

where

$$\bar{y} = \sum_{i=1}^{n_1} \frac{\log_e x_i}{n}$$

and

$$s_y = \sqrt{\sum_{i=1}^{n_1} \frac{(\log_e x_i - \bar{y})^2}{n - 1}}$$

are the mean and standard deviation of the natural logarithms of the detected values, and

$$\Delta_{n_1}(z) = 1 + \frac{n_1 - 1}{n_1}z + \frac{(n_1 - 1)^3}{n_1^2 2!}\frac{z^2}{n_1 + 1}$$

$$+ \frac{(n_1 - 1)^5}{n_1^3 3!}\frac{z^3}{(n_1 + 1)(n_1 + 3)} + \cdots \tag{10.17}$$

is a Bessel function with argument z, which can take on the values $z = 2s_y^2$ or $z = s_y^2/2$ or $z = s_y^2[(n_1 - 2)/(n_1 - 1)]$ as shown in the previous equations. In practice, this function generally converges in less than 10 steps, the first three of which are shown previously.

Aitchison's (1955) results also apply to a normal distribution with some probability mass at zero. In this case the adjusted mean concentration is computed as

$$\hat{\mu} = \left(1 - \frac{n_0}{n}\right)\bar{x}' \tag{10.18}$$

where $\bar{x}'$ is the average of the detected values, n is the total number of samples, and n_0 is the number of samples in which the compound is not present. The standard deviation is

$$\hat{\sigma} = \sqrt{\left(1 - \frac{n_0}{n}\right)s^{2\prime} + \frac{n_0}{n}\left(1 - \frac{n_0 - 1}{n - 1}\right)\bar{x}^{2\prime}} \tag{10.19}$$

where s' is the standard deviation of the detected measurements. In the case of a normal distribution, however, these results are identical to substituting zero for the nondetects in the usual calculation for the mean and variance.

Example 10.5

For the TOC example, we have $n = 10$, $n_0 = 3$, $n_1 = 7$, $\bar{y} = 1.60$, and $s_y = 0.29$. After 10 iterations,

$$\Delta_{n_1}\left(\frac{s_y^2}{2}\right) = 1.0365$$

$$\Delta_{n_1}\left(2s_y^2\right) = 1.1522$$

and

$$\Delta_{n_1}\left(\frac{n_1 - 2}{n_1 - 1}s_y^2\right) = 1.0614$$

The mean is therefore

$$\hat{\mu} = \tfrac{7}{10}(e^{1.6})1.0365 = 3.59$$

and the standard deviation is

$$\hat{\sigma} = \tfrac{7}{10}(e^{2(1.6)})\left[1.1522 - \tfrac{6}{9}(1.0614)\right] = 2.76$$

These estimates are, in fact, quite similar to the ML estimates for the censored normal distribution despite the fact that they assume an underlying lognormal distribution and place probability mass at zero rather than at the censoring point, which in this case is 1.0 mg/l.

Using Aitchison's (1955) estimator for the normal distribution, we obtain

$$\hat{\mu} = \left(1 - \tfrac{3}{10}\right)5.14 = 3.60$$

and

$$\hat{\sigma} = \left[\left(1 - \frac{3}{10}\right)1.35^2 + \frac{3}{10}\left(1 - \frac{3 - 1}{10 - 1}\right)5.14^2\right]^{1/2} = 2.73$$

which are remarkably similar to the lognormal results. Note that these estimates are identical to what would have been obtained had we simply substituted zero for the censored values. Recall that substitution of MDL/2 = .5 instead of zero yielded 3.75 and 2.50 for the mean and standard deviation, respectively.

10.8 REGRESSION METHODS

Hashimoto and Trussell (1983) and Gilliom and Helsel (1986) have suggested a method by which values for the censored observations can be imputed based on a linear regression of order statistics on measured concentrations for the uncensored data. Following Gilliom and Helsel (1986), normal scores are computed as

$$z = \Phi^{-1} \frac{r}{n + 1} \tag{10.20}$$

where r is the rank of the measurement ($r = n_0 + 1, \ldots, n$), n is the total number of measurements, and Φ^{-1} is the inverse normal cumulative distribution function. A least squares regression of concentration on normal scores for all uncensored data can be used to extrapolate values to the censored observations (ranks $r = 1, \ldots, n_0$). Estimated values below zero are set equal to zero. The estimated values are treated as observed, and the mean and variance of all measurements (i.e., estimated and observed concentrations) are used.

The least squares estimates of the intercept and slope of the regression line for the uncensored values are

$$b_1 = \frac{\sum_{i=n_0+1}^{n} (x_i - \bar{x})(z_i - \bar{z})}{\sum_{i=n_0+1}^{n} (z_i - \bar{z})^2} \tag{10.21}$$

and

$$b_0 = \bar{x} - b_1 \bar{z} \tag{10.22}$$

The prediction equation for a new concentration given the inverse normal probability (z) associated with rank r is therefore

$$\hat{x} = b_0 + b_1 z \tag{10.23}$$

which is computed for ranks 1 through n_0 (i.e., the n_0 lowest values). It is important to note that nothing prevents the estimated concentration for a censored value (i.e., $\hat{x}$) from being less than zero or greater than the censoring value. Estimated values less than zero should be set to zero and values greater than the censoring point (i.e., the MDL or PQL) should be set to that value.

Example 10.6

For the TOC data, the ordered values and inverse normal values associated with their rank are as follows:

Rank	Value	z
1	< 1	−1.34
2	< 1	−0.91
3	< 1	−0.60
4	3	−0.35
5	4	−0.11
6	5	0.11
7	5	0.35
8	6	0.60
9	6	0.91
10	7	1.34

The mean uncensored values are $\bar{x} = 5.14$ and $\bar{z} = 0.41$. The least squares estimates are

$$b_1 = \frac{4.59}{2.09} = 2.20$$

and

$$b_0 = 5.14 - 2.20(0.41) = 4.24$$

The estimated values for the three censored values are

Rank	Estimated Concentration
1	1.30
2	2.25
3	2.92

which are all greater than the MDL of 1.0 mg/l. In light of this, we would substitute the MDL for all three censored values and obtain

$$\hat{\mu} = 3.9$$

and

$$\hat{\sigma} = 2.28$$

However, had we used the imputed values, the estimates would have been

$\hat{\mu} = 4.25$ and $\hat{\sigma} = 1.85$, which are quite similar to the linear estimates. Gilliom and Helsel (1986) have also indicated that best overall success is obtained by first transforming the uncensored measurements to a natural log scale, which in our case yields

$$1.10, 1.39, 1.61, 1.61, 1.79, 1.79, 1.95$$

and a censoring value of $\log_e(1) = 0$. Here, $b_0 = 1.42$, $b_1 = 0.45$, and the three uncensored values are

Rank	$\log_e$ Estimated Censored Value	Estimated Raw Concentration
1	0.82	2.27
2	1.01	2.75
3	1.15	3.16

which again are all above the censoring limit of $\log_e(1) = 0$. If, however, we use these estimated values, we obtain $\hat{\mu}_{\log_e} = 1.42$ and $\hat{\sigma}_{\log_e} = 0.38$. These estimates may then be used to estimate the mean and standard deviation of the lognormal distribution as

$$\hat{\mu} = \exp\left(\hat{\mu}_{\log_e} + \frac{\hat{\sigma}^2_{\log_e}}{2} \right) = 4.45$$

and

$$\hat{\sigma} = \left[\hat{\mu}^2 \left(\exp\left(\hat{\sigma}^2_{\log_e} \right) - 1 \right) \right]^{1/2} = 1.75$$

which are quite similar to the normal regression results.

10.9 SUBSTITUTION OF EXPECTED VALUES OF ORDER STATISTICS

Gleit (1985) suggested an iterative procedure for obtaining improved estimators based on simple substitution methods. The basic idea is to replace the censored observations with the expected values of the order statistics for the n_0 censored observations conditional on a provisional estimate of the mean and standard deviation (e.g., begin by substituting MDL/2 for the censored values). The algorithm is as follows:

1. Compute a provisional estimate of the mean and standard deviation $\hat{\mu}_1$ and $\hat{\sigma}_1$ by substituting MDL/2 for the censored observations.
2. On the basis of $\hat{\mu}_1$ and $\hat{\sigma}_1$, calculate the expected values for the first n_0 order statistics. To do this, select the expected values of the first n_0

order statistics for a sample of size n from a standard normal distribution [i.e., $N(0, 1)$] from Table 10.5 and transform them to the current $N(\hat{\mu}_1, \hat{\sigma}_1)$ scale by

$$u'_i = (u_i \hat{\sigma}_1 + \hat{\mu}_1) \qquad (10.24)$$

where u_i is the expected value of the order statistic on the $N(0, 1)$ scale and u'_i is the expected value of the normal order statistic on the $N(\hat{\mu}_1, \hat{\sigma}_1)$ scale.

3. Now that all n data points are "known," compute the mean and standard deviation and call them $\hat{\mu}_2$ and $\hat{\sigma}_2$.
4. Continue steps 2 and 3 until the differences between $\hat{\mu}_t$ and $\hat{\mu}_{t-1}$ and $\hat{\sigma}_t$ and $\hat{\sigma}_{t-1}$ are less than 10^{-6}.

Example 10.7

In the TOC example, the first $n_0 = 3$ samples were censored and the corresponding $N(0, 1)$ order statistics are

$$u'_1 = -1.5388$$
$$u'_2 = -1.0014$$
$$u'_3 = -0.6561$$

Beginning with the simple substitution estimates (i.e., substituting MDL/2), we have

$$\hat{\mu}_1 = 3.75$$
$$\hat{\sigma}_1 = 2.50$$

Therefore

$$\hat{\mu}'_1 = -1.5388(2.50) + 3.75 = -.0970$$
$$\hat{\mu}'_2 = -1.0014(2.50) + 3.75 = 1.2465$$
$$\hat{\mu}'_3 = -0.6561(2.50) + 3.55 = 1.9098$$

Repeated iterations yielded

$$\hat{\mu} = 4.37$$

and

$$\hat{\sigma} = 1.70$$

The final values for the three censored measurements were 1.743 mg/l, 2.658 mg/l, and 3.247 mg/l. Note that all imputed values are greater than the detection limit but consistent with the detected concentrations. As such, this approach yields values similar to the linear estimates that ignore information regarding the actual censoring point (i.e., Type II censoring). The imputed values are somewhat larger than those obtained by the normal regression method, probably due to the normality assumption which is not present in the unweighted least squares interpolation. However, estimates of the mean and standard deviation of both methods are virtually identical.

10.10 COMPARISON OF ESTIMATORS

There have been several studies comparing the statistical properties of the various estimators, both analytically and via Monte Carlo simulation. In general, the studies have compared the ability of these estimators to recover the true mean and standard deviation of the uncensored parent distribution. In addition, the analytical studies have found that the modified maximum likelihood estimators perform nearly as well as the maximum likelihood estimator and that the alternative linear estimators perform nearly as well as the best linear unbiased estimator (Schneider, 1986). As expected, the efficiency of the estimators decreases with the degree of censoring, and the effect is more pronounced on the estimate of the variance versus the estimate of the mean (Sarhan and Greenberg, 1962).

Somewhat more relevant are Monte Carlo simulation studies which compared various estimators in the context of singly left-censored samples that are observed in groundwater monitoring applications. Again, these studies focus on the ability to recover the mean and variance of the total distribution and provide little information regarding the more important question in the present context of what effect use of these estimators has on the overall false positive rate of a groundwater monitoring program. The two most relevant studies in this area were conducted by Gilliom and Helsel (1986) and Haas and Scheff (1990). Gilliom and Helsel (1986) considered the following eight estimators.

1. ZE: Censored observations set to zero.
2. DL: Censored observations set equal to the MDL.
3. UN: Censored observations set equal to MDL/2.
4. NR: Censored observations were imputed using the normal regression method.
5. LR: Same as NR but data log transformed.
6. NM: Maximum likelihood estimates.
7. LM: Maximum likelihood estimates based on log-transformed data followed by reverse transformation due to Aitchison and Brown (1957).
8. DT: The lognormal delta distribution.

Evaluation of the reliability of the methods was based on root mean square errors (RMSEs) computed from the actual parameters of the underlying distribution used to generate the simulated data. For example,

$$\text{RMSE} = \left[\sum_{i=1}^{N} \left(\frac{\bar{x}_i - \mu}{\mu} \right)^2 \bigg/ N \right]^{1/2} \tag{10.25}$$

where $\bar{x}_i$ is the estimate of the mean for the ith of N datasets and μ is the true mean value used to generate the data.

Results of their study revealed that, overall, the lognormal regression (LR) method of imputing the censored values performed best. The maximum likelihood method computed on log-transformed data performed best for estimating the median and interquartile range.

Haas and Scheff (1990) compared the maximum likelihood, bias-corrected maximum likelihood, restricted maximum likelihood, one-half the detection limit, and normal regression methods in terms of recovering the mean of the parent distribution. The parent distributions were standard normal, a normal mixture with common variance, and a normal mixture with different variances. The authors noted that all methods yield biased results as the censoring level approaches 0.5. The normal regression method is positively biased (i.e., overestimates the true mean) over the entire range of censoring values, and for small amounts of censoring was the most strongly biased estimator. Overall, the bias-corrected restricted maximum likelihood estimator performed the best. Both one-half the detection limit and normal regression methods were shown to have "substantial deficiencies" with respect to bias and/or RMSE. The restricted maximum likelihood method also performed best for the normal mixture distributions which are more characteristic of the bimodal, heavy-tailed, and skewed distributions commonly found in environmental data.

The discrepancy between the two studies in terms of the utility of normal regression models is puzzling. Perhaps the emphasis on lognormally distributed generating distributions in the Gilliom and Helsel (1986) study versus the normal or normal mixture generating distributions used in the Haas and Scheff (1990) study may account for some of the difference.

Several other relevant studies have been conducted and have been summarized by Haas and Scheff (1990). Perhaps most interesting is a study by Gleit (1985) in which substituting the expected values of the normal order statistics for the censored values greatly outperformed both maximum likelihood and constant fill-in procedures (e.g., MDL/2) for samples drawn from normal distributions. Other studies (Hashimoto and Trussell, 1983; Helsel and Cohn, 1988; El-Shaarawi, 1989) generally found similar results for the regression and maximum likelihood methods.

McNichols and Davis (1988) performed a limited study on the effect of the type of censored data estimator on the overall false positive rates for the prediction limits of Gibbons (1987a) and Davis and McNichols (1987). The prediction limit factors were selected to provide an overall significance level of 5% across eight downgradient monitoring wells. Background sample sizes of 4 and 12 were considered with censoring levels ranging from 20% to 90%. Data were generated from normal and skewed normal distributions. Four methods were compared:

1. Nondetects = 0
2. Nondetects = MDL/2
3. Nondetects = MDL
4. Maximum likelihood estimators

Results of the study revealed that for high levels of censoring (i.e., > 80%), none of the methods achieved its intended nominal false positive rate; however, the Davis and McNichols (1987) prediction limits, which incorporate a verification resample, dramatically decreased the overall false positive rates for all methods, but not to the nominal level. The effects of high degrees of censoring were worse for the maximum likelihood methods than for the simple substitution methods; however, this may in part be due to the limited background sample sizes (i.e., 4 and 12) and the way in which they were drawn (i.e., each sample consisted of 4 aliquots and the maximum likelihood estimator was applied first to the aliquots and then to the set of 4 or 12 samples).

In terms of the TOC example, Table 10.6 displays a summary of the estimates. Inspection of Table 10.6 reveals the following. First, the MLE and RMLE typically have higher standard deviations and lower means due to their dependence on the censoring point which, in this example, is relatively far from the measured values. The simple substitution method (i.e., MDL/2) and both the normal and lognormal delta distributions are quite similar probably because the censoring point is close to zero. The linear estimators

TABLE 10.6   Summary of Estimates for TOC Example

Estimator	Mean	Standard Deviation
MLE (adjusted)	3.50	3.22
RMLE (adjusted)	2.91	3.60
Delta normal	3.60	2.73
Delta lognormal	3.59	2.76
BLUE	4.07	2.20
Alternative linear estimator	4.26	1.90
Regression method	4.25	1.85
Substitution of order statistics	4.37	1.70
Substitution of MDL/2	3.75	2.50

and the two more sophisticated substitution methods (i.e., linear regression and expected values of order statistics) all yield quite similar results due to the fact that they all treat the problem as Type II censoring; that is, they are not dependent on the value of the censoring point.

10.11 SOME SIMULATION RESULTS

Although far from conclusive, a limited Monte Carlo simulation was performed to compare the previously described methods for a typical ground-water monitoring problem. As in the examples, $n = 10$ background measurements were simulated. First, root mean square errors were compared for the mean and standard deviation of the various estimators over 1000 replications. Second, the new monitoring measurements were drawn from the same distribution and compared to a normal prediction limit for one of two samples in bounds at each of 10 monitoring wells (see Table 1.5). The resulting limit was

$$\hat{\mu} + \hat{\sigma}1.925$$

In the event that the new sample exceeded the limit, a second sample was generated, and failure was indicated if both samples exceeded the prediction limit (i.e., false positive result). The generating distribution was normal with $\mu = 5$ and $\sigma = 1$ [i.e., $N(5, 1)$]. With only 10 background measurements, censoring was restricted in probability to 20% and 50%. Background samples with fewer than three detected measurements were discarded.

Results of the simulation are displayed in Tables 10.7 and 10.8. Results of this very limited study indicate that the MLE is the best overall estimator even in a sample of only 10 measurements, in terms of minimizing both false positive rates and recovering the true population parameters. Interestingly, the normal and lognormal forms of the delta distribution were most effective at minimizing false positive rates, yet they were the worst at recovering the

TABLE 10.7 Comparison of Estimators for 20% Type I Censoring

Estimator	False Positive Rate	RMSE ($\hat{\mu}$)	RMSE ($\hat{\sigma}$)
MLE	.074	.074	.292
RMLE	.292	.065	.297
Linear	.385	.072	.312
Normal delta	.046	.213	1.325
Lognormal delta	.075	.213	1.329
Regression	.343	.073	.298
Order statistics	.377	.066	.290
MDL/2	.132	.119	.559

TABLE 10.8   Comparison of Estimators for 50% Type I Censoring

Estimator	False Positive Rate	RMSE ($\hat{\mu}$)	RMSE ($\hat{\sigma}$)
MLE	.057	.074	.390
RMLE	.284	.055	.385
Linear	.356	.085	.441
Normal delta	.032	.438	1.972
Lognormal delta	.023	.438	1.971
Regression	.367	.088	.434
Order statistics	.436	.079	.419
MDL/2	.132	.188	.739

true population mean and standard deviation. The linear estimator and the two more sophisticated substitution methods (i.e., linear regression and expected values of order statistics) were not effective at controlling the overall false positive rate. The simple substitution method (i.e., MDL/2) actually outperformed the more sophisticated substitution methods in terms of false positive rates (but not in terms of recovering the population parameters), but it still had an overall false positive rate almost three times the nominal level. As pointed out by McNichols and Davis (1988), use of verification resampling seems to help ensure that at least some methods can achieve their intended nominal Type I error rates.

10.12   SUMMARY

There are a great many methods for handling nondetectable or nonquantifiable samples in environmental data. Historically, these estimators have been compared on the basis of their ability to recover the true population parameters. In groundwater monitoring applications, a more relevant criterion is the overall false positive rate that results from use of a particular method. In many cases the method that best recovers the parameters of the distribution fails miserably at predicting future individual measurements from that distribution. Overall, the MLE appears to work best for small normally distributed samples, and lognormal versions of the estimator can be obtained simply by taking natural logarithms of the data and censoring point. The delta distributions also performed well in this simple example in terms of minimizing overall false positive rates. None of the other approaches provided adequate protection from false positive results even with verification resampling.

11 Tests for Departure from Normality

11.1 OVERVIEW

An assumption of many of the methods described in this book is that the measurements are continuous and normally distributed or can be suitably transformed to approximate a normal distribution [e.g., $\log_e(x) \sim N(\mu_x, \sigma_x^2)$]. There are several approaches to testing this assumption varying from graphical methods such as normal probability plots to inferential statistical approaches based on normal order statistics (e.g., Shapiro and Wilk, 1965). In this chapter attention is focused on statistical tests of the hypothesis that the data are normally distributed in the population of constituent measurements. In the context of groundwater monitoring applications, there are two special problems. First, measurements are nested within monitoring wells (e.g., a series of four upgradient wells). Due to spatial variability, the intrawell measurements may all be normally distributed; however, each well may have a different mean, offsetting the measurements from one well to another. A test of normality for the composite will generally yield a rejection of the null hypothesis that the data are normally distributed when no such rejection is warranted.

Second, the presence of censored measurements (i.e., measurements either not detected or not quantifiable) will generally produce rejection of the normality hypothesis regardless of whether or not the quantifiable measurements are normally distributed. One solution is to simply ignore the nonquantifiable measurements and test the assumption of normality in the measured samples. When the detection frequency is high (i.e., 90% or more), this may produce reasonable results. Alternatively, modifications of some normality tests, which incorporate the censored observations, have also been proposed and are preferable to simply ignoring the censored observations.

In the following sections several commonly used tests of normality are described, and generalizations to joint assessment of normality in several wells and extensions to censored normal distributions are discussed.

11.2 A SIMPLE GRAPHICAL APPROACH

In Chapter 10 the expected values of normal order statistics were used to impute quantitative values for the censored measurements. These same

TABLE 11.1   Ordered Chloride Measurements in mg / l with and without Transformation

Ordered Measurement i	Original Value $x_{(i)}$	Transformed Value $\log_{10}(10x_{(i)})$	Expected Value of Order Statistic $E(i \mid 15)$
1	0.200	0.301	−1.736
2	0.330	0.519	−1.248
3	0.450	0.653	−0.948
4	0.490	0.690	−0.715
5	0.780	0.892	−0.516
6	0.920	0.964	−0.335
7	0.950	0.978	−0.165
8	0.970	0.987	0.000
9	1.040	1.017	0.165
10	1.710	1.233	0.335
11	2.220	1.346	0.516
12	2.275	1.357	0.715
13	3.650	1.562	0.948
14	7.000	1.845	1.248
15	8.800	1.944	1.736

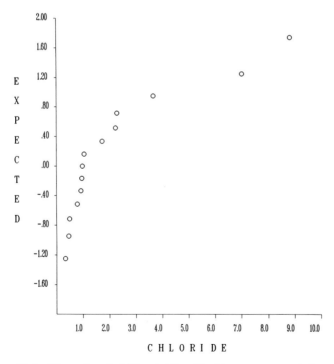

Figure 11.1   Normal probability plot for chloride measurements in mg/l.

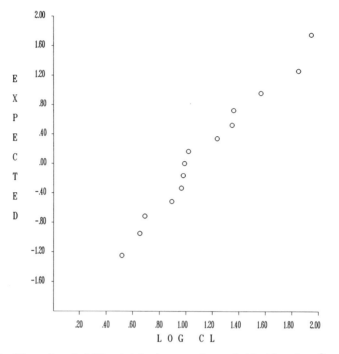

Figure 11.2   Normal probability plot for log-transformed chloride values [$\log_{10}(10x)$].

expected values can be used to produce a simple graphical test of normality. The observed measurements are first ordered from lowest to highest value and then plotted against the expected values of the normal order statistics (i.e., x axis). If the data are normally distributed, the points should lie close to a straight line, except for chance sampling fluctuations. The disadvantage of this type of graphical examination is that it provides no means of judging the significance of departure from linearity. It is often good to plot the ordered measurements in their original metric and following transformation [e.g., $x_{(i)}$ and $\log_{10}(x_{(i)})$].

Example 11.1

Consider the ordered chloride measurements in Table 11.1 obtained from a single upgradient well over 15 quarterly monitoring events. The normal probability plots are displayed graphically in Figures 11.1 and 11.2. Inspection of the figures reveals that the log-transformed values are more nearly normally distributed than the original values. The statistical significance of the departure from linearity observed in Figure 11.2 is, however, unknown using the graphical approach.

11.3 SHAPIRO–WILK TEST

In an attempt to formally summarize information contained in normal probability plots, Shapiro and Wilk (1965) proposed a test of normality based on normal order statistics. Their statistic W is proportional to the ratio of the slope of the normal probability plot to the usual mean square estimate, that is,

$$W = \left(\sum_{i=1}^{n} a_{i,n} x_{(i)} \right)^2 \Bigg/ \sum_{i=1}^{n} (x_i - \bar{x})^2 \qquad (11.1)$$

TABLE 11.2 Coefficients a_i for the Shapiro–Wilk W Test of Normality

	2	3	4	5	6	7	8	9	10	
1	.7071	.7071	.6872	.6646	.6431	.6233	.6052	.5888	.5739	
2		.0000	.1677	.2413	.2806	.3031	.3164	.3244	.3291	
3				.0000	.0875	.1401	.1743	.1976	.2141	
4						.0000	.0561	.0947	.1224	
5								.0000	.0399	

	11	12	13	14	15	16	17	18	19	20
1	.5601	.5475	.5359	.5251	.5150	.5056	.4968	.4886	.4808	.4734
2	.3315	.3325	.3325	.3318	.3306	.3290	.3273	.3253	.3232	.3211
3	.2260	.2347	.2412	.2460	.2495	.2521	.2540	.2553	.2561	.2565
4	.1429	.1586	.1707	.1802	.1878	.1939	.1988	.2027	.2059	.2085
5	.0695	.0922	.1099	.1240	.1353	.1447	.1524	.1587	.1641	.1686
6	.0000	.0303	.0539	.0727	.0880	.1005	.1109	.1197	.1271	.1334
7			.0000	.0240	.0433	.0593	.0725	.0837	.0932	.1013
8					.0000	.0196	.0359	.0496	.0612	.0711
9							.0000	.0163	.0303	.0422
10									.0000	.1040

	21	22	23	24	25	26	27	28	29	30
1	.4643	.4590	.4542	.4493	.4450	.4407	.4366	.4328	.4291	.4254
2	.3185	.3156	.3126	.3098	.3069	.3043	.3018	.2992	.2968	.2944
3	.2578	.2571	.2563	.2554	.2543	.2533	.2522	.2510	.2499	.2487
4	.2119	.2131	.2139	.2145	.2148	.2151	.2152	.2151	.2150	.2148
5	.1736	.1764	.1787	.1807	.1822	.1836	.1848	.1857	.1864	.1870
6	.1399	.1443	.1480	.1512	.1539	.1563	.1584	.1601	.1616	.1630
7	.1092	.1150	.1201	.1245	.1283	.1316	.1346	.1372	.1395	.1415
8	.0804	.0878	.0941	.0997	.1046	.1089	.1128	.1162	.1192	.1219
9	.0530	.0618	.0696	.0764	.0823	.0876	.0923	.0965	.1002	.1036
10	.0263	.0368	.0459	.0539	.0610	.0672	.0728	.0778	.0822	.0862
11	.0000	.0122	.0228	.0321	.0403	.0476	.0540	.0598	.0650	.0697
12			.0000	.1007	.0200	.0284	.0358	.0424	.0483	.0537
13					.0000	.0094	.0178	.0253	.0320	.0381
14							.0000	.0084	.0159	.0227
15									.0000	.0076

TABLE 11.2 (*Continued*)

	31	32	33	34	35	36	37	38	39	40
1	.4220	.4188	.4156	.4127	.4096	.4068	.4040	.4015	.3989	.3964
2	.2921	.2898	.2876	.2854	.2834	.2813	.2794	.2774	.2755	.2737
3	.2475	.2463	.2451	.2439	.2427	.2415	.2403	.2391	.2380	.2368
4	.2145	.2141	.2137	.2132	.2127	.2121	.2116	.2110	.2104	.2098
5	.1874	.1878	.1880	.1882	.1883	.1883	.1883	.1881	.1880	.1878
6	.1641	.1651	.1660	.1667	.1673	.1678	.1683	.1686	.1689	.1691
7	.1433	.1449	.1463	.1475	.1487	.1496	.1505	.1513	.1520	.1526
8	.1243	.1265	.1284	.1301	.1317	.1331	.1344	.1356	.1366	.1376
9	.1066	.1093	.1118	.1140	.1160	.1179	.1196	.1211	.1225	.1237
10	.0899	.0931	.0961	.0988	.1013	.1036	.1056	.1075	.1092	.1108
11	.0739	.0777	.0812	.0844	.0873	.0900	.0924	.0947	.0967	.0986
12	.0585	.0629	.0669	.0706	.0739	.0770	.0798	.0824	.0848	.0870
13	.0435	.0485	.0530	.0572	.0610	.0645	.0677	.0706	.0733	.0759
14	.0289	.0344	.0395	.0441	.0484	.0523	.0559	.0592	.0622	.0651
15	.0144	.0206	.0262	.0314	.0361	.0404	.0444	.0481	.0515	.0546
16	.0000	.0068	.0131	.0187	.0239	.0287	.0331	.0372	.0409	.0444
17		.0000	.0062	.0119	.0172	.0220	.0264	.0305	.0343	
18			.0000	.0057	.0110	.0158	.0203	.0244		
19				.0000	.0053	.0101	.0146			
20					.0000	.0049				

Note: columns for rows 17–20 above align as follows — row 17: .0000 (32), .0062 (33), .0119 (34), .0172 (35), .0220 (36), .0264 (37), .0305 (38), .0343 (39); row 18: .0000 (33), .0057 (34), .0110 (35), .0158 (36), .0203 (37), .0244 (38); row 19: .0000 (34), .0053 (35), .0101 (36), .0146 (37); row 20: .0000 (35), .0049 (36).

	41	42	43	44	45	46	47	48	49	50
1	.3940	.3917	.3894	.3872	.3850	.3830	.3808	.3789	.3770	.3751
2	.2719	.2701	.2684	.2667	.2651	.2635	.2620	.2604	.2589	.2574
3	.2357	.2345	.2334	.2323	.2313	.2302	.2291	.2281	.2271	.2260
4	.2091	.2085	.2078	.2072	.2065	.2058	.2052	.2045	.2038	.2032
5	.1876	.1874	.1871	.1868	.1865	.1862	.1859	.1855	.1851	.1847
6	.1693	.1694	.1695	.1695	.1695	.1695	.1695	.1693	.1692	.1691
7	.1531	.1535	.1539	.1542	.1545	.1548	.1550	.1551	.1553	.1554
8	.1384	.1392	.1398	.1405	.1410	.1415	.1420	.1423	.1427	.1430
9	.1249	.1259	.1269	.1278	.1286	.1293	.1300	.1306	.1312	.1317
10	.1123	.1136	.1149	.1160	.1170	.1180	.1189	.1197	.1205	.1212
11	.1004	.1020	.1035	.1049	.1062	.1073	.1085	.1095	.1105	.1113
12	.0891	.0909	.0927	.0943	.0959	.0972	.0986	.0998	.1010	.1020
13	.0782	.0804	.0824	.0842	.0860	.0876	.0892	.0906	.0919	.0932
14	.0677	.0701	.0724	.0745	.0765	.0783	.0801	.0817	.0832	.0846
15	.0575	.0602	.0628	.0651	.0673	.0694	.0713	.0731	.0748	.0764
16	.0476	.0506	.0534	.0560	.0584	.0607	.0628	.0648	.0667	.0685
17	.0379	.0411	.0442	.0471	.0497	.0522	.0546	.0568	.0588	.0608
18	.0283	.0318	.0352	.0383	.0412	.0439	.0465	.0489	.0511	.0532
19	.0188	.0227	.0263	.0296	.0328	.0357	.0385	.0411	.0436	.0459
20	.0094	.0136	.0175	.0211	.0245	.0277	.0307	.0335	.0361	.0386
21	.0000	.0045	.0087	.0126	.0163	.0197	.0229	.0259	.0288	.0314
22		.0000	.0042	.0081	.0118	.0153	.0185	.0215	.0244	
23			.0000	.0039	.0076	.0111	.0143	.0174		
24				.0000	.0037	.0071	.0104			
25					.0000	.0035				

Note: columns for rows 22–25 above align as follows — row 22: .0000 (42), .0042 (43), .0081 (44), .0118 (45), .0153 (46), .0185 (47), .0215 (48), .0244 (49); row 23: .0000 (43), .0039 (44), .0076 (45), .0111 (46), .0143 (47), .0174 (48); row 24: .0000 (44), .0037 (45), .0071 (46), .0104 (47); row 25: .0000 (45), .0035 (46).

Reproduced with permission from Shapiro and Wilk, *Biometrika*, **52** (1965): 591–611.

The coefficients $a_{i,n}$ are given in Table 11.2 for $n = 2$ to 50. Note that since the distribution is symmetric

$$a_{n-1+1,n} = -a_{i,n} \qquad (11.2)$$

The numerator in (11.1) can be written as

$$A^2 = \left\{ \sum_{i=1}^{h} a_{i,n} \left(x_{(n-i+1)} - x_i \right) \right\}^2 \qquad (11.3)$$

where $h = n/2$ if n is even or $(n-1)/2$ if n is odd. For ties [i.e., $x_{(i)} = x_{(i+1)}$] the authors suggest multiplying both by

$$\tfrac{1}{2}\left(a_{i,n} + a_{i+1,n} \right)$$

which can be extended to cover more than two ties. Critical values of the W-statistic are given in Table 11.3. The larger the value of W (i.e., closer to

TABLE 11.3   Lower 1% and 5% Critical Values for the Shapiro–Wilk Test Statistic W Used in Testing Normality

Sample Size	Value of $W_{.01}$	Value of $W_{.05}$	Sample Size	Value of $W_{.01}$	Value of $W_{.05}$
3	0.753	0.767	27	0.894	0.923
4	0.687	0.748	28	0.896	0.924
5	0.686	0.762	29	0.898	0.926
6	0.713	0.788	30	0.900	0.927
7	0.730	0.803	31	0.902	0.929
8	0.749	0.818	32	0.904	0.930
9	0.764	0.829	33	0.906	0.931
10	0.781	0.842	34	0.908	0.933
11	0.792	0.850	35	0.910	0.934
12	0.805	0.859	36	0.912	0.935
13	0.814	0.866	37	0.914	0.936
14	0.825	0.874	38	0.916	0.938
15	0.835	0.881	39	0.917	0.939
16	0.844	0.887	40	0.919	0.940
17	0.851	0.892	41	0.920	0.941
18	0.858	0.897	42	0.922	0.942
19	0.863	0.901	43	0.923	0.943
20	0.868	0.905	44	0.924	0.944
21	0.873	0.908	45	0.926	0.945
22	0.878	0.911	46	0.927	0.945
23	0.881	0.914	47	0.928	0.946
24	0.884	0.916	48	0.929	0.947
25	0.886	0.918	49	0.929	0.947
26	0.891	0.920	50	0.930	0.947

Reproduced with permission from Shapiro and Wilk, *Biometrika*, **52**(1965):591–611.

one), the greater is the support for the normality assumption. The assumption of normality is rejected if the computed value of W is less than the critical value.

Example 11.2

Using the chloride data in Table 11.1, we obtain

$$\sum (x_i - \bar{x})^2 = 90.41 \qquad A = 8.036 \qquad W(x) = .714$$

for the original data, and, letting $y = \log_{10}(10x)$,

$$\sum (y_i - \bar{y})^2 = 3.062 \qquad A = 1.723 \qquad W(y) = .970$$

Table 11.3 reveals that for $n = 15$ the lower one percentage point is 0.835; therefore we reject normality for the original data, but not for the log-transformed data.

11.4   SHAPIRO–FRANCIA TEST

When $n > 50$ the Shapiro–Francia test can be used in place of the Shapiro–Wilk test (Shapiro and Francia, 1972). The modified W-statistic is computed as

$$W' = \frac{\left[\sum_{i=1}^{n} b_{i,n} x_{(i)}\right]^2}{\sum_{i=1}^{n} (x_{(i)} - \bar{x})^2 \sum_{i=1}^{n} b_{i,n}^2} \qquad (11.4)$$

where the $x_{(i)}$ represent the ordered observations, the b_i represent the expected values of the normal order statistics $b_i = E(i|n)$. For large n (i.e., $n > 50$), the expected value of the normal order statistic can be approximated as

$$b_i = \Phi^{-1}\left(\frac{i}{n+1}\right) \qquad (11.5)$$

where Φ^{-1} is the inverse standard normal distribution function described in the previous chapter. Critical values of W' are provided in Table 11.4.

TABLE 11.4   Lower 1% and 5% Critical Values of the Shapiro–Francia Test of Normality for $n > 50$

Sample Size	Lower 1% Value	Lower 5% Value
35	.919	.943
50	.935	.953
51	.935	.954
53	.938	.957
55	.940	.958
57	.944	.961
59	.945	.962
61	.947	.963
63	.947	.964
65	.948	.965
67	.950	.966
69	.951	.966
71	.953	.967
73	.956	.968
75	.956	.969
77	.957	.969
79	.957	.970
81	.958	.970
83	.960	.971
85	.961	.972
87	.961	.972
89	.961	.972
91	.962	.973
93	.963	.973
95	.965	.974
97	.965	.975
99	.967	.976

11.5   D'AGOSTINO'S TEST

An alternative and even easier test to compute when $n > 50$ was developed by D'Agostino (1971). To test if the underlying distribution is normal, compute the statistic

$$D = \frac{\sum_{i=1}^{n}\left[i - \frac{1}{2}(n + 1)\right]x_{(i)}}{n^2 s} \tag{11.6}$$

where s is the sample standard deviation and the $x_{(i)}$ are the sample order

TABLE 11.5   Critical Values of the D'Agostino Test of Normality for $n > 50$

Sample Size	$Y_{.01}$	$Y_{.05}$	$Y_{.95}$	$Y_{.99}$
50	−3.442	−2.220	0.923	1.140
60	−3.360	−2.179	0.986	1.236
70	−3.293	−2.146	1.036	1.312
80	−3.237	−2.118	1.076	1.374
90	−3.100	−2.095	1.109	1.426
100	−3.150	−2.075	1.137	1.470
150	−3.009	−2.004	1.233	1.623
200	−2.922	−1.960	1.290	1.715
250	−2.861	−1.926	1.328	1.779
300	−2.816	−1.906	1.357	1.826
350	−2.781	−1.888	1.379	1.863
400	−2.753	−1.873	1.396	1.893
450	−2.729	−1.861	1.411	1.918
500	−2.709	−1.850	1.423	1.938
550	−2.691	−1.841	1.434	1.957
600	−2.676	−1.833	1.443	1.972
650	−2.663	−1.826	1.451	1.986
700	−2.651	−1.820	1.458	1.999
750	−2.640	−1.814	1.465	2.010
800	−2.630	−1.809	1.471	2.020
850	−2.621	−1.804	1.476	2.029
900	−2.613	−1.800	1.481	2.037
950	−2.605	−1.796	1.485	2.045
1000	−2.599	−1.792	1.489	2.052

Reproduced with permission from D'Agostino, *Biometrika*, **58**(1971):341–348.

statistics (i.e., the observed measurements ranked from lowest to highest). To test the significance of the statistic, compute

$$Y = \frac{D - 0.28209479}{0.02998598/\sqrt{n}} \tag{11.7}$$

The values of Y can be compared to the critical values computed by D'Agostino (1971) and provided here in Table 11.5, for selected values of $n = 50$ to 1000 and percentage points of 1% to 99%. The null hypothesis of normality is rejected if Y is less than $Y_{\alpha/2}$ or greater than $Y_{1-\alpha/2}$.

11.6   METHODS BASED ON MOMENTS OF A NORMAL DISTRIBUTION

There are three general methods for testing departure from normality based on sample moments. The sample moments are defined as

$$\bar{x} = \sum_{i=1}^{n} \frac{x_i}{n} \tag{11.8}$$

and

$$m_r = \sum_{i=1}^{n} \frac{(x_i - \bar{x})^r}{n} \quad \text{for } r \geq 2 \tag{11.9}$$

The relevant moment ratios are

$$\sqrt{b_1} = \frac{m_3}{m_2^{3/2}} \tag{11.10}$$

and

$$b_2 = \frac{m_4}{m_2^2} \tag{11.11}$$

Departure of $\sqrt{b_1}$ from the normal value of zero is an indication of skewness

TABLE 11.6   Critical Values for the Distribution of $\sqrt{b_1} = m_3 / m_2^{3/2}$ in Samples from a Normal Distribution

Sample Size	Upper (or Lower) 5%	Upper (or Lower) 1%
20	.772	1.150
25	.711	1.059
30	.662	.986
35	.621	.923
40	.588	.871
45	.559	.826
50	.534	.788
60	.492	.724
70	.459	.673
80	.432	.632
90	.409	.597
100	.390	.567
125	.351	.508
150	.322	.465
175	.299	.430
200	.280	.403
250	.251	.361
300	.230	.329
350	.213	.305
400	.200	.285
450	.188	.269
500	.179	.255

Reproduced with permission from Pearson and Hartley, *Biometrika Tables for Statisticians*, 1(1976):207–208.

TABLE 11.7   Critical Values for the Distribution of $b_2 = m_4 / m_2^2$ in Sample from a Normal Population

Sample Size	Upper 1.0%	Upper 5.0%	Lower 5.0%	Lower 1.0%
20	5.38	4.18	1.83	1.64
30	5.20	4.12	1.98	1.79
40	5.04	4.06	2.07	1.89
50	4.88	4.00	2.15	1.95
75	4.59	3.87	2.27	2.08
100	4.39	3.77	2.35	2.18
125	4.24	3.70	2.40	2.24
150	4.13	3.65	2.45	2.29
175	4.04	3.61	2.48	2.34
200	3.98	3.57	2.51	2.37
250	3.87	3.52	2.55	2.42
300	3.79	3.47	2.59	2.46
400	3.67	3.41	2.64	2.52
500	3.60	3.37	2.67	2.57
600	3.54	3.34	2.70	2.60
700	3.50	3.31	2.72	2.62
800	3.46	3.29	2.74	2.65
900	3.43	3.28	2.75	2.66
1000	3.41	3.26	2.76	2.68
2000	3.28	3.18	2.83	2.77

Reproduced with permission from Pearson and Hartley, *Biometrika Tables for Statisticians*, 1(1976):207–208.

in the frequency distribution, whereas departure of b_2 from the normal value of three is an indication of kurtosis. Pearson and Hartley (1976) suggest that for large samples (e.g., $n > 50$) a "rough" test of normality can be obtained by comparing $\sqrt{b_1}$ and $b_2 - 3$ with the approximate values of their standard errors which are $\sqrt{6/n}$ and $\sqrt{24/n}$, respectively. The ratio of the estimate to its standard error is approximately normal; therefore rejection is indicated by comparing the ratio to the corresponding normal tail probability. Fisher (1929, 1930) first developed methods for calculating higher sample moments of $\sqrt{b_1}$ and b_2 which led to the critical values in Tables 11.6 and 11.7. Shapiro, Wilk, and Chen (1968) have shown that combined use of $\sqrt{b_1}$ and b_2 is slightly less powerful than the W-statistic.

An alternative statistic, better suited to small samples, was proposed by Geary (1935, 1936). The statistic is the ratio of mean deviation to standard deviation, that is,

$$a = \frac{\sum_{i=1}^{n}|x_i - \bar{x}|}{\left\{n\sum_{i=1}^{n}(x_i - \bar{x})^2\right\}^{1/2}} \tag{11.12}$$

For a normally distributed random variable, the ratio has value $\sqrt{2/\pi} = .7979$. For platykurtic distributions the ratio will be higher and for leptokurtic

TABLE 11.8   Critical Values for the Distribution of
$a = $ **(Mean Deviation)/(Standard Deviation)**

Sample Size	Upper 1%	Upper 5%	Lower 5%	Lower 1%
11	.9359	.9073	.7153	.6675
16	.9137	.8884	.7236	.6829
21	.9001	.8768	.7304	.6950
26	.8901	.8686	.7360	.7040
31	.8827	.8625	.7404	.7110
36	.8769	.8578	.7440	.7167
41	.8722	.8540	.7470	.7216
46	.8682	.8508	.7496	.7256
51	.8648	.8481	.7518	.7291
61	.8592	.8434	.7554	.7347
71	.8549	.8403	.7583	.7393
81	.8515	.8376	.7607	.7430
91	.8484	.8353	.7626	.7460
101	.8460	.8344	.7644	.7487
201	.8322	.8229	.7738	.7629
301	.8260	.8183	.7781	.7693
401	.8223	.8155	.7807	.7731
501	.8198	.8136	.7825	.7757
601	.8179	.8123	.7838	.7776
701	.8164	.8112	.7848	.7791
801	.8152	.8103	.7857	.7803
901	.8142	.8096	.7864	.7814
1001	.8134	.8090	.7869	.7822

Reproduced with permission from Pearson and Hartley, *Biometrika Tables for Statisticians,* **1**(1976):207–208.

distributions the ratio will be lower. Table 11.8 gives upper and lower 5% and 1% points for a.

Example 11.3

Consider the 20 chloride measurements from four upgradient wells listed in Table 11.9. Assuming that samples from all four upgradient wells were drawn from the same population, we obtain the following three statistics:

$$\sqrt{b_1} = \frac{m_3}{m_2^{3/2}} = \frac{.0040}{1.3475^{3/2}} = .0025$$

$$b_2 = \frac{m_4}{m_2^2} = \frac{5.200}{1.3475^2} = 2.8640$$

$$a = \frac{\sum_{i=1}^{n}|x_i - \bar{x}|}{\left\{ n\sum_{i=1}^{n}(x_i - \bar{x})^2 \right\}^{1/2}} = \frac{19.0000}{\left\{ 20(26.9500) \right\}^{1/2}} = .8184$$

TABLE 11.9   Chloride Measurements from Four Upgradient Wells in mg/l

Sample	Well 1	Well 2	Well 3	Well 4
1	6	4	5	5
2	5	6	6	8
3	6	5	4	5
4	7	6	7	5
5	4	5	3	6

Comparison of these estimates to the critical values in Tables 11.6 to 11.8 reveal that in all three cases normality is not rejected. In fact, the estimated value of $a = .8184$ is virtually identical to the mean value of .8181 (see Table 11.8).

11.7   MULTIPLE INDEPENDENT SAMPLES

In the context of groundwater monitoring, it is hard to imagine a case in which normality would be assessed in a single sample. More typical is the case described in Example 11.3 in which the collection of upgradient wells is used to establish background and multiple measurements are available in each well. Due to spatial variability, the wells may have different means and variances, so that the simple pooling of measurements as in Example 11.3 is not justifiable. Wilk and Shapiro (1968) have suggested a generalization of their original test that is suitable for the joint assessment of normality in K independent samples. The idea is to compute individual values of W for each well, denoted W_t, obtain a normal equivalent G_t, and obtain an overall test by referring the normalized mean

$$G = \sum_{i=1}^{K} \frac{G_t}{\sqrt{K}} \qquad (11.13)$$

to a standard table of the normal integral. Values of G_t are given by

$$G_t = \gamma(n) + \delta(n) \log\left\{ \frac{W_t - \varepsilon(n)}{1 - W_t} \right\} \qquad (11.14)$$

The coefficients γ, δ, and ε as functions of n are given in Table 11.10 for $n = 7$ to 50. For values of $n = 3$ to 6, values of G_t can be obtained in terms of the transformed function

$$\gamma_t = \log_e\left[\frac{\{W_t - \varepsilon(n)\}}{1 - W_t} \right] \qquad (11.15)$$

where G_t is given as a function of γ_t in Table 11.11 [values of $\varepsilon(n)$ are given in the column heads in Table 11.11].

TABLE 11.10 Coefficients γ, δ, and ε as Functions of $n = 7(1)50$ for Assessing Normality Jointly in Several Independent Samples

Sample Size	$\gamma(n)$	$\delta(n)$	$\varepsilon(n)$	Sample Size	$\gamma(n)$	$\delta(n)$	$\varepsilon(n)$
7	−2.356	1.245	0.4533	31	−6.248	1.965	0.1840
8	−2.696	1.333	0.4186	32	−6.324	1.976	0.1811
9	−2.968	1.400	0.3900	33	−6.402	1.988	0.1781
10	−3.262	1.471	0.3660	34	−6.480	2.000	0.1755
				35	−6.559	2.012	0.1727
11	−3.485	1.515	0.3451	36	−6.640	2.024	0.1702
12	−3.731	1.571	0.3270	37	−6.721	2.037	0.1677
13	−3.936	1.613	0.3111	38	−6.803	2.049	0.1656
14	−4.155	1.655	0.2969	39	−6.887	2.062	0.1633
15	−4.373	1.695	0.2842	40	−6.961	2.075	0.1612
16	−4.567	1.724	0.2727	41	−7.035	2.088	0.1591
17	−4.713	1.739	0.2622	42	−7.111	2.101	0.1572
18	−4.885	1.770	0.2528	43	−7.188	2.114	0.1552
19	−5.018	1.786	0.2440	44	−7.266	2.128	0.1534
20	−5.153	1.802	0.2359	45	−7.345	2.141	0.1516
21	−5.291	1.818	0.2264	46	−7.414	2.155	0.1499
22	−5.413	1.835	0.2207	47	−7.484	2.169	0.1482
23	−5.508	1.848	0.2157	48	−7.555	2.183	0.1466
24	−5.605	1.862	0.2106	49	−7.615	2.198	0.1451
25	−5.704	1.876	0.2063	50	−7.677	2.212	0.1436
26	−5.803	1.890	0.2020				
27	−5.905	1.905	0.1980				
28	−5.988	1.919	0.1943				
29	−6.074	1.934	0.1907				
30	−6.150	1.949	0.1872				

Reproduced with permission from Pearson and Hartley, *Biometrika Tables for Statistics*, 2(1976):221.

Example 11.4

Returning to the data from Example 11.3 in Table 11.9 in which there were five samples in each of four upgradient wells, we have $K = 4$, $n_t = n = 5$. Using the coefficients in Table 11.2, we obtain the following values of W_t, γ_t, and G_t:

	Well			
	1	2	3	4
W_t	.961	.881	.987	.833
γ_t	2.342	1.015	3.478	0.516
G_t	0.869	−0.493	1.717	−1.058

TABLE 11.11 Values of G_t for Argument γ_t for Normal Conversion of W [$n = 3(1)6$]

γ_t	$n = 3$ (0.7500)	$n = 4$ (0.6297)	$n = 5$ (0.5521)	$n = 6$ (0.4963)
−7.0	−3.29	—	—	—
−5.4	−2.81	—	—	—
−5.0	−2.68	—	—	—
−4.6	−2.54	—	—	—
−4.2	−2.40	—	—	—
−3.8	−2.25	−3.50	—	—
−3.4	−2.10	−3.27	—	—
−3.0	−1.94	−3.05	−4.01	—
−2.6	−1.77	−2.84	−3.70	—
−2.2	−1.59	−2.64	−3.38	—
−1.8	−1.40	−2.44	−3.11	—
−1.4	−1.21	−2.22	−2.87	—
−1.0	−1.01	−1.96	−2.56	−3.72
−0.6	−0.80	−1.66	−2.20	−2.88
−0.2	−0.60	−1.31	−1.81	−2.27
0.2	−0.39	−0.94	−1.41	−1.85
0.6	−0.19	−0.57	−0.97	−1.38
1.0	−0.00	−0.19	−0.51	−0.84
1.4	0.18	0.15	−0.06	−0.33
1.8	0.35	0.45	0.37	0.18
2.2	0.52	0.74	0.75	0.64
2.6	0.67	1.00	1.09	1.06
3.0	0.81	1.23	1.40	1.45
3.4	0.95	1.44	1.67	1.83
3.8	1.07	1.65	1.91	2.17
4.2	1.19	1.85	2.15	2.50
4.6	1.31	2.03	2.47	2.77
5.0	1.42	2.19	2.85	3.09
5.4	1.52	2.34	3.24	3.54
5.8	1.62	2.48	3.64	—
6.2	1.72	2.62	—	—
6.6	1.81	2.75	—	—
7.0	1.90	2.87	—	—
7.4	1.98	2.97	—	—
7.8	2.07	3.08	—	—
8.2	2.15	3.22	—	—
8.6	2.23	3.36	—	—
9.0	2.31	—	—	—
9.4	2.38	—	—	—
9.8	2.45	—	—	—

Reproduced with permission from Pearson and Hartley, *Biometrika Tables for Statisticians*, **2**(1976):207–208.

TABLE 11.12   1% and 5% Critical Values for the Normal Probability Plot Correlation Coefficient Z

Sample Size	Upper 1% Point	Upper 5% Point	Sample Size	Upper 1% Point	Upper 5% Point
3	.869	.879	32	.949	.966
4	.822	.868	33	.950	.967
5	.822	.879	34	.951	.967
6	.835	.890	35	.952	.968
7	.847	.899	36	.953	.968
8	.859	.905	37	.955	.969
9	.868	.912	38	.956	.970
10	.876	.917	39	.957	.971
11	.883	.922	40	.958	.972
12	.889	.926	41	.958	.972
13	.895	.931	42	.959	.973
14	.901	.934	43	.959	.973
15	.907	.937	44	.960	.973
16	.912	.940	45	.961	.974
17	.916	.942	46	.962	.974
18	.919	.945	47	.963	.974
19	.923	.947	48	.963	.975
20	.925	.950	49	.964	.975
21	.928	.952	50	.965	.977
22	.930	.954	55	.967	.978
23	.933	.955	60	.970	.980
24	.936	.957	65	.972	.981
25	.937	.958	70	.974	.982
26	.939	.959	75	.975	.983
27	.941	.960	80	.976	.984
28	.943	.962	85	.977	.985
29	.945	.962	90	.978	.985
30	.947	.964	95	.979	.986
31	.948	.965	100	.981	.987

Reprinted with permission from Filliben, *Technometrics*, **17**(1975):113. © 1975 by the American Statistical Association and the American Society for Quality Control.

From the last row of figures, we find $G = \Sigma G_t / \sqrt{4} = 0.518$ which has an associated probability of .302, indicating that we cannot reject the assumption of normality.

11.8   TESTING NORMALITY IN CENSORED SAMPLES

When data are censored (e.g., reported as less than the detection limit), the previously described distributional tests do not apply. Ignoring the nondetects can be misleading because it eliminates the lower tail of the distribution and can therefore falsely reject the null hypothesis of normality. Including the nondetects at the MDL introduces a spike in the distribution that can also incorrectly cause the rejection of normality.

To develop a test of normality, we begin by noting that an alternative view of the Shapiro–Wilk test is as a correlation coefficient. Ryan and Joiner (1973) noted that the Shapiro–Francia test could be written as the square of the correlation coefficient Z:

$$Z^2 = \frac{\sum_{i=1}^n \left[(x_{(i)} - \bar{x})(z_{i,n} - \bar{z}) \right]}{\left[\sum_{i=1}^n (x_i - \bar{x})^2 \sum_{i=1}^n (z_{i,n} - \bar{z})^2 \right]^{1/2}}$$

where the $z_{i,n}$ are approximately $\Phi^{-1}[i/(n+1)]$. Filliben (1975) suggested substituting the inverse normal transform of the median of the ith order statistic (m_i) from a sample of n standard normal random variables. The median order statistics (m_i) are given by

$$m_i = \begin{cases} 1 - m_n & i = 1 \\ \dfrac{i - .3175}{n + .365} & i = 2, 3, \ldots, n-1 \\ .5^{1/n} & i = n \end{cases}$$

Therefore, $z_i = \Phi^{-1}(m_i)$. Filliben (1975) developed percentage points of the normal probability plot correlation coefficient ρ for $n = 3$ to 100 which are reproduced in Table 11.12 for the 1% and 5% points. Major advantages of this interpretation include that it is more general in terms of sample size than any of the other tests, it is easier to compute, and it is essentially indistinguishable in terms of power.

Smith and Bain (1976) obtained percentage points of the statistic $1 - \rho$ for complete and censored samples for $n = 8$ to 80 and censoring of 0%, 25%, and 50%. The critical 1% and 5% values of ρ^2 are displayed in Table 11.13.

Smith and Bain (1976) note that the choice of z_i is not critical and suggest the approximation

$$z_i = t - \frac{2.30753 + 0.27061t}{1.0 + 0.99229t + 0.04481t^2}$$

where

$$t = \left\{ \log_e \left[\frac{(n+1)^2}{i^2} \right] \right\}^{1/2}$$

Note that this is an approximation for $\Phi^{-1}[i/(n+1)]$; therefore there are

TABLE 11.13   1% and 5% Critical Values for the Censored Normal Probability Plot Correlation Coefficient Z

r/n	Sample Size	Upper 95%	Upper 99%
0.5	8	.268	.369
	20	.170	.268
	40	.115	.175
	60	.085	.128
	80	.066	.101
0.75	8	.227	.315
	20	.128	.197
	40	.078	.125
	60	.052	.078
	80	.043	.062
1.0	8	.187	.286
	20	.100	.146
	40	.059	.086
	60	.040	.060
	80	.033	.045

Reprinted with permission from Smith and Bain, *Communications in Statistical Theory and Methods*, **A5**(1976):119–132. Courtesy of Marcel Dekker, Inc.

slight differences between Tables 11.12 and 11.13 for the complete data case. To apply their test, the correlation of the uncensored data with the corresponding largest order statistics is computed.

Example 11.5

Using the chloride data from Table 11.1, we obtain the various estimates of z_i given in Table 11.14. The three estimates of z_i lead to values of $Z = .917$, .915, and .917, respectively, further indicating that the correlation estimate is robust to choice of z_i. Comparison of these estimates to the critical values in Table 11.12 reveals a 1% critical value of .907 and a 5% critical value of .937, indicating that the data fit a normal distribution marginally at best. Recall that for the Shapiro–Wilk test normality was rejected at both the 5% and 1% levels.

To illustrate application to censored data, assume that the seven lowest values were not detected. In this case summation was from 8 to 15 and the resulting values of $Z = .766$, .757, and .765 for expected values of normal order statistics, $\Phi^{-1}[i/(n + 1)]$, and $\Phi^{-1}(m_i)$, respectively. For 50% censoring and $n = 15$, the 1% critical value found by interpolation is .834, indicating rejection of normality.

TABLE 11.14   Three Methods for Computing z_i Using Chloride Data from Table 11.1

Ordered Measurement	Value	Expected Value of Order Statistics	$\Phi^{-1}[i/(n + 1)]$	$\Phi^{-1}(m_i)$
1	0.200	−1.736	−1.538	−1.698
2	0.330	−1.248	−1.152	−1.232
3	0.450	−0.948	−0.887	−0.937
4	0.490	−0.715	−0.673	−0.706
5	0.780	−0.516	−0.487	−0.509
6	0.920	−0.335	−0.317	−0.330
7	0.950	−0.165	−0.156	−0.162
8	0.970	0.000	0.000	0.000
9	1.040	0.165	0.156	0.162
10	1.710	0.335	0.317	0.330
11	2.220	0.516	0.487	0.509
12	2.275	0.715	0.673	0.706
13	3.650	0.948	0.887	0.937
14	7.000	1.248	1.152	1.232
15	8.800	1.736	1.538	1.698

11.9   KOLMOGOROV–SMIRNOV TEST

The Kolmogorov–Smirnov (KS) test is a nonparametric test that can be used to evaluate the fit of any hypothesized distribution. These tests are described in detail by Conover (1980) and also discussed by Gilbert (1987) for environmental applications. In general, the KS test is considered more powerful than alternative goodness-of-fit chi-square tests. The three general limitations are: (1) the method is computationally complex; (2) it requires large sample sizes (i.e., 50 or more); and (3) the parameters of the hypothesized distribution (e.g., mean and variance of a normal distribution) are assumed to be known. Lilliefors (1967, 1969) generalized the test to the case of a normal or lognormal distribution with unknown mean and variance, although the method is still computationally complex and requires large samples.

11.10   SUMMARY

Although several methods for testing normality have been described in this chapter, no single method is suitable for generic use because none of the methods incorporates both multiple groups and censoring. From the perspective of testing normality in multiple independent samples, the test described by Wilk and Shapiro (1968) appears to be the most rigorous. From the perspectives of computational simplicity, generality with respect to sample size, and censoring, tests based on normal probability plot correlation coefficients appear to be the most useful. Certainly, one possibility is to compute censored versions of Z separately in each well and compute the average value of ρ weighted by the number of measurements available in each well (in the event the number of intrawell measurements is unequal). This average value of ρ could then be compared to the corresponding critical value corresponding to the total number of measurements. While not exact, this approach which combines both multiple wells and censoring should be sufficiently accurate for most practical purposes.

12 Variance Component Models

12.1 OVERVIEW

For those constituents that occur naturally in the environment, the most critical task of statistical modeling is the characterization of background variability. It is critical to the analysis to preserve the context in which the measurements are obtained. For example, the usual estimator for the sample standard deviation

$$s = \left(\frac{\sum_{i=1}^{n}(x_i - \bar{x})^2}{n - 1} \right)^{1/2} \qquad (12.1)$$

assumes that observations are drawn from a single population where measurements are nested within monitoring wells with potentially different means and possibly collected on different sampling events. Thus the traditional estimator is biased and will produce consistent underestimates of the true population value σ. In the following sections a series of methods ranging from simple to complex will be described for decomposing variation into its component parts. From these variance component estimates an unbiased estimate of σ can be obtained and used in constructing statistical limit estimates. The reader should note that an analogous problem exists for MDL and PQL studies where measurements are nested within laboratories, analysts, and instruments.

12.2 LEAST SQUARES ESTIMATORS

When data are balanced (i.e., each monitoring well has the same number of measurements), least squares estimators can be used to obtain variance component estimates. The principal advantage is that these estimators are unbiased and are computationally simple relative to maximum likelihood estimators. These estimators assume that the data are normally distributed or can be transformed to be so. Principally, the variance components analysis is performed on background wells only; however, in those cases where predisposal data are available or it can be documented that no previous groundwater impacts have occurred, all monitored wells can be used.

236

The unbiased variance estimate for the case of multiple upgradient wells and multiple samplings of each well (selected at sufficient intervals to ensure independence, for example, quarterly sampling) can be obtained from the *random effects* analysis of variance model (see Gibbons, 1987a; Winer, 1971, page 427, for a similar example). Here, we are modeling the background measurements in terms of two random factors corresponding to wells and time points. The general form of this model is given by

$$x_{ij} = \mu + \omega_i + \tau_j + \varepsilon_{ij} \tag{12.2}$$

where x_{ij} is the measurement on background (upgradient) well i on sampling event j, μ is the unknown mean level for background measurements (i.e., averaged across wells and time), ω_i is a random variable distributed $N(0, \sigma_\omega^2)$ that describes the deviation of well i from μ, τ_j is a random variable distributed $N(0, \sigma_\tau^2)$ that describes the deviation at time j from μ, and ε_{ij} is a random residual distributed $N(0, \sigma_\varepsilon^2)$. Assuming independence of the x_{ij}, the expected value of a measurement is

$$E(x_{ij}) = \mu \tag{12.3}$$

with variance

$$\mathrm{var}(x_{ij}) = \sigma_\omega^2 + \sigma_\tau^2 + \sigma_\varepsilon^2 \tag{12.4}$$

This model corresponds to what is termed a *randomized block design* in the statistical literature, and the expected mean squares for the wells, time points, and residuals are given, respectively, as

$$\mathrm{MS}_\omega = t\sigma_\omega^2 + \sigma_\varepsilon^2$$

$$= t \sum_{i=1}^{w} \frac{(\bar{x}_i - \bar{x})^2}{w - 1} \tag{12.5}$$

$$\mathrm{MS}_\tau = w\sigma_\tau^2 + \sigma_\varepsilon^2$$

$$= w \sum_{j=1}^{t} \frac{(\bar{x}_j - \bar{x})^2}{t - 1} \tag{12.6}$$

$$\mathrm{MS}_\varepsilon = \sigma_\varepsilon^2$$

$$= \sum_{i=1}^{w} \sum_{j=1}^{t} \frac{(x_{ij} - \bar{x}_i - \bar{x}_j + \bar{x})^2}{(w - 1)(t - 1)} \tag{12.7}$$

where

x_{ij} = the measurement for well i on occasion j
w = the number of wells
t = the number of quarterly measurements per well
$\bar{x}_i$ = the mean for all measurements from well i
$\bar{x}_j$ = the mean for all measurements from time j
$\bar{x}$ = the mean of all measurements across wells and times

Solving for the individual variance components σ_ω^2, σ_τ^2, and σ_ε^2, we find that

$$\sigma_\omega^2 = \frac{\mathrm{MS}_\omega - \mathrm{MS}_\varepsilon}{t} \tag{12.8}$$

$$\sigma_\tau^2 = \frac{\mathrm{MS}_\tau - \mathrm{MS}_\varepsilon}{\omega} \tag{12.9}$$

and

$$\sigma_\varepsilon^2 = \mathrm{MS}_\varepsilon \tag{12.10}$$

The unbiased estimate of the total variance is therefore

$$\mathrm{var}(x_{ij}) = \sigma_\omega^2 + \sigma_\tau^2 + \sigma_\varepsilon^2$$
$$= \frac{\mathrm{MS}_\omega - \mathrm{MS}_\varepsilon}{t} + \frac{\mathrm{MS}_\tau - \mathrm{MS}_\varepsilon}{\omega} + \mathrm{MS}_\varepsilon \tag{12.11}$$

The preceding model assumes that all wells are measured at the same time points, and thus the two factors are referred to as "crossed" factors. When the time points corresponding to the different wells cannot be assumed to be equal, then the time-point factor is said to be "nested" within the well factor (see Winer, 1971, page 360, for a detailed description of nested factor designs). This would occur, for instance, if the background observations for well 1 were made quarterly in 1972, while the observations for well 2 were made quarterly in 1976, and those for well 3 were made quarterly in 1975, and so on. This nested model is given by

$$x_{ij} = \mu + \omega_i + \tau_{ij} + \varepsilon_{ij} \tag{12.12}$$

where x_{ij}, μ, ω_i, and ε_{ij} are as before, while τ_{ij} is a random variable distributed $N(0, \sigma_\tau^2)$ that describes the deviation of well i at time j from the overall mean value for well i $(\mu + \omega_i)$ averaged over time.

If sampling events are "nested" within wells, the expected mean squares must be modified to reflect this design. Specifically, the expected mean square for the wells is

$$\mathrm{MS}_\omega = t\sigma_\omega^2 + \sigma_\tau^2 + \sigma_\varepsilon^2$$

$$= t \sum_{i=1}^{w} \frac{\left(\bar{x}_i - \bar{x}\right)^2}{w - 1} \tag{12.13}$$

while the mean square for time is

$$\mathrm{MS}_\tau = \sigma_\tau^2 + \sigma_\varepsilon^2$$

$$= \sum_{i=1}^{w} \sum_{j=1}^{t} \frac{\left(x_{ij} - \bar{x}_i\right)^2}{w(t - 1)} \tag{12.14}$$

With time nested within wells the error term σ_ε^2 cannot be separately estimated since there is only one observation per well $\times$ time cell. Thus, solving for the individual variance components σ_ω^2 and σ_τ^2, we find that

$$\sigma_\omega^2 = \frac{\mathrm{MS}_\omega - \mathrm{MS}_\tau}{t} \tag{12.15}$$

and

$$\sigma_\tau^2 + \sigma_\varepsilon^2 = \mathrm{MS}_\tau \tag{12.16}$$

The unbiased estimate of the total variance is now

$$\mathrm{var}(x_{ij}) = \sigma_\omega^2 + \sigma_\tau^2 + \sigma_\varepsilon^2$$

$$= \frac{\mathrm{MS}_\omega - \mathrm{MS}_\tau}{t} + \mathrm{MS}_\tau$$

$$= \frac{t\sum_{i=1}^{w}\left(\bar{x}_i - \bar{x}\right)^2/(w - 1) - \sum_{i=1}^{w}\sum_{j=1}^{t}\left(x_{ij} - \bar{x}_i\right)^2/(w(t - 1))}{t}$$

$$+ \sum_{i=1}^{w} \sum_{j=1}^{t} \frac{\left(x_{ij} - \bar{x}_i\right)^2}{w(t - 1)} \tag{12.17}$$

It is of considerable interest to compare the biased variance estimate s^2 with the unbiased variance estimate $\mathrm{var}(x_{ij})$ given previously. The simple variance estimate that assumes equal spatial and temporal variability has expectation

$$E[\sigma^2] = E\left[\sum_{i=1}^{w}\sum_{j=1}^{t}\frac{(x_{ij}-\bar{x}_i)^2}{n-1}\right]$$

$$= \frac{n-t}{n}\sigma_\omega^2 + \sigma_\tau^2 + \sigma_\varepsilon^2 \qquad (12.18)$$

where n is the total number of measurements. This equation reveals that s^2 will always underestimate $var(x_{ij})$. For example, with two wells and four quarterly measurements, the contribution of the well variance σ_ω^2 is underestimated by 43%. The addition of two upgradient wells decreases this bias to 20%, and the addition of a second year's background increases the bias only slightly to 23%.

When multiple upgradient wells comprise the background sample, the unbiased estimate $var(x_{ij})$ should *always* be used in place of s^2. The importance of using multiple upgradient wells is due both to the large spatial variability that is commonly observed for naturally occurring constituents of groundwater and because it eliminates the confound between spatial variability and contamination. In the case of a single upgradient well, an upgradient versus downgradient difference may reflect either spatial variability or contamination (i.e., there is no estimate of σ_ω^2). In contrast, when multiple upgradient wells are sampled, an estimate of spatial variability σ_ω^2 is available, and the resulting prediction limit will provide for spatial effects. A critical assumption, however, is that both spatial variability and temporal variability are the same for upgradient and downgradient locations. This is a complex problem and the feasibility of this assumption should always be evaluated on a site-specific basis to bring into consideration hydrogeology as well as problems with previous conditions and sources of off-site contamination.

Example 12.1

Consider the following data for pH obtained from two upgradient wells each with four quarterly measurements:

Quarter	Upgradient Well 1	Upgradient Well 2
1	7.1	7.7
2	7.4	7.8
3	7.7	7.9
4	8.0	8.0

The summary statistics for the nested design are

$$\bar{x} = (7.1 + 7.4 + 7.7 + 8.0 + 7.7 + 7.8 + 7.9 + 8.0)/8 = 7.70$$

$$\bar{x}_{well\ 1} = (7.1 + 7.4 + 7.7 + 8.0)/4 = 7.55$$

$$\bar{x}_{well\ 2} = (7.7 + 7.8 + 7.9 - 8.0)/4 = 7.85$$

$$MS_\omega = 4\left[(7.55 - 7.70)^2 + (7.85 - 7.70)^2\right]/(2 - 1) = 0.180$$

$$MS_\tau = \left[(7.1 - 7.55)^2 + (7.4 - 7.55)^2 + (7.7 - 7.55)^2 + (8.0 - 7.55)^2\right.$$
$$+ (7.7 - 7.85)^2 + (7.8 - 7.85)^2 + (7.9 - 7.85)^2$$
$$\left. + (8.0 - 7.85)^2\right]/(2(4 - 1))$$
$$= 0.083$$

The resulting variance estimates are therefore

$$\sigma_\omega^2 = \frac{MS_\omega - MS_\tau}{t}$$

$$= \frac{.180 - .083}{4} = 0.024$$

$$\sigma_\tau^2 = MS_\tau = 0.083$$

$$var(x_{ij}) = 0.108$$

As expected, the biased variance estimate $s^2 = .097$ underestimates $var(x_{ij})$ since, in this example, $s^2 = (4/7)\sigma_\omega^2 + \sigma_\tau^2 + \sigma_\varepsilon^2$. The variance estimate for the randomized block design $var(x_{ij}) = .120$ is somewhat larger than the result obtained for the nested model, because a unique estimate of the residual error is available ($\sigma_\omega^2 = .037$, $\sigma_\tau^2 = .050$, and $\sigma_\varepsilon^2 = .033$).

12.3   MAXIMUM LIKELIHOOD ESTIMATORS

Recent statistical research in the area of variance component models has been in the area of maximum likelihood/empirical Bayesian methods (Laird and Ware, 1982; Searle, 1987; Aitkin and Longford, 1986; Longford, 1987; Donner, 1985; Gibbons et al., 1988b; Hedeker et al., 1989; Goldstein, 1987). This work has proceeded along parallel lines for the analysis of longitudinal and clustered data. In the context of groundwater monitoring data, we have both a longitudinal component (i.e., measurements repeatedly made over time within the same well) and a clustered component (multiple monitoring wells within which temporal measurements are clustered). When the numbers of measurements are unequal or missing, the traditional least squares estimators described previously are no longer valid. The maximum likelihood estimators, however, have no such limitation. A complete description of these methods is beyond the scope of this book but would include discussion of autocorrelated errors of measurement (Chi and Reinsel, 1989) and three-level models in which both longitudinal and clustered components of the data are jointly modeled (Hedeker et al., 1991). In the following discussion maximum likelihood estimators for the interwell and residual components of variability are examined.

In the present context, consider measurements clustered within monitoring wells. Our interest is in modeling the measurement x_{ij} in well i on occasion j. Consider the following model for measurement x_{ij} (where $j = 1, 2, \ldots, n_i$ occasions) within well i ($i = 1, \ldots, N$ wells in the sample):

$$x_{ij} = \mu + \omega_i + \varepsilon_{ij} \tag{12.19}$$

where

x_{ij} = the measurement for sampling event j in well i

μ = the overall mean concentration

ω_i = the effect due to well i

ε_{ij} = an independent residual distributed normally, $N(0, \sigma_\varepsilon^2)$

As sampled wells are thought to be representative of a larger population of possible wells, ω_i is considered a random parameter and the assumed distribution for ω is a normal distribution with mean 0 and variance σ_ω^2, that is, $N(0, \sigma_\omega^2)$. To the degree that clustering within wells has little effect on concentration data, estimates of ω_i will all be near 0, and the estimate of well variance (σ_ω^2) will approach 0. If, on the other hand, clustering within wells has a strong effect on concentration levels, estimates of ω_i will deviate from 0 and differ for each well i, and thus well variance (σ_ω^2) will increase in value.

The x_{ij} are distributed as independent normals with mean μ and variance $\sigma_\omega^2 + \sigma_\varepsilon^2$ (by "independent" it is meant that x_{ij} is independent of $x_{i'j}$, where $i \neq i'$). To the extent that wells are placed close together and there is spatial correlation, this assumption may be violated. The variance–covariance structure of the measurements assumed by the model, however, explicitly takes into account the clustering of measurements within wells and is the same form as that corresponding to the previously described randomized block analysis of variance model (also see Kirk, 1982, pages 253–257). As such, the interdependency of the measurements within a well is expressed by the variance–covariance structure. Specifically, for a given well i, the variance associated with each of the $j = 1, \ldots, n_i$ measurements is assumed to be the same with form $\sigma_\omega^2 + \sigma_\varepsilon^2$, while the covariance of any two measurements within a well is assumed to be σ_ω^2. This structure is sometimes referred to as the "compound symmetry" form, since, within a well, variances are assumed to be homogeneous ($\sigma_\omega^2 + \sigma_\varepsilon^2$), as are all covariances (σ_ω^2). The ratio of the well, or cluster, variance σ_ω^2 to the total variance $\sigma_\omega^2 + \sigma_\varepsilon^2$ is the intraclass correlation, which indicates the proportion of variance in the data attributable to the well, or cluster.

In terms of parameter estimation, a combination of two complementary methods has been proposed (Laird and Ware, 1982; Bock, 1989). For estimation of well effects ω_i, empirical Bayes (EB) methods have been recommended, while maximum marginal likelihood (MML) methods are recommended for estimation of variance parameters, σ_ε^2 and σ_ω^2 and the overall concentration mean μ.

Empirical Bayes estimates of well effects are sometimes termed EAP (expected a posteriori) estimates, since they are derived as the mean of the posterior distribution of ω, given the x_{ij}. Denoting the EAP estimate of ω_i as $\tilde{\omega}_i$ to distinguish it from subsequent MML estimates, and given the preceding assumptions, we get the following EAP estimator of the well effect:

$$\tilde{\omega}_i = \rho_{n_i n_i} \frac{1}{n_i} \sum_{j=1}^{n_i} (x_{ij} - \mu) \qquad (12.20)$$

where

$$\rho_{nn} = \frac{nr}{1 + (n-1)r} \qquad (12.21)$$

and r is the intraclass correlation

$$r = \frac{\sigma_\omega^2}{\sigma_\varepsilon^2 + \sigma_\omega^2} \qquad (12.22)$$

A property of EB estimation is that $\tilde{\omega}_i$ is a function of both the actual data and the empirical prior distribution specified for ω_i. As information about a well increases (i.e., the reliability $\rho_{n_i n_i}$ increases toward 1), by either increasing data interdependency within the well (increasing r) and/or increasing well sample size (n_i), the well estimate approaches the average measurement deviation for that well, $\sum_{j=1}^{n_i}(x_{ij} - \mu)/n_i$. Alternatively, as information about a well decreases (i.e., $\rho_{n_i n_i}$ decreases toward 0), by either decreasing data interdependency within the well and/or decreasing well sample size, the well estimate approaches the posited mean of the empirical prior distribution of ω_i, namely 0.

In addition to the EB estimate of the posterior mean, the variance of the posterior distribution of ω is given as

$$\sigma_{\omega|\mathbf{x}_i}^2 = \sigma_\omega^2 \left(1 - \rho_{n_i n_i}\right) \qquad (12.23)$$

Again, the form reveals the nature of this EB estimator of the posterior variance: As information about the well increases, the posterior variance becomes a fraction of the empirical prior variance (σ_ω^2), while as information about the well decreases, this variance approaches the empirical prior variance.

To estimate the overall concentration mean μ and the variance parameters σ_ω^2 and σ_ε^2, MML estimation is recommended. The MML estimation procedure can be implemented using two numerical algorithms: the EM algorithm solution (Dempster et al., 1981) and the Fisher scoring solution (Longford, 1987). As an illustration, the following equations are used in the iterative EM algorithm solution:

$$\hat{\mu} = \frac{1}{N} \sum_{i=1}^{N} \sum_{j=1}^{n_i} \left(x_{ij} - \tilde{\omega}_i \right) \tag{12.24}$$

$$\hat{\sigma}_\omega^2 = \frac{1}{N} \sum_{i}^{N} \tilde{\omega}_i^2 + \sigma_{\omega|x_i}^2 \tag{12.25}$$

$$\hat{\sigma}_\varepsilon^2 = \frac{1}{N} \sum_{i}^{N} \sum_{j}^{n_i} \left(x_{ij} - \mu - \tilde{\omega}_i \right)^2 + n_i \hat{\sigma}_{\omega|x_i}^2 \tag{12.26}$$

where x_i denotes the vector of n_i measurements from well i. The solution is obtained by iterating between EB equations (12.20) and (12.21) and MML equations (12.24) to (12.26) until convergence.

Example 12.2

To compare the two approaches to variance component estimators, the data in Example 12.1 were used to obtain the marginal maximum likelihood estimates of σ_ω^2 and σ_ε^2. The computer program MIXREG (Hedeker, 1993) was used, which is distributed by the National Institute of Health Division of Services Research. Results of the analysis yielded $\sigma_\omega^2 = .002$ and $\sigma_\varepsilon^2 = .083$. Note that in this model the variance component due to time (σ_τ^2) is confounded with the residual variance σ_ε^2; therefore $\sigma^2 = \sigma_\varepsilon^2 + \sigma_\tau^2$. The marginal maximum likelihood estimate of σ_ε^2 is identical to σ_τ^2 from the least squares analysis, but σ_ω^2 is smaller than the corresponding least squares estimate. This type of downward bias is characteristic of empirical Bayes estimators, and therefore they should only be used in cases where the least squares estimators are inappropriate (e.g., unequal numbers of measurements within wells). Note that use of the maximum likelihood/empirical Bayesian estimator in a sample of this small size is also ill advised and is done here simply as a comparative illustration.

12.4 SUMMARY

The variance component model attempts to partition the overall variability into component parts within which the assumption of independence is viable. In groundwater monitoring applications the two potential sources are temporal variability and spatial variability. In general, the spatial variance component can be estimated, but the temporal variance component requires several years of background measurements. In this chapter both traditional least squares and maximum likelihood/empirical Bayes estimators have been proposed. When similar numbers of measurements are available for all wells, the least squares approach is the more robust and easier to calculate; however, when the data are unbalanced (i.e., varying numbers of measurements per well), the maximum likelihood/empirical Bayes approach may be the only viable alternative. The latter approach can be greatly expanded to include the effects of covariates (e.g., water level, temperature, etc.) and autocorrelation (i.e., short-run association between measurement occasions that are close in time). Further research in this emerging area is strongly encouraged.

While extremely useful, the variance component approach is not without substantial limitations. In general, these methods assume normality, and it is unclear how censored data could be incorporated. Ignoring the variance components, however, will lead to underestimates of the true sampling variability and will produce groundwater detection monitoring programs with false positive rates larger than anticipated.

13 Detecting Outliers

13.1 OVERVIEW

It is common to find outliers in environmental data. Outliers, those values that do not conform to the pattern established by other observations (see Hunt, et al., 1981; Gilbert, 1987), can arise from errors in transcription, data-coding, analytical instrument failure and calibration errors, or underestimation of inherent spatial or temporal variability in constituent concentrations. Despite the cause, including outliers to establish statistical limit estimates will often lead to biased and misleading results. As an example, consider the TOC data in Table 13.1 from a single upgradient well.

The mean concentration is 4.13 mg/l and the standard deviation is 2.03 mg/l. A 99% normal upper prediction limit for the next single measurement from that population is

$$\bar{x} + t_{7, .01} s \left(1 + \frac{1}{n} \right)^{1/2} = 10.59 \text{ mg/l}$$

Now assume that the 1992 fourth-quarter observation was incorrectly recorded as 50 mg/l. In this case the mean is 9.75 mg/l, the standard deviation is 16.39 mg/l, and the 99% normal upper prediction limit is 61.87 mg/l. Inclusion of the outlier will have profound effects on the false negative rate of the statistical test. Including the outlier will cause us to accept potentially contaminated downgradient measurements as consistent with upgradient water quality levels. This is also true for low outliers since the increase in s increases the prediction limit $t\sqrt{1 + 1/n}$ times more than the decrease in $\bar{x}$.

In practice, however, we rarely know the source of an outlying observation. Is the TOC value of 50 mg/l really due to error or is it simply natural variability that one might observe in a downgradient well? Deleting the observation from the upgradient or background database may cause us to incorrectly reject a new downgradient measurement at that level (i.e., false positive). Should we remove the outlier or not? Fortunately, there is a simple solution to this problem which is another virtue of verification resampling. By the very definition of an outlier, its frequency must be low. The probability of observing a rare event twice in a row in a downgradient well, even if it is real and not an error, is remote. Note that no verification resampling is or should be performed on the upgradient or background data; therefore excluding outliers is good practice as long as new downgradient measurements that exceed background limits can be verified on or between the next scheduled

TABLE 13.1 Hypothetical TOC Data in mg/l

Year	Quarter	TOC in MW01
1991	1	1
	2	7
	3	3
	4	6
1992	1	5
	2	4
	3	2
	4	5

sampling event. In the following sections several methods for detecting outliers in background data are described. There are numerous approaches to this problem and the interested reader is referred to Beckman and Cook (1983) and the book by Barnett and Lewis (1984) for excellent reviews. Since there is no reason to assume a specific number of outliers (e.g., one or two) or direction (i.e., outliers can be either too small or too large), attention is focused on two-tailed outlier detection procedures for up to m outliers.

13.2 ROSNER'S TEST

Gilbert (1987) suggests using Rosner's test (Rosner, 1983) with environmental data. Rosner's test is a generalization of the extreme Studentized deviate (ESD) originally suggested by Grubbs (1969) and tabulated by Grubbs and Beck (1972) for testing a single outlier. Rosner's generalization is for multiple outliers (i.e., $m = 2$ to 10). To use Rosner's test, an upper limit m must be specified on the number of potential outliers present. The test is valid for samples of size 25 or more; therefore, in general, it can only be applied to a pooled background consisting of multiple upgradient wells. The danger here is that spatial variability (i.e., variability from well to well) will either mask or produce outliers depending on the number of measurements available for each well.

Let $\bar{x}^{(i)}$ and $s^{(i)}$ be the mean and standard deviation of the $m - i$ measurements that remain after the i most extreme observations have been deleted, that is,

$$\bar{x}^{(i)} = \frac{1}{n-i} \sum_{j=1}^{n-i} x_j \tag{13.1}$$

$$s^{(i)} = \left[\frac{1}{n-i} \sum_{j=1}^{n-i} \left(x_j - \bar{x}^{(i)} \right)^2 \right]^{1/2} \tag{13.2}$$

Let $x^{(i)}$ denote the furthest remaining observation from the mean $\bar{x}^{(i)}$. After

TABLE 13.2 Critical Values for Rosner's Test Statistic

u	$i+1$	0.05	0.01	u	$i+1$	0.05	0.01
25	1	2.82	3.14	32	1	2.94	3.27
	2	2.80	3.11		2	2.92	3.25
	3	2.78	3.09		3	2.91	3.24
	4	2.76	3.06		4	2.89	3.22
	5	2.73	3.03		5	2.88	3.20
	10	2.59	2.85		10	2.78	3.09
26	1	2.84	3.16	33	1	2.95	3.29
	2	2.82	3.14		2	2.94	3.27
	3	2.80	3.11		3	2.92	3.25
	4	2.78	3.09		4	2.91	3.24
	5	2.76	3.06		5	2.89	3.22
	10	2.62	2.89		10	2.80	3.11
27	1	2.86	3.18	34	1	2.97	3.30
	2	2.84	3.16		2	2.95	3.29
	3	2.82	3.14		3	2.94	3.27
	4	2.80	3.11		4	2.92	3.25
	5	2.78	3.09		5	2.91	3.24
	10	2.65	2.93		10	2.82	3.14
28	1	2.88	3.20	35	1	2.98	3.32
	2	2.86	3.18		2	2.97	3.30
	3	2.84	3.16		3	2.95	3.29
	4	2.82	3.14		4	2.94	3.27
	5	2.80	3.11		5	2.92	3.25
	10	2.68	2.97		10	2.84	3.16
29	1	2.89	3.22	36	1	2.99	3.33
	2	2.88	3.20		2	2.98	3.32
	3	2.86	3.18		3	2.97	3.30
	4	2.84	3.16		4	2.95	3.29
	5	2.82	3.14		5	2.94	3.27
	10	2.71	3.00		10	2.86	3.18
30	1	2.91	3.24	37	1	3.00	3.34
	2	2.89	3.22		2	2.99	3.33
	3	2.88	3.20		3	2.98	3.32
	4	2.86	3.18		4	2.97	3.30
	5	2.84	3.16		5	2.95	3.29
	10	2.73	3.03		10	2.88	3.20
31	1	2.92	3.25	38	1	3.01	3.36
	2	2.91	3.24		2	3.00	3.34
	3	2.89	3.22		3	2.99	3.33
	4	2.88	3.20		4	2.98	3.32
	5	2.86	3.18		4	2.97	3.30
	10	2.76	3.06		10	2.89	3.22

TABLE 13.2   (*Continued*)

u	$i + 1$	0.05	0.01	u	$i + 1$	0.05	0.01
39	1	3.03	3.37	46	1	3.09	3.45
	2	3.01	3.36		2	3.09	3.44
	3	3.00	3.34		3	3.08	3.43
	4	2.99	3.33		4	3.07	3.41
	5	2.98	3.32		5	3.06	3.40
	10	2.91	3.24		10	3.00	3.34
40	1	3.04	3.38	47	1	3.10	3.46
	2	3.03	3.37		2	3.09	3.45
	3	3.01	3.36		3	3.09	3.44
	4	3.00	3.34		4	3.08	3.43
	5	2.99	3.33		5	3.07	3.41
	10	2.92	3.25		10	3.01	3.36
41	1	3.05	3.39	48	1	3.11	3.46
	2	3.04	3.38		2	3.10	3.46
	3	3.03	3.37		3	3.09	3.45
	4	3.01	3.36		4	3.09	3.44
	5	3.00	3.34		5	3.08	3.43
	10	2.94	3.27		10	3.03	3.37
42	1	3.06	3.40	49	1	3.12	3.47
	2	3.05	3.39		2	3.11	3.46
	3	3.04	3.38		3	3.10	3.46
	4	3.03	3.37		4	3.09	3.45
	5	3.01	3.36		5	3.09	3.44
	10	2.95	3.29		10	3.04	3.38
43	1	3.07	3.41	50	1	3.13	3.48
	2	3.06	3.40		2	3.12	3.47
	3	3.05	3.39		3	3.11	3.46
	4	3.04	3.38		4	3.10	3.46
	5	3.03	3.37		5	3.09	3.45
	10	2.97	3.30		10	3.05	3.39
44	1	3.08	3.43	60	1	3.20	3.56
	2	3.07	3.41		2	3.19	3.55
	3	3.06	3.40		3	3.19	3.55
	4	3.05	3.39		4	3.18	3.54
	5	3.04	3.38		5	3.17	3.53
	10	2.98	3.32		10	3.14	3.49
45	1	3.09	3.44	70	1	3.26	3.62
	2	3.08	3.43		2	3.25	3.62
	3	3.07	3.41		3	3.25	3.61
	4	3.06	3.40		4	3.24	3.60
	5	3.05	3.39		5	3.24	3.60
	10	2.99	3.33		10	3.21	3.57

TABLE 13.2   (*Continued*)

u	$i + 1$	0.05	0.01	u	$i + 1$	0.05	0.01
80	1	3.31	3.67	250	1	3.67	4.04
	2	3.30	3.67		5	3.67	4.04
	3	3.30	3.66		10	3.66	4.03
	4	3.29	3.66				
	5	3.29	3.65	300	1	3.72	4.09
	10	3.26	3.63		5	3.72	4.09
					10	3.71	4.09
90	1	3.35	3.72				
	2	3.34	3.71	350	1	3.77	4.14
	3	3.34	3.71		5	3.76	4.13
	4	3.34	3.70		10	3.76	4.13
	5	3.33	3.70				
	10	3.31	3.68	400	1	3.80	4.17
					5	3.80	4.17
100	1	3.38	3.75		10	3.80	4.16
	2	3.38	3.75				
	3	3.38	3.75	450	1	3.84	4.20
	4	3.37	3.74		5	3.83	4.20
	5	3.37	3.74		10	3.83	4.20
	10	3.35	3.72				
				500	1	3.86	4.23
150	1	3.52	3.89		5	3.86	4.23
	2	3.51	3.89		10	3.86	4.22
	3	3.51	3.89				
	4	3.51	3.88	750	1–10	3.95	4.30
	5	3.51	3.88				
	10	3.50	3.87	1000	1–10	4.02	4.37
200	1	3.61	3.98	2000	1–10	4.20	4.54
	2	3.60	3.98				
	3	3.60	3.97	3000	1–10	4.29	4.63
	4	3.60	3.97				
	5	3.60	3.97	4000	1–10	4.36	4.70
	10	3.59	3.96	5000	1–10	4.41	4.75

Reprinted with permission from R. O. Gilbert, *Statistical Methods for Environmental Pollution Monitoring*, Van Nostrand Reinhold (1987):268–270.

i more extreme values (large or small) have been detected,

$$R_{i+1} = \frac{|x^{(i)} - \bar{x}^{(i)}|}{s^{(i)}} \tag{13.3}$$

is a test statistic for deciding whether the $i + 1$ most extreme values in the complete dataset are outliers from a normal distribution. Critical values for this statistic are provided in Table 13.2.

TABLE 13.3   Critical Values for Extreme Studentized Deviate

Size of Sample	Upper 5% Level	Upper 1% Level
5	.858	.882
6	.844	.882
7	.825	.873
8	.804	.860
9	.783	.844
10	.763	.827
11	.745	.811
12	.727	.795
13	.711	.779
14	.695	.764
15	.681	.750
16	.668	.737
17	.655	.724
18	.643	.711
19	.632	.700
20	.621	.688
22	.602	.668
24	.584	.649
26	.568	.632
28	.554	.616
30	.540	.601

Gilbert (1987) suggests that a test for outliers from a lognormal distribution can be constructed by simply log-transforming the original data (e.g., $y_i = \log_e[x_i]$). FORTRAN IV source code that iteratively rejects up to m outliers is given by Rosner (1983) and reproduced by Gilbert (1987). Illustrative examples are given by Gilbert (1987).

For samples of less than 25, critical values for the ESD test for a single outlier can be used (see Table 13.3). When testing outliers, a conservative approximation can be obtained by assuming independence and setting $\alpha^* = \alpha/m$ and interpolating to find α^* in Table 13.3.

13.3   SKEWNESS TEST

The skewness test can be used to test discordance for one or more upper or lower outliers in a normal sample with μ and σ^2 unknown. The sample skewness statistic is

$$\left[\frac{\sum_{i=1}^{n}(x_i - \bar{x})^3}{ns^3} \right]^{1/2} \tag{13.4}$$

To use the test consecutively, observations are ordered from lowest to highest and the value of the statistic is compared to the critical value in Table 11.6. If the sample skewness statistic exceeds the critical value, the lowest or highest value is rejected depending on the sign of the statistic (note that the square root must be taken on the absolute value of the statistic). The procedure is repeated until the sample skewness statistic does not exceed the critical value. Note that this test is essentially a data screening procedure that ends when the data are consistent with a normal distribution. If the data are not normally distributed, it is possible to falsely reject values. Comparison of results for raw and log-transformed values is strongly recommended.

13.4   KURTOSIS TEST

An analogous test to the skewness test is based on the sample kurtosis statistic

$$\frac{\sum_{i=1}^{n}(x_i - \bar{x})^4}{ns^4} \tag{13.5}$$

Discordancy is indicated by high values of the statistic. The smallest or largest value is rejected based on whichever is further from $\bar{x}$. Critical values for this statistic are provided in Table 11.7. As in the skewness test, the kurtosis test can be repeatedly applied until the sample value of the statistic is less than the critical value.

13.5   SHAPIRO–WILK TEST

As in the two prior outlier tests, the Shapiro–Wilk test of normality (W-statistic)

$$W = \frac{\left(\sum_{i=1}^{n/2}a_{(n,\,n-i+1)}[x_{(n-i+1)} - x_{(i)}]\right)^2}{s^2} \tag{13.6}$$

can be used to consecutively reject outliers until the fit of a normal distribution can no longer be rejected. Critical values of W are given in Table 11.2 and the coefficients (a) are given in Table 11.3. Note that for the skewness test, kurtosis test, and Shapiro–Wilk test, critical values underestimate their intended nominal levels as the number of repeated evaluations increases. For example, with five outlier tests, the 1% critical value of these tests approximates the 5% critical value. It is only approximate because the repeated tests are not independent.

TABLE 13.4 **Critical Values for** E_m **Test Statistic**

n	1	2	3	4	5	6	7	8	9	10
3	.001	—	—	—	—	—	—	—	—	—
4	.025	.001	—	—	—	—	—	—	—	—
5	.081	.010	—	—	—	—	—	—	—	—
6	.146	.034	.004	—	—	—	—	—	—	—
7	.208	.065	.016	—	—	—	—	—	—	—
8	.265	.099	.034	.010	—	—	—	—	—	—
9	.314	.137	.057	.021	—	—	—	—	—	—
10	.356	.172	.083	.037	.014	—	—	—	—	—
11	.386	.204	.107	.055	.026	—	—	—	—	—
12	.424	.234	.133	.073	.039	.018	—	—	—	—
13	.455	.262	.156	.092	.053	.028	—	—	—	—
14	.484	.293	.179	.112	.068	.039	.021	—	—	—
15	.509	.317	.206	.134	.084	.052	.030	—	—	—
16	.526	.340	.227	.153	.102	.067	.041	.024	—	—
17	.544	.362	.248	.170	.116	.078	.050	.032	—	—
18	.562	.382	.267	.187	.132	.091	.062	.041	.026	—
19	.581	.398	.287	.203	.146	.105	.074	.050	.033	—
20	.597	.416	.302	.221	.163	.119	.085	.059	.041	.028
25	.652	.493	.381	.298	.236	.186	.146	.114	.089	.068
30	.698	.549	.443	.364	.298	.246	.203	.166	.137	.112
35	.732	.596	.495	.417	.351	.298	.254	.214	.181	.154
40	.758	.629	.534	.458	.395	.343	.297	.259	.223	.195
45	.778	.658	.567	.492	.433	.381	.337	.299	.263	.233
50	.797	.684	.599	.529	.468	.417	.373	.334	.299	.268

n	1	2	3	4	5	6	7	8	9	10
3	.000	—	—	—	—	—	—	—	—	—
4	.004	.000	—	—	—	—	—	—	—	—
5	.029	.002	—	—	—	—	—	—	—	—
6	.068	.012	.001	—	—	—	—	—	—	—
7	.110	.028	.006	—	—	—	—	—	—	—
8	.156	.050	.014	.004	—	—	—	—	—	—
9	.197	.078	.026	.009	—	—	—	—	—	—
10	.235	.101	.018	.006	—	—	—	—	—	—
11	.274	.134	.064	.030	.012	—	—	—	—	—
12	.311	.159	.083	.042	.020	.008	—	—	—	—
13	.337	.181	.103	.056	.031	.014	—	—	—	—
14	.374	.207	.123	.072	.042	.022	.012	—	—	—
15	.404	.238	.146	.090	.054	.032	.018	—	—	—
16	.422	.263	.166	.107	.068	.040	.024	.014	—	—
17	.440	.290	.188	.122	.079	.052	.032	.018	—	—
18	.459	.306	.206	.141	.094	.062	.041	.026	.014	—
19	.484	.323	.219	.156	.108	.074	.050	.032	.020	—
20	.499	.339	.236	.170	.121	.086	.058	.040	.026	.017
25	.571	.438	.320	.245	.188	.146	.110	.087	.066	.050
30	.624	.482	.386	.308	.250	.204	.166	.132	.108	.087
35	.669	.533	.435	.364	.299	.252	.211	.177	.149	.124
40	.704	.574	.480	.408	.347	.298	.258	.220	.190	.164
45	.728	.607	.518	.446	.386	.336	.294	.258	.228	.200
50	.748	.636	.550	.482	.424	.376	.334	.297	.264	.235

Reprinted with permission from Tietjen and Moore, *Technometrics*, **14**(1972). ©1972 by the American Statistical Association and the American Society for Quality Control.

13.6 E_m-STATISTIC

Tietjen and Moore (1972) developed a two-sided test for m outliers in a normal sample with μ and σ^2 unknown. Their statistic is

$$E_m = \frac{\sum_{i=1}^{n-m}\left(r_{(i)} - \bar{r}_{n-m}\right)^2}{\sum_{i=1}^{n}\left(r_{(i)} - \bar{r}\right)^2} \tag{13.7}$$

where $r_i = |x_i - \bar{x}|$ is the absolute deviation of x_i from the sample mean, $\{r_{(i)}\}$ are the values of r_i in ascending order, (i.e., $r_{(1)} < r_{(2)} \cdots < r_{(n)}$), $\bar{r}$ is the mean of all of the r_i, and $\bar{r}_{n-m}$ is the mean of the $(n-m)$ lowest r_i. Critical values for the statistic are provided in Table 13.4. Since E_m is a simultaneous test for m outliers, it is applied only once. Note that if the choice of m is too large (e.g., $m = 3$, when only $m = 2$ true outliers are present), then the third observation will often be rejected as well, or even worse, the third may mask the first two outliers.

13.7 DIXON'S TEST

Although not specifically designed for a general number of possible outliers, Dixon's test can be used for applications in which a small number of outliers are suspected. Arranging the sample in ascending order, the best test criterion for varying sample sizes is provided by Dixon (1953) as

n	Highest Value	Lowest Value
3–7	$\dfrac{x_n - x_{n-1}}{x_n - x_1}$	$\dfrac{x_2 - x_1}{x_n - x_1}$
8–10	$\dfrac{x_n - x_{n-1}}{x_n - x_2}$	$\dfrac{x_2 - x_1}{x_{n-1} - x_1}$
11–13	$\dfrac{x_n - x_{n-2}}{x_n - x_2}$	$\dfrac{x_3 - x_1}{x_{n-1} - x_1}$
14–25	$\dfrac{x_n - x_{n-2}}{x_n - x_3}$	$\dfrac{x_3 - x_1}{x_{n-2} - x_1}$

Critical values for Dixon's statistic are provided in Table 13.5. The test may be consecutively repeated by first testing the most extreme value and then testing the next most extreme value in a sample of size $(n - 1)$, found by omitting the most extreme value. If m outliers are suspected, all m tests must be performed no matter what the verdict of the first $m - 1$ test, since the former may mask the latter, particularly if they are on the same side. If the mth test exceeds the critical value, all m outliers must be rejected. Note

TABLE 13.5   Critical Values for Dixon's Statistic

n	5% Level	1% Level	n	5% Level	1% Level
3	.941	.988	14	.546	.641
4	.765	.889	15	.525	.616
5	.642	.780	16	.507	.595
6	.560	.698	17	.490	.577
7	.507	.637	18	.475	.561
8	.554	.683	19	.462	.547
9	.512	.635	20	.450	.535
10	.477	.597	21	.440	.524
11	.576	.679	23	.421	.505
12	.546	.642	24	.413	.497
13	.521	.615	25	.406	.489

Reproduced from W. J. Dixon, "Processing Data for Outliers," *Biometrics*, **9**(1953):74–89. With permission from the Biometric Society.

that the probability values in Table 13.5 are for a single test, and repeated application of the test to multiple suspected outliers increases the probability of a false rejection.

Example 13.1

Consider the 12 TOC measurements (see Table 13.6) obtained from a single background well over a period of 3 years. The values on the first quarter of 1990 and second quarter of 1991 appear to be possible outliers for this well and constituent.

TABLE 13.6   Hypothetical TOC Data in mg/l

Year	Quarter	TOC in MW01	$\log_e(x_i) + 1$
1990	1	< 1	0.693
	2	21	3.091
	3	22	3.135
	4	18	2.944
1991	1	19	2.996
	2	40	3.714
	3	21	3.091
	4	25	3.258
1992	1	17	2.890
	2	18	2.944
	3	19	2.996
	4	22	3.135

TABLE 13.7   Raw TOC Data Sorted by $|x - \bar{x}|$ and Associated Test Statistics for $m = 2$ Outliers

Raw Data	Sorted	ESD	Skew	Kurt	SWT	E_m	Dixon
< 1	40	3.456**	0.287	4.413*	0.819*	0.542	0.783**
21	< 1	7.868**	− 1.358**	5.719**	0.706**	0.026**	0.708**
22	25						
18	17						
19	18						
40	18						
21	22						
25	22						
17	19						
18	19						
19	21						
22	21						

$^{*}p < .05.$
$^{**}p < .01.$

Table 13.7 lists data sorted by maximum deviation from the mean and test statistics for tests of $m = 2$ outliers. Table 13.8 lists the same results for $\log_e(x)$.

In general, each test rejected the second largest deviation for both raw and log-transformed data. Each method would have resulted in rejection of the two suspected outliers. Some variation in result was found for the most extreme deviation, particularly for raw data. For the skewness test and

TABLE 13.8 $\log_e(x) + 1$ Transformed Data Sorted by $|x - \bar{x}|$ and Associated Test Statistics for $m = 2$ Outliers

$\log_e(x) = y$	Sorted	ESD	Skew	Kurt	SWT	E_m	Dixon
0.693	0.693	10.619**	− 1.481**	7.175**	0.592**	0.120**	0.878**
3.091	3.714	5.888**	− 1.251**	4.701**	0.779**	0.025**	0.702**
3.135	3.258						
2.944	3.135						
2.996	3.135						
3.714	3.091						
3.091	3.091						
3.258	2.996						
2.890	2.996						
2.944	2.944						
2.996	2.944						
3.135	2.890						

$^{*}p < .05.$
$^{**}p < .01.$

E_m-statistic, the most extreme value was masked by the second most extreme value. For the kurtosis test and Shapiro–Wilk test, the most extreme value was significant at the 5% level, but the second most extreme value was significant at the 1% level, suggesting small masking effect. The most consistent results in the example appeared to have been for Rosner's extension of the ESD test and for Dixon's test. Also note that the ESD test and the E_m-statistic are designed for simultaneous tests of all m outliers and should not be interpreted as sequential tests. Nonetheless, that the largest outlier was masked by the second largest for the E_m-statistic is of little consequence since with $m = 2$ we would have computed the second test only. However, if we had failed to recognize the second outlier and set $m = 1$, we would have obtained a misleading result.

Finally, it is of some interest to investigate what would have happened if we had set $m = 3$. The third most deviant value is 25 mg/l which is not particularly deviant. Nevertheless, this value would have been rejected by Rosner's test for $m = 3$ and by the E_m-statistic for $m = 3$. These results indicate that the two simultaneous tests are influenced by having a subset of the m outliers as true outliers. Setting m too large will result in false rejection of valid measurements for the simultaneous tests.

13.8   SUMMARY

Statistical approaches to outlier detection can be broadly classified into simultaneous and sequential methods. Simultaneous methods are attractive because they control the false positive rate for tests of all m suspected outliers; however, the presence of some outliers leads to rejection of all m suspected outliers. In contrast, sequential methods appear to work reasonably well despite the lack of explicitly controlled multiple comparisons through the adjustment of critical values. Skewness and kurtosis tests should probably be used jointly to eliminate potential masking produced by the presence of high and low outliers. The Shapiro–Wilk test is computationally complex in that separate sets of coefficients must be used at each step (i.e., for each suspected outlier). In addition, the test is also sensitive to nonnormality (i.e., the test may confuse nonnormality with the presence of outliers from a normal distribution). We are then left with the one test really not designed for testing multiple outliers, namely Dixon's test. As described here, the test can be applied easily to the the two highest and two lowest values in a sample, which may be sufficient for most practical applications where historical measurements from each well are examined separately.

14 Applications to Regulatory Issues

14.1 REGULATORY STATISTICS

Waste disposal practices in the United States have evolved significantly since the EPA promulgated the initial set of RCRA regulations in 1982 (EPA, 1982). These regulations established a "cradle-to-grave" system for managing the generation, transportation, treatment, storage, and disposal of hazardous wastes. More recently, RCRA regulations were promulgated to control the siting, construction, and operation of municipal solid waste landfills (EPA, 1991). These detailed technical regulations represent the public commitment to protect human health and the environment from waste handling practices, and to preclude the creation of additional Superfund sites. For a variety of reasons, groundwater is the principal medium of concern for which protection is sought. Consequently, a key component of the EPA regulations for Subtitle C and D landfills is the reliance on periodic detection monitoring of groundwater at these facilities, and the subsequent statistical analysis of these groundwater monitoring data, to determine whether a release has occurred. In the past 10 years, the EPA has issued a series of rules and guidelines containing statistical decision rules to assist in making objective determinations whether a release to groundwater has occurred (EPA, 1982, 1987a, 1988, 1989, 1991, 1992). Unfortunately, the implementation of these statistical rules has been highly varied and often poorly conceptualized. Many of these methods are described in the following section. Unfortunately, they are methods to be avoided.

14.2 METHODS TO BE AVOIDED

14.2.1 Analysis of Variance (ANOVA)

Application of ANOVA procedures to groundwater detection monitoring programs, both parametric and nonparametric, is inadvisable for the following reasons.

First, univariate ANOVA procedures do not adjust for multiple comparisons due to multiple constituents which can be devastating to the sitewide false positive rate. As such, a site with 10 indicator constituents will have a 40% chance of failing at least one on every monitoring event.

Second, ANOVA is more sensitive to spatial variability than contamination. Spatial variability affects mean concentrations but typically not the variance; hence small yet consistent differences will achieve statistical significance. In contrast, contamination affects both variability and mean concentration; therefore a much larger effect is required to achieve statistical significance. In fact, application of ANOVA methods to predisposal groundwater monitoring data can result in statistically significant differences between upgradient and downgradient wells, despite the fact that there is no waste in between. The reasons for this are: (1) The overall F-statistic tests the null hypothesis of no differences among any of the wells regardless of gradient (i.e., it will be significant if two downgradient wells are different); and (2) the distribution of the mean of four measurements (i.e., four measurements collected from the same well within a six-month period) is normal with mean μ and variance $\sigma^2/4$, whereas the distribution of each of the individual measurements is normal with mean μ and variance σ^2. This means that the standard deviation of the mean of four measurements is one-half the size of the standard deviation of the individual measurements themselves. As a result, small but consistent geochemical differences that are invariably observed naturally across a waste disposal facility will be attributed to contamination. To make matters worse, since there are far more downgradient than upgradient wells at these facilities, spatial variation has a far greater chance of occurrence downgradient than upgradient, further increasing the likelihood of falsely concluding that contamination is present. While spatial variation is also a problem for prediction limits and tolerance limits for single future measurements, it is not nearly as severe a problem as for ANOVA since the distribution of the individual measurement is considered and not the more restrictive distribution of the sample mean.

Third, nonparametric ANOVA is often presented by the EPA as if it protects the user from all of the weaknesses of its parametric counterpart. This is *not* the case. Both methods assume identical distributions for the analyte in *all* monitoring wells. The only difference is that the parametric ANOVA assumes that the distribution is normal and the nonparametric ANOVA is indifferent to what the distribution is. Both parametric and nonparametric ANOVA assume homogeneity of variance, a condition that almost never occurs in practice. This is not a weakness of methods for single future samples (i.e., prediction and tolerance limits) since the variance estimates rely solely on the background data. Why would anyone want to use downgradient data from an existing site (which could be affected by the site) to characterize natural variability? Yet this is exactly what ANOVA does. Furthermore, ANOVA is not a good statistical technique for detecting a narrow plume that might affect only 1 of 10 or 20 monitoring wells.

Fourth, ANOVA requires the pooling of downgradient data. Specifically, the EPA has suggested that four samples per semiannual monitoring event be collected (i.e., eight samples per year). As such, on average, it will never most rapidly detect a release, since only a subset of the required four semiannual samples will be affected by a site impact. This heterogeneity will decrease the mean concentration and dramatically increase the variance for the affected well thereby limiting the ability of the statistical test to detect contamination when it occurs. This is not true for tolerance limits, prediction limits, and control charts, which can and *should* be applied to individual measurements. The EPA may like ANOVA because it will appear to be more powerful than prediction and tolerance limits for single future values. The increased power, however, is only realized when all four measurements from a single well are equally affected by the site impact which, on average, will only occur 25% of the time (i.e., if four semiannual sampling events are evenly spaced, all four will be impacted by a new release only one in four times). For these reasons, when applied to groundwater detection monitoring, ANOVA will maximize both the false positive and false negative rates and double the cost of monitoring (i.e., ANOVA requires four samples per semiannual event or eight per year versus a maximum of four quarterly samples per year for prediction or tolerance limits that test each new individual measurement).

To illustrate, consider the data in Table 14.1 which were obtained from a facility in which no disposal of waste has yet occurred (Gibbons, 1994a).

The results of applying both parametric and nonparametric ANOVA to these predisposal data yielded an effect that approached significance for chemical oxygen demand (COD) ($p < .072$ parametric and $p < .066$ nonparametric) and a significant difference for alkalinity (ALK) ($p < .002$ parametric and $p < .009$ nonparametric). In terms of individual comparisons, significantly increased COD levels were found for well MW05 ($p < .026$) and significantly increased ALK was found for wells MW06 ($p < .026$) and P14 ($p < .003$) relative to upgradient wells. Of course, these results represent false positives due to spatial variability, since there is no garbage. What is perhaps most remarkable, however, is the absence of any significant results for TOC, where some of the values are as much as 20 times higher than the others. The reason, of course, is that these extreme values tremendously increase the within-well variance estimate, rendering the ANOVA powerless to detect any differences regardless of magnitude. This is yet another testimonial to why it is environmentally negligent to average measurements from downgradient monitoring wells, a problem that is inherent to ANOVA-type analyses when applied to dynamic groundwater quality measurements.

14.2.2   Risk-Based Compliance Determinations: Comparisons to ACLs and MCLs

Under EPA regulations different types of groundwater compliance criteria are applied to different types of facilities. For instance, consider a RCRA

TABLE 14.1   Raw Data for All Detection Monitoring Wells and Constituents in (mg/l) (This Facility Has No Garbage in It)

Well	Event	TOC	TKN	COD	ALK
MW01	1	5.2000	0.8000	44.0000	58.0000
MW01	2	6.8500	0.9000	13.0000	49.0000
MW01	3	4.1500	0.5000	13.0000	40.0000
MW01	4	15.1500	0.5000	40.0000	42.0000
MW02	1	1.6000	1.6000	11.0000	59.0000
MW02	2	6.2500	0.3000	10.0000	82.0000
MW02	3	1.4500	0.7000	10.0000	54.0000
MW02	4	1.0000	0.2000	13.0000	51.0000
MW03	1	1.0000	1.8000	28.0000	39.0000
MW03	2	1.9500	0.4000	10.0000	70.0000
MW03	3	1.5000	0.3000	11.0000	42.0000
MW03	4	4.8000	0.5000	26.0000	42.0000
MW04	1	4.1500	1.5000	41.0000	54.0000
MW04	2	1.0000	0.3000	10.0000	40.0000
MW04	3	1.9500	0.3000	24.0000	32.0000
MW04	4	1.2500	0.4000	45.0000	28.0000
MW05	1	2.1500	0.6000	39.0000	51.0000
MW05	2	1.0000	0.4000	26.0000	55.0000
MW05	3	19.6000	0.3000	31.0000	60.0000
MW05	4	1.0000	0.2000	48.0000	52.0000
MW06	1	1.4000	0.8000	22.0000	118.0000
MW06	2	1.0000	0.2000	23.0000	66.0000
MW06	3	1.5000	0.5000	25.0000	59.0000
MW06	4	20.5500	0.4000	28.0000	63.0000
P14	1	2.0500	0.2000	10.0000	79.0000
P14	2	1.0500	0.3000	10.0000	96.0000
P14	3	5.1000	0.5000	10.0000	89.0000

Subtitle C (hazardous waste) facility subject to the agency's groundwater monitoring provisions, and assume that the facility is in detection monitoring (40 CFR Part 264.98). If a statistically significant increase (or decrease in the case of pH) in one or more indicator parameters occurs, the facility owner/operator enters a compliance monitoring program (40 CFR Part 264.99). Under such a program, the facility must analyze samples from its groundwater monitoring wells for the entire 40 CFR Part 264, Appendix IX list of 222 hazardous constituents, and recompute groundwater statistical comparisons. If the statistically significant adverse effect on downgradient groundwater quality is confirmed, the facility owner/operator is afforded two options beyond the standard of no degradation of downgradient groundwater quality beyond background groundwater quality (i.e., the compliance limit is the concentration in upgradient wells). The two options allow the facility owner/operator to attempt to demonstrate that either:

1. The alleged release is statistical artifact; or

2. There has been a release, but it poses negligible present or potential hazard to human health or the environment at the point of compliance (i.e., the edges or boundary of a regulated unit).

Both demonstrations involve comparisons with chemical-specific, statistically derived, compliance limits. The EPA provides two types of statistical methods for making such compliance demonstrations: tolerance limits and parametric or nonparametric confidence limits.

In the first option, downgradient sample results are compared to a groundwater protection standard (GWPS) representing the average on-site background data. The GWPS serves as an alternate concentration limit (ACL) for the facility or unit. In this case a lower one-sided 99% confidence limit is calculated for each measured hazardous constituent in each individual compliance well. Essentially, the normal confidence limit is equivalent to computing a one-sample t-statistic to test if the mean of the n measurements for the chemical in the compliance well exceeds the ACL, which in this case is the upgradient mean concentration for the chemical. If the lower confidence limit for any compliance well exceeds an ACL, "...there is significant evidence that the true mean concentration at the compliance well exceeds the GWPS and that the facility permit has been violated" (EPA, 1992, page 61). Note that uncertainty in the background mean value is completely ignored as is the problem of multiple comparisons (i.e., repeated comparisons of different compliance wells and constituents to the background mean).

For the second option, the EPA wants an owner/operator to use a tolerance limit. "Specifically, the tolerance interval approach is recommended for comparison with a Maximum Contaminant Level (MCL) or with an ACL if the ACL is derived from health-based data" (EPA, 1992, page 50). In this case an upper one-sided tolerance limit for coverage of 95% with 95% confidence is calculated for each measured hazardous constituent in each compliance monitoring well. If the upper tolerance limit from any compliance monitoring well exceeds an MCL or ACL, "...there is significant evidence that as much as 5 percent or more of all the compliance well measurements will exceed the limit and consequently that the compliance point wells are in violation of the facility permit" (EPA, 1992, page 52). The use of tolerance intervals is a complete mystery. Here, the entire tolerance interval must be less than the MCL or ACL, indicating that we can have 95% confidence that 95% of all possible measurements from the distribution are below the ACL or MCL. These are very different approaches to the same problem and will yield dramatically different results as will be shown.

As an example, consider a normally distributed random variable in which a sample of size 7 yielded a mean of 5, a standard deviation of 1, and a minimum value of 3. The lower nonparametric 99% confidence limit is 3, the lower normal 99% confidence limit is 3.81, and the upper 95% tolerance limit is 8.40. Using the confidence limits, we are in compliance as long as the ACL or MCL is greater than 3.81; however, using the tolerance limit, the ACL or MCL must be greater than 8.40 in order to be in compliance. This is a difference of almost five standard deviation units. Something is clearly wrong here.

The confidence limits appear to be the correct method to determine if a well has significantly exceeded an ACL or MCL. In contrast to detection monitoring, assessment or compliance monitoring involves a single well and constituent, conditional on prior demonstration of a significant increase over background. In this case it makes perfect sense to pool measurements and compare their mean value to a standard using Student's t-test or a nonparametric alternative. This is a *very* different problem than detection monitoring. Note, however, that when the ACL is the background mean, this is a two-sample t-statistic and not a one-sample t-statistic or normal confidence limit. The reason is that the one-sample test or confidence limit only pertains when the ACL is a population value known with certainty. The background mean, however, is a statistical estimate based on a sample of n upgradient measurements and is subject to normal sampling variability. A more appropriate test here would be Dunnett's test (see Chapter 1) for the repeated comparison of multiple compliance well means to a single pooled background mean. The confidence level associated with each test should be $1 - .05/k$, where k represents the number of chemicals being tested.

The motivation for the selection between these two disparate approaches is described by the EPA (1992) in its Addendum to the Statistical Guidance Document for Groundwater Monitoring. The addendum explains that the confidence limit approach is to be used for comparisons to ACLs based on background data, whereas the tolerance limit approach is to be used for comparisons to MCLs or health-based ACLs. Presumably, the more stringent approach (i.e., tolerance limit) is to be used for those cases in which a known and "quantifiable" health risk is present (i.e., the MCL). It should be noted, however, that since at most four independent samples will be available on a semiannual monitoring event, the 95% confidence, 95% coverage tolerance limit is approximately five standard deviation units above the mean concentration. In light of this, even if all four semiannual measurements for a given compliance well are below the MCL, the tolerance limit will invariably exceed the MCL or health-based ACL and never-ending corrective action will be required. The net result is, of course, corrective action due to nothing more than natural variability and results that are consistent with chance expectations.

14.2.3   Cochran's Approximation to the Behrens–Fisher t-test

Although no longer required, for years the EPA RCRA regulation was based on application of Cochran's approximation to the Behrens–Fisher (CABF) t-test. The test was incorrectly implemented by requiring that four quarterly

TABLE 14.2   Illustration of pH Data Used in Computing the CABF t-test

Date	Replicate 1	2	3	4	Average
Background					
11/81	7.77	7.76	7.78	7.78	7.77
2/82	7.74	7.80	7.82	7.85	7.80
5/82	7.40	7.40	7.40	7.40	7.40
8/82	7.50	7.50	7.50	7.50	7.50
$\bar{x}_B$		7.62			7.62
s_B		0.18			0.20
n_B		16			4
Monitoring					
9/83	7.39	7.40	7.38	7.42	7.40
$\bar{x}_B$		7.40			7.40
s_B		0.02			
n_B		4			1

upgradient samples from a single well and single samples from a minimum of three downgradient wells each be divided into four aliquots and treated as if there were $4n$ independent measurements. The net result was that every hazardous waste disposal facility regulated under RCRA was declared "leaking." As an illustration consider the data in Table 14.2.

Note that the aliquots are almost perfectly correlated and add virtually no independent information, yet they are assumed to be completely independent by the statistic. The CABF t-test is computed as

$$t = \frac{\bar{x}_B - \bar{x}_M}{\sqrt{\dfrac{s_B^2}{n_B} + \dfrac{s_M^2}{n_M}}} = \frac{7.62 - 7.40}{\sqrt{\dfrac{.032}{16} + \dfrac{.0004}{4}}} = \frac{.22}{.05} = 4.82$$

The associated probability of this test statistic is 1 in 10,000, indicating that the chance that the new monitoring measurement came from the same population as the background measurements is 1 in 10,000. Note that, in fact, the mean concentration of the four aliquots for the new monitoring measurement is identical to one of the four mean values for background, suggesting that intuitively the probability is closer to 1 in 4 rather than 1 in 10,000. Averaging the aliquots, which should have never been split in the first place, yields the statistic

$$t = \frac{\bar{x}_B - \bar{x}_M}{s_B\sqrt{\dfrac{1}{n_B} + 1}} = \frac{7.62 - 7.40}{.20\sqrt{\dfrac{1}{4} + 1}} = \frac{.22}{.22} = 1.0$$

which has an associated probability of 1 in 2. Had the sample size been increased to $n_B = 20$, the probability would have decreased to 1 in 3. It took the EPA 6 years to recognize this flaw and to change this regulation.

14.2.4 Control of False Positive Rate by Constituent

Sitewide false positive and false negative rates are more important than choice of statistic; nonetheless, certain statistics make it impossible to control the sitewide false positive rate because the rate is controlled separately for each constituent (e.g., parametric and nonparametric ANOVA). The only important false positive rate is the one that includes all monitoring wells and all constituents, since any single exceedance can trigger an assessment. This criterion impacts greatly on the selection of statistical method. These error rates are dependent on the number of wells, number of constituents, number of background measurements, type of comparison (i.e., intrawell versus interwell), distributional form of the constituents, detection frequency of the constituents, and the individual comparison false positive rate of the statistic being used. Invariably, this leads to a problem in interval estimation, the solution of which is typically a prediction limit that incorporates the effects of verification resampling as well as multiple comparisons introduced by both multiple monitoring wells and multiple monitoring constituents.

14.2.5 EPA's 40 CFR Computation of MDLs and PQLs

The role of the laboratory is critical in the detection monitoring program because compounds not detected or quantified in background can cause as much or even more grief as those that are detected. Practical quantitation limits—PQLs (not method detection limits—MDLs) are appropriate statistical triggers for constituents not detected in background. The EPA requires that the MDL be computed as

$$MDL = 3.14s \tag{14.1}$$

where s is the standard deviation of seven replicate spiked samples at some concentration in the range of two to five times the suspected MDL, and 3.14 is the 99% point of Student's t-distribution on 6 degrees of freedom. This MDL estimator is referred to by the EPA as the "40 CFR" method and was originally suggested by Glaser et al. (1981) (see Chapter 5). Figure 5.3 presents calibration data for benzene and illustrates that variability is proportional to concentration. While this is true for many constituents, it is always true for volatile organic priority pollutant compounds which form the basis of most detection monitoring programs. Note that the relative standard deviation (i.e., standard deviation/concentration) is fairly constant in this example. However, it is the absolute standard deviation that is used in computing the MDL per 40 CFR Part 136, Appendix B. In light of this, had we spiked at 10 μg/l, we would have obtained $s = 0.50$ and an MDL of

1.58 μg/l. Alternatively, had we spiked at 1 μg/l, we would have obtained $s = 0.16$ and an MDL of 0.5 μg/l. Finally, had we spiked at 0.5 μg/l, we would have obtained $s = 0.11$ and an MDL of 0.35 μg/l. To make matters worse, the EPA defines the PQL as 5 to 10 times the MDL; therefore the PQL is just as arbitrary as the MDL. Using information from the entire calibration line and the method described by Gibbons et al. (1991c—see Chapter 5), the MDL is estimated as 0.75 μg/l.

The important point here is that when variability is not constant over concentration, we must model both concentration and variability in order to obtain a valid estimate of the MDL. Such an estimate is *only* available from calibration data (i.e., multiple concentrations in the range of the MDL) and cannot be obtained from data at a single spiking concentration, regardless of the method. As discussed in Chapter 5, not only is the 40 CFR Part 136, Appendix B method statistically flawed, but it is based on a single spiking concentration which is unwisely left to the discretion of the user and never published by method developers who report MDLs and PQLs for their methods. This is the worst of all possible choices and use of these estimates to make groundwater monitoring decisions will lead to false assessment and corrective action.

To illustrate, consider the data in Table 14.3 for a new facility that has not yet disposed of any waste (Gibbons, 1994a). The measurements in Table 14.3 are reported regardless of whether or not they were above or below the MDL. In each well a list of 31 volatile organic compounds (VOCs) was analyzed. Table 14.3 reports the detected concentrations only, assuming an MDL of zero. Clearly, using this logic, the trip blank has been consistently impacted by the facility (despite the fact that it never made it out of the truck), and there is clear evidence of contamination at all wells despite the fact that there has been nothing disposed of at this facility. Note that these

TABLE 14.3   Uncensored VOC Detections in μg/l (All Wells Were Analyzed for 31 VOCs)

Compound	MW01	MW02	MW03	MW04	MW05	MW06	MW07	TRIP
PCE	.1	.1		.1	.2	.2	.1	.1
TOL	.4						.4	.4
TCFM	.2			.2	.2	.3	.7	.5
VC		.1						
CBENZ			.1					
EBENZ				.1			.1	.1
BENZ							.2	.2
12CDB								.1

TRIP refers to a trip blank; PCE-tetrachloroethene; TOL = toluene; TCFM = trichlorofluoromethane; VC = vinyl chloride; CBENZ = chlorobenzene; EBENZ = ethylbenzene; BENZ = benzene; 12DCB = 1,2-dichlorobenzene.

"detected" levels are, in fact, above some published MDLs and easily achievable using the 40 CFR methodology.

14.3 VERIFICATION RESAMPLING

Without verification resampling, all is hopeless. Regulations have been quite vague on this topic, often requiring some form of assessment monitoring prior to verification resampling. Also, it is unclear how verification resampling would be performed for those methods such as ANOVA, which require multiple samples at each downgradient monitoring well. Obviously, to recompute the statistic, multiple samples must also be included in the verification. However, ANOVA uses the downgradient variance from each well in computing a pooled estimate of the total variance; therefore the same time frame for the original set of samples (e.g., four samples in six months) must also be used in collecting the verification samples. It seems unlikely that environmentally conscious regulators would wait six months to verify an initial statistical exceedance.

It should also be noted that requiring passage of two verification resamples is less powerful than requiring passage of a single resample, or one of two resamples, since in the parametric case lower limits can be used. Regulators, particularly in California, claim the reverse is true, arguing that if two verification resamples must be passed, twice as much evidence in favor of no impact is available. Use of lower limits for verification resampling plans requiring only one of two, or one of three, samples to pass (e.g., pass only one of two verification resamples) will, in fact, result in lower false negative rates than corresponding plans in which all resamples must pass.

In the nonparametric case the limit is generally fixed at the highest order statistic, and false positive and false negative rates become a function of background sample size and the number of future comparisons (i.e., both number of downgradient wells and constituents). In most cases the number of background measurements required to provide a sitewide false positive of 5% is prohibitive, but this is not true for verification resampling plans that require passage of one of two or one of three verification samples.

14.4 INTERWELL VERSUS INTRAWELL COMPARISONS

If the wells are not contaminated by on-site impacts, whenever possible intrawell comparisons should be used since they completely eliminate spatial variability, and thus are more powerful than their interwell counterparts. It is unreasonable to assume that all of the spatial variability at a facility will be represented by a small number of upgradient monitoring wells. In light of this, spatial variability and contamination are invariably confounded when using the traditional upgradient versus downgradient comparison study. Often, an owner/operator will have to perform upgradient versus downgradient comparisons, identify which downgradient wells exceed upgradient limits for which constituents, provide evidence that these exceedances are not due to the facility, and use an intrawell strategy for these cases. In the absence of predisposal data, this may be a "hard sell"; however, without it, most sites in this country will eventually find themselves in a site assessment program and possibly corrective action whether they have impacted the environment or not.

14.5 COMPUTER SOFTWARE

To date, little useful software has been developed for implementing statistical detection monitoring programs. Most notable among the handful of packages available is the GRITS/STAT computer program produced by the EPA. This program (as well as the other available programs) implements the approaches advocated in the original EPA guidance document (EPA, 1989) which, as previously described, is severely flawed and of little practical value. As the unit of analysis, GRITS/STAT considers each sample point and constituent separately; therefore the amount of required user input is enormous (i.e., the user must test distributional assumptions, screen outliers, and select an appropriate statistical method for each well and constituent individually). With 30 or 40 constituents and 10 or 20 monitoring wells, this could take a lifetime. Statistical weaknesses of the program include but are not limited to:

1. Replacement of nondetects with half the detection limit regardless of detection frequency.
2. Computation of tests of normality and other statistical procedures even if all data were nondetects.
3. Use of differences in detection limits as estimates of natural variability.
4. No control over the effects of multiple comparisons (e.g., only 95% confidence normal prediction limits for the next single sample or mean of m samples are computed regardless of the number of monitoring wells and constituents).
5. For normal tolerance limits, individual wells are only compared to a 95% coverage 95% confidence tolerance limit. The tolerance limit has a built-in failure rate; that is, 5 out of every 100 measurements are *expected* to fail with 95% confidence. Using GRITS/STAT, a site with 20 monitoring wells would be expected to fail at one compliance well on every monitoring event for every constituent subjected to this test. Clearly, this approach will lead owner/operators down a path of constant site assessment.
6. ANOVA, either parametric or nonparametric, is presented as the statistical method of choice.

7. There is no discussion of the role of verification resampling and its effects on balancing false positive and false negative rates in the manual, or methods by which the verification resampling strategy are explicitly considered in computing limit values and corresponding probability estimates.

8. Nonparametric prediction limits and Poisson-based prediction and tolerance limits which are discussed in great detail in the 1992 Addendum to the Statistical Guidance document for Groundwater Monitoring are not even mentioned in the program or the User Documentation Manual.

9. None of the methods in the program is suited to data in which detection frequencies are less than 50%.

Given these limitations, it is hard to imagine how the EPA expects owners and operators to comply with the new Subtitle D regulation without constantly triggering site assessments and possible corrective action regardless of whether the facility has contaminated groundwater or not.

A new computer program, entitled DUMPSTAT (Downgradient Upgradient Monitoring Program Statistics; Gibbons, 1994b), implements the methods described in this book, many of which are explicitly recommended in the EPA's 1992 Addendum to the Statistical Guidance Document for Groundwater Monitoring. The program is based on an artificial intelligence algorithm that automatically selects the appropriate form of the statistical prediction limit on the basis of outlier screening, distributional testing, and detection frequency. The program is capable of providing either interwell or intrawell comparisons or a mixture of the two, which will probably be the most frequent case. The user can interactively define fields in the dataset, create a database, and update that database as new data become available. Once the sample points, constituents, and type of comparison (i.e., interwell, intrawell, or some combination) are selected, a complete statistical evaluation of the data is automatically generated in both tabular and graphical formats. An important feature of the program is that via simulation it can compute expected false positive and false negative rates for the entire detection monitoring program based on existing conditions at the facility (i.e., number of background measurements, type of prediction limit used for each constituent, distributional form of the data, detection frequency, and number of monitoring wells and constituents). Based on the results of these power characteristics, overrides of system selections can be requested (e.g., use of normal and Poisson versus nonparametric prediction limits when the number of background measurements is too small to provide sufficient confidence for the site as a whole). The specific verification resampling plan is incorporated into all statistical computations and its effects can be examined via simulation (i.e., false positive and false negative rates for the site as a whole).

Summary

Having made it to this point in the book, I sincerely hope you have gained insight into the complexities and range of solutions to groundwater monitoring statistical problems. Indeed, to preserve the subtleties of natural phenomena, methodology must necessarily be complicated. I hope that you, as a groundwater scientist or engineer, have become convinced that statistical methods are useful in setting limits for both naturally occurring and anthropogenic substances. For synthetic substances, the setting of limits may be especially complex. The heart of the analytic process is variability, but routine laboratory measurements may encompass unknown or unconsidered sources which impact statistical inferences and groundwater monitoring decisions. If nothing else, one must be an informed consumer of data and statistical methodologies, and understand the basis for "limits" reported by the laboratory, for what they are useful, and for what they are not useful. More often than not, these "limits" are useless in making groundwater impact decisions.

To the statistician interested in environmental problems, my goal has been to raise the level of presentation such that it is obvious which areas need further study, and also to clearly explicate our knowledge of statistical foundations. Work is needed in detection limits and quantification (Chapters 5 and 6). New approaches to variance component estimation (Chapter 12) are also vital for accurate decisions in this field.

Use of prediction limits have been stressed throughout (Chapters 1 through 3) because they efficiently limit overall sitewide false positive rates without sacrificing false negative rates. Prediction limits for normal, lognormal, and Poisson random variables and their nonparametric alternatives have been considered. Applications to interwell (i.e., upgradient versus downgradient) and intrawell comparisons (Chapter 8) have been illustrated, and appropriate modifications of the limit estimates and overall tests of trend have been considered (Chapter 9). Methods for screening background data for outliers (Chapter 13) and tests of distributional assumptions to produce efficient prediction limit applications (Chapter 11) have been discussed in detail. In general, when detection frequency is less than 50%, nonparametric prediction limits should be used. However, an adequate number of background data must be available to provide the required sitewide confidence level. When detection frequency is greater than 50% but less than 100%, the distributional form must be assessed and an appropriate method for incorporating censored measurements (i.e., nondetects) must be selected (Chapter 10). Chapter 10 also calls for a censored data estimator for prediction limits quite different from that of an estimator to determine mean and variance of censored normal or censored lognormal distributions. Of course, when sufficient background data are available, nonparametric limits are desirable alternatives and do not require a method to recover underlying distribution statistics from a censored dataset. In some cases tolerance limits or tolerance and prediction limits combined may provide a useful alternative (Chapter 4).

A new application of multivariate statistics, sometimes termed "chemometric" or "pattern recognition methods," is used to determine and differentiate contaminant sources (see Chapter 7). These methods contrast with the rest of the book because they are more useful in assessment monitoring than detection monitoring. Potential impacts are geochemically differentiated, then the classification function determines the probability of impact source, helping to determine the source of statistically significant exceedance over background when multiple contaminant sources may be present.

Finally, applicability to regulatory issues is presented (Chapter 14). Historically, environmental regulatory agencies have made poor use of statistical methods, especially for groundwater monitoring applications. Lack of attention to sitewide false positive rates, prediction limits, nonparametric approaches, censored data, verification resampling plans, spatial variability, and laboratory-specific limits of detection and quantification have led to inefficient and costly regulation and guidance. Recent developments at the EPA are encouraging though and should be supported.

In the future statistically rigorous groundwater detection monitoring programs should become the rule, not the exception. The methods described in this book provide an underpinning for the future and presage the foundation of more complex efforts to upgrade environmental statistical methodology.

Bibliography

Adams, P. B., Passmore, W. O., and Campbell, D. E. (1966). *Symposium on Trace Characterization—Chemical and Physical*. National Bureau of Standards, Washington, DC.

Aitchison, J. (1955). On the distribution of a positive random variable having a discrete probability mass at the origin. *J. Amer. Statist. Assoc.*, **50** 901–908.

Aitchison, J., and Brown, J. A. C. (1957). *The Lognormal Distribution*. Cambridge Univ. Press, Cambridge.

Aitkin, M., and Longford, N. (1986). Statistical modelling issues in school effectiveness studies. *J. Roy. Statist. Soc.*, *Ser. A* **149** 1–43.

Aroian, L. A. (1941). A study of Fisher's z distribution and the related F distribution. *Ann. Math. Statist.* **12** 429–448.

Barnett, V., and Lewis, T. (1984). *Outliers in Statistical Data* 2nd ed. Wiley, New York.

Beckman, R. J., and Cook, R. D. (1983). Outlier.........s. *Technometrics* **25** 119–152.

Bock, R. D. (1975). *Multivariate Statistical Methods in Behavioral Research*. McGraw-Hill, New York.

Bock, R. D. (1989). *Multilevel Analysis of Educational Data*. Academic, San Diego.

Bortkiewicz, L. von (1898). *Das Gesetz der kleinen Zahlen* (The Law of Small Numbers). Teubner, Leipzig.

Box, G. E. P., and Cox, D. R. (1964). An analysis of transformations (with discussion). *J. Roy. Statist. Soc. Ser. B*. **26** 211–252.

Box, G. E. P., and Jenkins, G. M. (1976). *Time Series Analysis: Forecasting and Control* 2nd ed. Holden-Day, San Francisco.

Brown, K. W., and Donnelly, K. C. (1988). An estimation of the risk associated with the organic constituents of hazardous and municipal waste leachates. *Hazardous Waste and Hazardous Materials* **5** 1–30.

Carlson, R. F., MacCormick, A. J. A., and Watts, D. G. (1970). Applications of linear random models to four annual streamflow series. *Water Resources Res.*, **6** 1070–1078.

Chatfield, C. (1984). *The Analysis of Time Series: An Introduction* 3rd ed. Chapman and Hall, London.

Chew, V. (1968). Simultaneous prediction intervals. *Technometrics* **10** 323–331.

Chi, E. M., and Reinsel, G. C. (1989). Models of longitudinal data with random effects and AR(1) errors. *J. Amer. Statist. Assoc.*, **84** 452–459.

Chou, Y. M., and Owen, D. B. (1986). One-sided distribution-free simultaneous prediction limits for p future samples. *J. Quality Technol.* **18** 96–98.

Clayton, C. A., Hines, J. W., and Elkins, P. D. (1987). Detection limits with specified assurance probabilities. *Anal. Chem.* **59** 2506–2514.

Cohen, A. C. (1959). Simplified estimators for the normal distribution when samples are singly censored or truncated. *Technometrics* **1** 217–237.

Cohen, A. C. (1961). Tables for maximum likelihood estimates: singly truncated and singly censored samples. *Technometrics* **3** 535–541.

Cohen, A. C. (1991). *Truncated and Censored Samples: Theory and Applications.* Marcel Dekker, New York.

Coleman, D. (1993). Comparison of different types of method detection limits (MDL's). *Alcoa Memorandum*, March 26.

Conover, W. J. (1980). *Practical Nonparametric Statistics* 2nd ed. Wiley, New York.

Cox, D. R., and Hinkley, D. V. (1974). *Theoretical Statistics.* Chapman and Hall, London.

Currie, L. A. (1968). Limits for qualitative detection and quantitative determination: application to radiochemistry. *Anal. Chem.* **40** 586–593.

Currie, L. A. (1988). *Detection in Analytical Chemistry: Importance, Theory and Practice.* American Chemical Society, Washington, DC.

D'Agostino, R. B. (1971). An omnibus test of normality for moderate and large size samples. *Biometrika* **58** 341–348.

Dahiya, R. C., and Guttman, I. (1982). Shortest confidence and prediction intervals for the lognormal. *Canad. J. Statist.* **10** 277–291.

Davis, C. B. (1994). Environmental regulatory statistics. In *Handbook of Statistics: Environmental Statistics* **12** Chapter 26, G. P. Patil and C. R. Rao (eds.). Elsevier.

Davis, C. B., and McNichols, R. J. (1987). One-sided intervals for at least p of m observations from a normal population on each of r future occasions. *Technometrics* **29** 359–370.

Davis, C. B., and McNichols, R. J. (1993). Nonparametric simultaneous prediction limits. Technical Report, Environmetrics and Statistics LTD, Henderson, NV.

Dempster, A. P., Rubin, D. B., and Tsutakawa, R. K. (1981). Estimation in covariance component models, *J. Amer. Statist. Assoc.* **76** 341–353.

Dixon, W. J. (1953). Processing data for outliers. *Biometrics* **9** 74–89.

Donner, A. (1985). A regression approach to the analysis of data arising from cluster randomization. *Internat. J. Epidemiology* **14** 322–326.

Dunnett, C. W. (1955). A multiple comparisons procedure for comparing several treatments with a control. *J. Amer. Statist. Assoc.* **50** 1096–1121.

Dunnett, C. W., and Sobel, M. (1955). Approximations to the probability integral and certain percentage points of a multivariate analogue of Student's t-distribution. *Biometrika* **42** 258–260.

El-Shaarawi, A. H. (1989). Inferences about the mean from censored water quality data. *Water Resources Res.,* **25** 685–690.

El-Shaarawi, A. H., and Niculescu, S. P. (1992). On Kendall's tau as a test of trend in time. *Environmetrics* **3** 385–412.

EPA (1980). Standards for owners and operators of hazardous waste treatment, storage, and disposal facilities. *Federal Register* **45**(98) 33154–33258 (May 19, 1980).

EPA (1982). Hazardous waste management system; permitting requirements for land disposal facilities. *Federal Register* **47**(143) 32274–32373 (July 26, 1982).

EPA (1985). National primary drinking water regulations; volatile synthetic organic chemicals; proposed rule. *Federal Register* **50**(219) 46902–46933.

EPA (1986). Hazardous waste land disposal facilities, statistical procedures for detecting ground-water contamination. *Federal Register* **51**(161) 29812–29814 (August 20, 1986).

EPA (1987a). 40CFR Part 264: statistical methods for evaluating ground-water monitoring data from hazardous waste facilities: proposed rule. *Federal Register* **52**(163) 31948–31956 (August 24, 1987).

EPA (1987b). List (Phase 1) of hazardous constituents for ground-water monitoring; final rule. *Federal Register* **52**(131) 25942–25953 (July 9, 1987).

EPA (1988). 40CFR Part 264: statistical methods for evaluating ground-water monitoring data from hazardous waste facilities: final rule. *Federal Register* **53**(196) 39720–39731 (October 11, 1988).

EPA (1989). Statistical analysis of ground-water monitoring data at RCRA facilities. Interim Final Guidance, April 1989.

EPA (1991). Solid waste disposal facility criteria: final rule. *Federal Register* **56**(196) 50978–51119 (October 9, 1991).

EPA (1992). Statistical analysis of ground-water monitoring data at RCRA facilities. Addendum to Interim Final Guidance. Office of Solid Waste, July 1992.

Everitt, B. S., and Hand, D. J. (1981). *Finite Mixture Distributions*. Chapman and Hall, London.

Fieller, E. C. (1940). The biological standardization of insulin. *J. Roy. Statist. Soc.* **S7** 1–54.

Filliben, J. J. (1975). The probability plot correlation coefficient test for normality. *Technometrics* **17** 111–117.

Fisher, R. A. (1929). Moments and product-moments of sampling distributions. *Proc. London Math. Soc.* **30**(2) 199.

Fisher, R. A. (1930). The moments of the distribution for normal samples of measures of departure from normality. *Proc. Roy. Soc.* **A 130** 16.

Fisher, R. A. (1936). The use of multiple measurements in taxonomic problems. *Ann. Eugenics* **7** 179–188.

Fuller, F. C., and Tsokos, C. P. (1971). Time series analysis of water pollution data. *Biometrics* **27** 1017–1034.

Geary, R. C. (1935). The ratio of the mean deviation to the standard deviation as a test of normality. *Biometrika* **27** 310–332.

Geary, R. C. (1936). Moments of the ratio of the mean deviation to the standard deviation for normal samples. *Biometrika* **28** 295–307.

Gibbons, R. D. (1987a). Statistical prediction intervals for the evaluation of ground-water quality. *Ground Water* **25** 455–465.

Gibbons, R. D. (1987b). Statistical models for the analysis of volatile organic compounds in waste disposal sites. *Ground Water* **25** 572–580.

Gibbons, R. D., Jarke, F. H., and Stoub, K. P. (1988a). Method detection limits. *Proc. Fifth Annual USEPA Waste Testing and Quality Assurance Symp.*, **2** 292–319.

Gibbons, R. D., Hedeker, D. R., Waternaux, C., and Davis, J. M. (1988b). Random regression models: a comprehensive approach to the analysis of the longitudinal psychiatric data. *Psychopharmacol. Bull.* **24** 438–443.

Gibbons, R. D. (1990). A general statistical procedure for ground-water detection monitoring at waste disposal facilities. *Ground Water* **28** 235–243.

Gibbons, R. D. (1991a). Some additional nonparametric prediction limits for ground-water detection monitoring at waste disposal facilities. *Ground Water* **29** 729–736.

Gibbons, R. D. (1991b). Statistical tolerance limits for ground-water monitoring. *Ground Water* **29** 563–570.

Gibbons, R. D., Jarke, F. H., and Stoub, K. P. (1991c). Detection limits for linear calibration curves with increasing variance and multiple future detection decisions. *Waste Testing and Quality Assurance* 3, **ASTM STP 1075** 377–390, D. Friedman (ed.). American Society for Testing and Materials, Philadelphia.

Gibbons, R. D., and Baker, J. (1991d). The properties of various statistical prediction intervals for ground-water detection monitoring. *J. Environ. Sci. Health* **A26** 535–553.

Gibbons, R. D., Grams, N. E., Jarke, F. H., and Stoub, K. P. (1992a). Practical quantitation limits. *Chemometrics and Intelligent Laboratory Systems* **12** 225–235.

Gibbons, R. D., Dolan, D., Keough, H., O'Leary, K., and O'Hara, R. (1992b). A comparison of chemical constituents in leachate from industrial hazardous waste and municipal solid waste landfills. *Proc. Fifteenth Annual Madison Waste Conf.*, Univ. Wisconsin, Madison, September 23–24.

Gibbons, R. D. (1994a). The folly of Subtitle D statistics: when green field sites fail. *Proc. Waste Technol. 94*. National Solid Waste Management Assoc., Washington, DC, January 13–15.

Gibbons, R. D. (1994b). DUMPSTAT: an artificially intelligent computer program for ground-water monitoring applications. *Discerning Systems*. Burnaby, British Columbia.

Gilbert, R. O. (1987). *Statistical Methods for Environmental Pollution Monitoring*. Van Nostrand Reinhold, New York.

Gilliom, R. J., and Helsel, D. R. (1986). Estimation of distributional parameters for censored trace level water quality data. I. Estimation techniques. *Water Resources Res.*, **22** 135–146.

Glaser, J. A., Foerst, D. L., McKee, G. D., Quane, S. A., and Budde, W. L. (1981). Trace analyses for wastewaters. *Environ. Sci. Technol.*, **15** 1426–1435.

Gleit, A. (1985). Estimation for small normal data sets with detection limits. *Environ. Sci. Technol.*, **19** 1201–1206.

Goldstein, H. (1987). *Multilevel Models in Educational and Social Research*. Oxford Univ. Press, New York.

Grubbs, F. E. (1969). Procedures for detecting outlying observations in samples. *Technometrics* **11** 1–21.

Grubbs, F. E., and Beck, G. (1972). Extension of sample sizes and percentage points for significance tests of outlying observations. *Technometrics* **14** 847–854.

Gupta, A. K. (1952). Estimation of the mean and standard deviation of a normal population from a censored sample. *Biometrika* **39** 260–273.

Gupta, S. S. (1963). Probability integrals of multivariate normal multivariate t. *Ann. Math. Statist.* **34** 792–828.

Guttman, I. (1970). *Statistical Tolerance Regions: Classical and Bayesian*. Hafner, Darien, CT.

Haas, C. N., and Scheff, P. A. (1990). Estimation of averages in truncated samples. *Environ. Sci. Technol.* **24** 912–919.

Hahn, G. J. (1970). Additional factors for calculating prediction intervals for samples from a normal distribution. *J. Amer. Statist. Assoc.*, **65** 1668–1676.

Hahn, G. J., and Meeker, W. Q. (1991). *Statistical Intervals: A Guide for Practitioners*. Wiley, New York.

Hall, I. J., Prarie, R. R., and Motlagh, C. K. (1975). Non-parametric prediction intervals. *J. Quality Technol.* **7** 109–114.

Hand, D. J. (1981). *Discrimination and Classification*. Wiley, New York.

Harman, H. (1970). *Modern Factor Analysis*. Univ. Chicago Press, Chicago.

Hartley, H. O., and Pearson, E. S. (1950). Tables of the chi-squared integral and of the cumulative Poisson distribution. *Biometrika* **37** 313–325.

Hashimoto, L. K., and Trussell, R. R. (1983). Evaluating water quality data near the detection limit. Paper presented at the American Water Works Assoc. Advanced Technology Conf. American Water Works Assoc., Las Vegas, NV., June 5–9.

Hedeker, D., Gibbons, R. D., Waternaux, C., and Davis, J. M. (1989). Investigating drug plasma levels and clinical response using random regression models. *Psychopharmacol. Bull.* **25** 227–231.

Hedeker, D., Gibbons, R. D., and Davis, J. M. (1991). Random regression models for multicenter clinical trials data. *Psychopharmacol. Bull.* **27** 73–77.

Hedeker, D. (1993). MIXREG: a Fortran program for mixed-effects linear regression models. Technical Report. Prevention Research Center, School of Public Health, Univ. Illinois at Chicago.

Helsel, D. R., and Gilliom, R. J. (1986). Estimation of distributional parameters for censored trace level water quality data. II. Verification and applications. *Water Resources Res.*, **22** 147–155.

Helsel, D. R., and Cohn, T. A. (1988). Estimation of descriptive statistics for multiply censored water quality data. *Water Resources Res.*, **24** 1997–2004.

Helsel, D. R. (1990). Less than obvious: statistical treatment of data below the detection limit. *Environ. Sci. Technol.*, **24** 1766–1774.

Hsu, D. A., and Hunter, J. S. (1976). Time series analysis and forecasting for air pollution concentrations with seasonal variations. *Proc. EPA Conf. on Environmental Modeling and Simulation* **PB-257** 142 673–677. National Technical Information Service, Springfield, VA.

Hubaux, A., and Vos, G. (1970). Decision and detection limits for linear calibration curves. *Anal. Chem.* **42** 849–855.

Hunt, W. F., Jr., Akland, G., Cox, W., Curran, T., Frank, N., Goranson, S., Ross, P., Sauls, H., and Suggs, J. (1981). *U.S. Environmental Protection Agency Intra-Agency Task Force Report on Air Quality Indicators*, EPA-450/4-81-015. Environmental Protection Agency, National Technical Information Service, Springfield, VA.

Johnson, N. L., and Kotz, S. (1969). *Discrete Distributions*. Wiley, New York.

Kaiser, H., and Specker, H. (1956). Bewertung und Vergleich von Analysenverfahren. *Fresenius' Z. Anal. Chem.*, **149** 46.

Kaiser, H. (1965). Zum Problem der Nachweisgrenze. *Fresenius' Z. Anal. Chem.* **209** 1.

Kaiser, H. (1966). Zur Definition der Nachweisgrenze, der Garantiegrenze und der dabei benutzen Begriffe: Fragen und Ergebnisse der Diskussion. *Fresenius' Z. Anal. Chem.* **216** 80.

Kendall, M. G. (1975). *Rank Correlation Methods* 4th ed. Charles Griffin, London.

Kirk, R. E. (1982). *Experimental Design* 2nd ed. Brooks/Cole, Monterey, CA.

Koorse, S. J. (1989). False positives, detection limits, and other laboratory imperfections: the regulatory implications. *Environ. Law Reporter* **19** 10211–10222.

Kushner, E. J. (1976). On determining the statistical parameters for pollution concentrations from a truncated data set. *Atmos. Environ.* **10** 975.

Laird, N. M., and Ware, J. H. (1982). Random-effects models for longitudinal data. *Biometrics* **38** 963–974.

Lieberman, G. J., and Miller, R. G. (1963). Simultaneous tolerance intervals in regression. *Biometrika* **50** 155–168.

Lilliefors, H. W. (1967). On the Kolmogorov–Smirnov test for normality with mean and variance unknown. *J. Amer. Statist. Assoc.*, **62** 399–402.

Lilliefors, H. W. (1969). Correction to the paper "On the Kolmogorov–Smirnov test for normality with mean and variance unknown." *J. Amer. Statist. Assoc.* **64** 1702.

Liteanu, C., and Rica, I. (1980). *Statistical Theory and Methodology of Trace Analysis*. Halsted, Chicago.

Longford, N. T. (1987). A fast scoring algorithm for maximum likelihood estimation in unbalanced mixed models with nested random effects. *Biometrika* **74** 817–827.

Lucas, J. M. (1982). Combined Shewart–CUSUM quality control schemes. *J. Quality Technol.*, **14** 51–59.

Lucas, J. M. (1985). Cumulative sum (CUSUM) control schemes. *Comm. Statist. A—Theory Methods* **14**(11) 2689–2704.

Maindonald, J. H. (1984). *Statistical Computation*. Wiley, New York.

Mann, H. B. (1945). Nonparametric tests against trend. *Econometrica* **13** 245–259.

McCleary, R., and Hay, R. A. (1980). *Applied Time Series Analysis for the Social Sciences*. Sage, Beverly Hills, CA.

McCollister, G. M., and Wilson, K. R. (1975). Linear stochastic models for forecasting daily maxima and hourly concentrations of air pollutants. *Atmos. Environ.*, **9** 417–423.

McMichael, F. C., and Hunter, J. S. (1972). Stochastic modeling of temperature and flow in rivers. *Water Resources Res.* **8** 87–98.

McNichols, R. J., and Davis, C. B. (1988). Statistical issues and problems in ground water detection monitoring at hazardous waste facilities. *Ground Water Monitoring Rev.* **8** 135–150.

Miller, R. G. (1966). *Simultaneous Statistical Inference*. McGraw-Hill, New York.

Montgomery, D. C., and Johnson, L. A. (1976). *Forecasting and Time Series Analysis*. McGraw-Hill, New York.

Nelson, W. (1982). *Applied Life Data Analysis*. Wiley, New York.

Ott, W. R. (1990). A physical explanation of the lognormality of pollutant concentrations. *J. Air Waste Management Assoc.* **40** 1378–1383.

Owen, D. B. (1962). *Handbook of Statistical Tables*. Addison-Wesley, Reading, MA.

Owen, W. J., and DeRouen, T. A. (1980). Estimation of the mean for lognormal data containing zeros and left-censored values, with applications to the measurement of worker exposure to air contaminants. *Biometrics* **36** 707–719.

Page, E. S. (1954). Continuous inspection schemes. *Biometrika* **41**(1) 100–114.

Paulson, E. (1942). A note on the estimation of some mean values for a bivariate distribution. *Ann. Math. Statist.*, **13** 440–445.

Pearson, E. S., and Hartley, H. O. (1976). *Biometrika Tables for Statisticians*. Biometrika Trust, London.

Persson, T., and Rootzen, H. (1977). Simple and highly efficient estimators for a type I censored normal sample. *Biometrika* **64** 123–128.

Rosner, B. (1983). Percentage points for a generalized ESD many-outlier procedure. *Technometrics* **25** 165–172.

Rutherford, E., and Geiger, H. (1910). The probability variations in the distribution of alpha particles. *Philos. Mag.* **20** 698–704.

Ryan, T., and Joiner, B. (1973). Normal probability plots and tests for normality. Technical Report, Pennsylvania State Univ.

Sarhan, A. E., and Greenberg, B. G. (1956). Estimation of location and scale parameters by order statistics from singly and doubly censored samples, Part I, The normal distribution up to samples of size 10. *Ann. Math. Statist.* **27** 427–457.

Sarhan, A. E., and Greenberg, B. G. (eds.) (1962). *Contributions to Order Statistics*. Wiley, New York.

Saw, J. G. (1961). Estimation of the normal population parameters given a type I censored sample. *Biometrika* **48** 367–377.

Schmee, J., Gladstein, D., and Nelson, W. (1985). Confidence limits of a normal distribution from singly censored samples using maximum likelihood. *Technometrics* **27** 119–128.

Schneider, H. (1986). *Truncated and Censored Samples from Normal Populations*. Marcel Dekker, New York.

Searle, S. R. (1987). *Linear Models for Unbalanced Data*. Wiley, New York.

Sen, P. K. (1968a). On a class of aligned rank order tests in two-way layouts. *Ann. Math. Statist.* **39** 1115–1124.

Sen, P. K. (1968b). Estimates of the regression coefficient based on Kendall's tau. *J. Amer. Statist. Assoc.*, **63** 1379–1389.

Shapiro, S. S., and Wilk, M. B. (1965). An analysis of variance test for normality (complete samples). *Biometrika* **52** 591–611.

Shapiro, S. S., Wilk, M. B., and Chen, H. J. (1968). A comparative study of various tests for normality. *J. Amer. Statist. Assoc.* **63** 1343–1372.

Shapiro, S. S., and Francia, R. S. (1972). An approximate analysis of variance test for normality. *J. Amer. Statist. Assoc.* **67** 215–216.

Shewart, W. A. (1931). *Economic Control of Quality of Manufactured Product*. Van Nostrand, Princeton, NJ.

Smith, R. A., Hirsch, R. M., and Slack, J. R. (1982). A study of trends in total phosphorus measurements at NASQAN stations. U.S. Geological Survey Water Supply Paper 2190, U.S. Geological Survey, Alexandria, VA.

Smith, R. M., and Bain, L. J. (1976). Correlation type goodness-of-fit statistics with censored sampling. *Comm. Statist. A—Theory Methods* **5**(2) 119–132.

Snedecor, G. W., and Cochran, W. G. (1980). *Statistical Methods.* Iowa State Univ. Press, Ames, IA.

Starks, T. H. (1988). Evaluation of control chart methodologies for RCRA waste sites. EPA Technical Report CR814342-01-3.

Stefansky, W. (1972). Rejecting outliers in factorial designs. *Technometrics* **14** 469–478.

Theil, H. (1950). A rank-invariant method of linear and polynomial regression analysis. *Proc. Koninalijke Nederlandse Akademie Van Wetenschatpen* **A53** part 3, 1397–1412.

Tietjen, G. L., and Moore, R. H. (1972). Some Grubbs-type statistics for the detection of several outliers. *Technometrics* **14**(3) 583–597.

Wald, A., and Wolfowitz, J. (1946). Tolerance limits for a normal distribution. *Ann. Math. Statist.* **17** 208–215.

Wilk, M. B., and Shapiro, S. S. (1968). The joint assessment of normality of several independent samples. *Technometrics* **10**(4) 825–839.

Wilks, S. S. (1941). Determination of sample sizes for setting tolerance limits. *Ann. Math. Statist.* **12** 91–96.

Willits, N. (1993). Personal communication, Univ. California at Davis.

Winer, B. J. (1971). *Statistical Principles in Experimental Design* 2nd ed. McGraw-Hill, New York.

Zacks, S. (1970). Uniformly most accurate upper tolerance limits for monotone likelihood ratio families of discrete distributions. *J. Amer. Statist. Assoc.,* **65** 307–316.

Index